AF620146

B. Zhilinskii

Introduction to Louis Michel's lattice geometry through group action

CURRENT NATURAL SCIENCES
EDP Sciences/CNRS Éditions

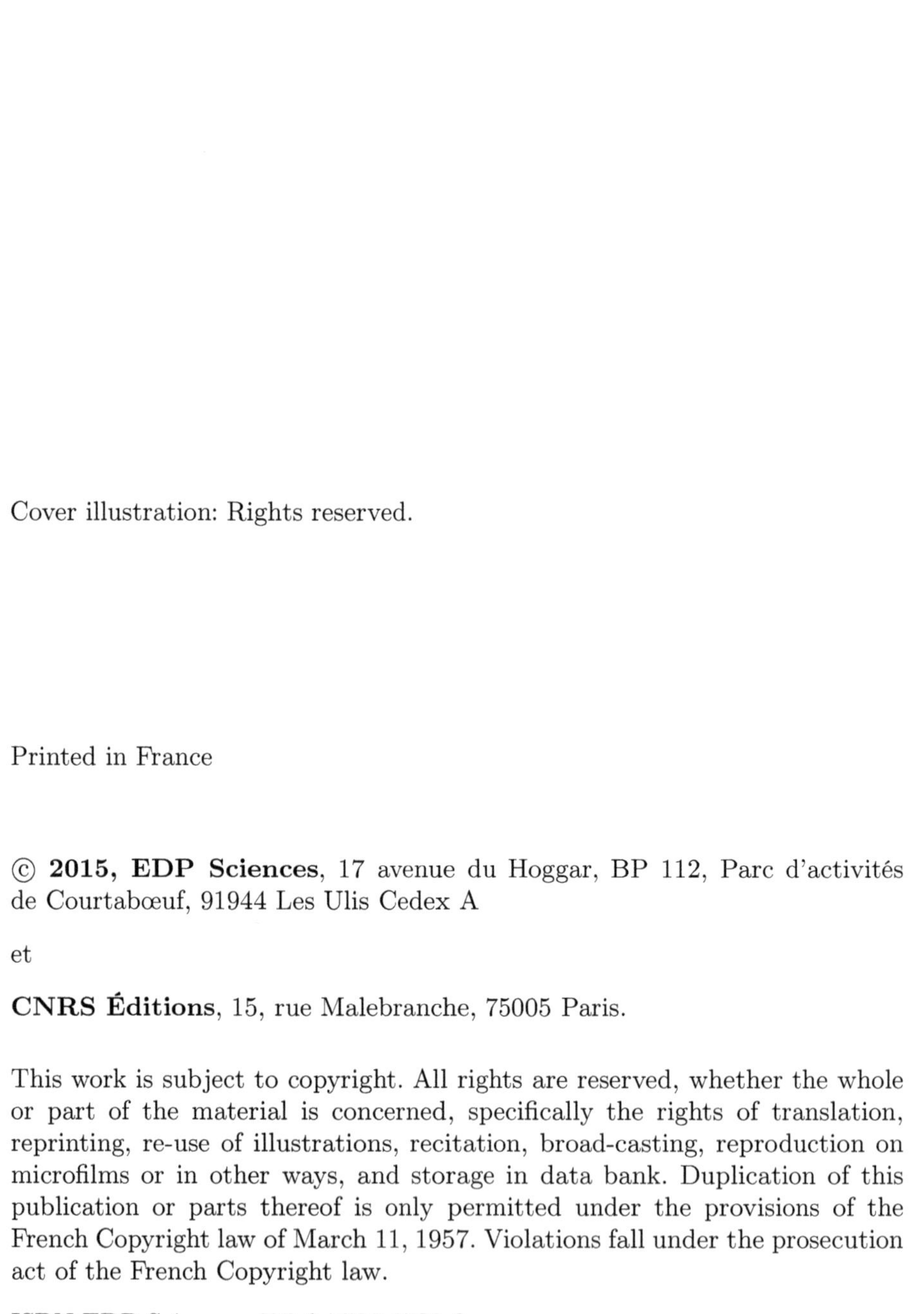

Cover illustration: Rights reserved.

Printed in France

© **2015, EDP Sciences**, 17 avenue du Hoggar, BP 112, Parc d'activités de Courtabœuf, 91944 Les Ulis Cedex A

et

CNRS Éditions, 15, rue Malebranche, 75005 Paris.

This work is subject to copyright. All rights are reserved, whether the whole or part of the material is concerned, specifically the rights of translation, reprinting, re-use of illustrations, recitation, broad-casting, reproduction on microfilms or in other ways, and storage in data bank. Duplication of this publication or parts thereof is only permitted under the provisions of the French Copyright law of March 11, 1957. Violations fall under the prosecution act of the French Copyright law.

ISBN EDP Sciences: 978-2-7598-1738-2
ISBN CNRS Éditions: 978-2-271-08739-3

Contents

Preface

This book has a rather long and complicated history. One of the authors, Louis Michel, passed away on the 30 December, 1999. Among a number of works in progress at that time there were a near complete series of big papers on "Symmetry, invariants, topology" published soon after in Physics Reports [75] and a project of a book "Lattice geometry", started in collaboration with Marjorie Senechal and Peter Engel [53]. The partially completed version of the "Lattice geometry" by Louis Michel, Marjorie Senechal and Peter Engel is available as a IHES preprint version of 2004. In 2011, while starting to work on the preparation of selected works of Louis Michel [19] it became clear that scientific ideas of Louis Michel developed over the last thirty years and related to group action applications in different physical problems are not really accessible to the young generation of scientists in spite of the fact that they are published in specialized reviews. It seems that the comment made by Louis Michel in his 1980's talk [70] remains valid till now:

"Fifty years ago were published the fundamental books of Weyl and of Wigner on application of group theory to quantum mechanics; since, some knowledge of the theory of linear group representations has become necessary to nearly all physicists. However the most basic concepts concerning group actions are not introduced in these famous books and, in general, in the physics literature."

After rather long discussions and trials to revise initial "Lattice geometry" text which require serious modifications to be kept at the current level of the scientific achievements, it turns out that probably the most wise solution is to restrict it to the basic ideas of Louis Michel's approach concentrated on the use of group actions. The present text is based essentially on the preliminary version of the "Lattice geometry" manuscript [53] and on relevant publications by Louis Michel [71, 76, 72, 73, 74], especially on reviews published in Physics Reports [75], but the accent is made on the detailed presentation of the two- and three-dimensional cases, whereas the generalization to arbitrary dimension is only outlined.

Chapter 1

Introduction

This chapter describes the outline of the book and explains the interrelations between different chapters and appendices.

The specificity of this book is an intensive use of group action ideas and terminology when discussing physical and mathematical models of lattices. Another important aspect is the discussion and comparison of various approaches to the characterization of lattices. Along with symmetry and topology ideas, the combinatorial description based on Voronoï and Delone cells is discussed along with classical characterization of lattices via quadratic forms.

We start by introducing in Chapter 2 the most important notions related to group action: orbit, stabilizer, stratum, orbifold, ... These notions are illustrated on several concrete examples of the group action on groups and on vector spaces. The necessary basic notions of group theory are collected in appendix A which should be considered as a reference guide for basic notions and notation rather than as an exposition of group theory.

Before starting description of lattices, chapter 3 deals with a more general concept, the Delone system of points. Under special conditions Delone sets lead to lattices of translations which are related to the fundamental physical notion of periodic crystals. The study of the Delone set of points is important not only to find necessary and sufficient conditions for the existence of periodic lattices. It allows discussion of a much broader mathematical frame and physical objects like aperiodic crystals, named also as quasicrystals.

Chapter 4 deals with symmetry aspects of periodic lattices. Point symmetry classification and Bravais classes of lattices are introduced using two-dimensional and three-dimensional lattices as examples. Stratification of the ambient space and construction of the orbifolds for the symmetry group action is illustrated again on many examples of two- and three-dimensional lattices. The mathematical concepts necessary for the description of point symmetry of higher dimensional lattices are introduced and the crystallographic restrictions imposed on the possible types of point symmetry groups by periodicity condition are explicitly introduced.

Chapter 5 introduces the combinatorial description of lattices in terms of their Voronoí and Delone cells. The duality aspects between Voronoí and Delone tesselations are discussed. Voronoí cells for two- and three-dimensional lattices are explicitly introduced along with their combinatorial classification as an alternative to the symmetry classification of lattices introduced in the previous chapter. Such notions as corona, facet, and shortest vectors are defined and their utility for description of arbitrary N-dimensional lattices is outlined.

Description of the lattices by using their symmetry or by their Voronoí cells does not depend on the choice of basis used for the concrete realization of the lattice in Euclidean space. At the same time practical calculations with lattices require the use of a specific lattice basis which can be chosen in a very ambiguous way. Chapter 6 discusses a very old subject: the description of lattices in terms of positive quadratic forms. The geometric representation of the cone of positive quadratic forms and choice of the fundamental domain of the cone associated with different lattices is discussed in detail for two-dimensional lattices. The reduction of quadratic forms is viewed through the perspective of the group action associated with lattice basis modification. The correspondence between the combinatorial structure of the Voronoí cell and the position of the point representing lattice on the cone of positive quadratic forms is carefully analyzed. The dimension of the cone of positive quadratic forms increases rapidly with the dimension of lattices. That is why the straightforward geometric visualization becomes difficult for three- and higher dimensional lattices. Nevertheless, for three-dimensional lattices the construction of the model showing the distribution of Bravais lattices and combinatorially different lattices by taking an appropriate section of the cone of positive quadratic forms is possible. This presentation is done on the basis of the very detailed analysis realized by Louis Michel during his lectures given at Smith College, Northampton, USA. Generalizations of the combinatorial description of lattices to arbitrary dimension requires introduction of a number of new concepts, which are shortly outlined in this chapter following mainly the fundamental works by Peter Engel and his collaborators. Symbolic visualization of lattices via graphs is introduced intuitively by examples of 3-, 4-, and partially 5-dimensional lattices without going into details of matroid theory.

Concrete examples of lattices in arbitrary dimensions related to reflection groups are studied in chapter 7. These examples allow us to see important correspondence between different mathematical domains, finite reflection groups, Lie groups and algebra, Dynkin diagrams, ...

Chapter 8 turns to discussion of the comparison between different classifications of lattices introduced in previous chapters and some other more advanced classifications suggested and used for specific physical and mathematical applications in the scientific literature. Among these different classifications we describe the correspondence between geometric and arithmetic classes of lattices and more general crystallographic classes necessary to classify the symmetry of the system of points which are more general than simple regular point lattices. Among the most important for physical applications aspects of lattice symmetry, the notion of enantiomorphism and of time reversal invariance are additionally discussed. The simultaneous use of symmetry and combinatorial classification for three-dimensional lattices is demonstrated by using the Delone approach.

Some physical and mathematical applications of lattices are discussed in chapter 9. These include analysis of sphere packing, covering, and tiling related mainly with specific lattices relevant for each type of problem. More physically related applications are the classification of the regular phases of matter and in particular the description of quasicrystals which are more general than regular crystals. Another generalization of regular lattices includes discussion of lattice defects. Description of different types of lattice defects is important not only from the point of view of classification of defects of periodic crystals. It allows also the study of defects of more formal lattice models, for example defects associated with lattices appearing in integrable dynamical models which are tightly related with singularities of classical dynamical integrable models and with qualitative features of quantum systems associated with lattices of common eigenvalues of several mutually commuting observables.

Appendices can be used as references for basic definitions of group theory (Appendix A), on graphs and partially ordered sets (Appendix B), and for comparison of notations (Appendix C) used by different authors. Also the complete list of orbifolds for 17 two-dimensional crystallographic groups (Appendix D) and for $3D$-irreducible Bravais groups (Appendix E) is given together with short explication of their construction and notation.

The bibliography includes a list of basic books for further reading on relevant subjects and a list of original papers cited in the text, which is obviously very partial and reflects the personal preferences of authors.

Chapter 2

Group action. Basic definitions and examples

This chapter is devoted to the definitions and short explanations of basic notions associated with *group actions*, which play a fundamental role in mathematics and in other fields of science as well. In physics group actions appear naturally in different domains especially when one discusses qualitative features of physical systems and their qualitative modifications.

We also introduce here much of the notation that will be used systematically in this book. Thus this section can be used as a dictionary.

Group action involves two "objects": a group G, and a mathematical structure M on which the group acts. M may be algebraic, geometric, topological, or combinatorial. Aut M, its automorphism group, is the group of one-to-one mappings of M to itself.

Definition: group action. An action of a group G on a mathematical structure M is a group morphism (homomorphism) $G \xrightarrow{\rho} \text{Aut } M$.

The examples we give are designed for the applications we need in this book. Let us start with a very simple mathematical object M, an equilateral triangle in the (two-dimensional) Euclidean plane R^2. The isometries of R^2 that leave this triangle invariant form a group consisting of 6 elements (identity, rotations through $2\pi/3$ and through $4\pi/3$, and reflections across the lines passing through its three vertices and the midpoints of the opposite sides). In the classical notation used by physicists and chemists, this group is denoted D_3. (Alternative notations of groups are discussed in Appendix C).

We can also consider the action of D_3 on other objects, for example on the entire plane (see Figure 2.1). In this case the group morphism $D_3 \xrightarrow{\rho} \text{Aut } R^2$ maps each group element to an automorphism (symmetry transformation) of $M = R^2$. This action is said to be *effective* because each $g \in G$ (other than the identity) effects the displacement of at least one point of the plane.

As another example of the action of D_3, we can take for M a single point, the center of the equilateral triangle. This point is left fixed by every element of D_3; thus this action is not effective.

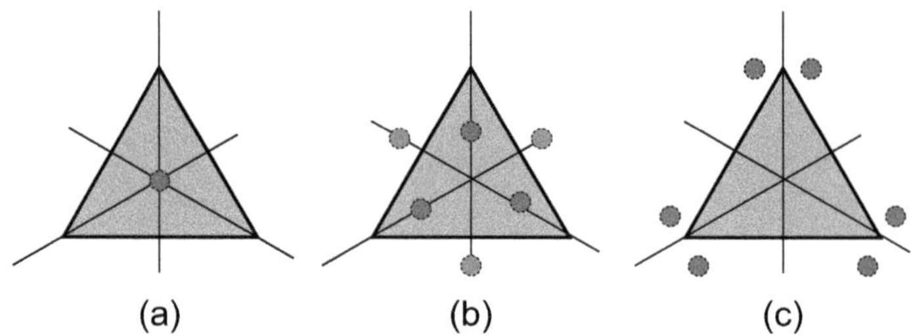

FIG. 2.1 – Orbits of the action of D_3 (the symmetry group of an equilateral triangle) on the 2D-plane. (*a*) The sole fixed point of the D_3 group action. The stabilizer of this one-point orbit is the whole group D_3. (*b*) Two examples of orbits consisting of three points. Each point of the orbit has one of the reflection subgroups r_i, $i = 1, 2, 3$, as a stabilizer. The three stabilizers r_i, $i = 1, 2, 3$ form the conjugacy class r of D_3 subgroups. (*c*) Example of an orbit consisting of six points. The stabilizer of each point of such an orbit and of the orbit itself is a trivial group $C_1 \equiv 1$.

We can also extend the action of D_3 from R^2 to R^3. The rotations through $2\pi/3$ and $4\pi/3$ about the axis passing through the center of the triangle and orthogonal to it generalize the plane rotations in a natural way.

There are two ways to generalize the reflections of D_3 to transformations of 3D-space.

First, we can replace reflection across a line ℓ by reflection in the plane orthogonal to the triangle and intersecting it in ℓ. This gives us a symmetry group whose symbol is C_{3v} (or $3m$ or $*33$). Alternatively, we can replace 2D-reflection across ℓ by rotation in space, through π, around the axis coinciding with that line. This group is denoted D_3 (or 32 [ITC]=[14], or 223 [Conway]=[31]). The groups D_3 and C_{3v} are isomorphic; thus one abstract group has two very different actions on R^3, while their actions on a 2D-dimensional subspace are identical.

We began this discussion with the example of an equilateral triangle in the plane. What is the symmetry group if the triangle is situated in three-dimensional space? Obviously, this group includes the six symmetry transformations forming the two-dimensional group D_3. But now the complete set of transformations leaving the triangle invariant also includes reflection in the plane of the triangle and the composition of this reflection with all the elements of D_3. Thus in R^3 the symmetry group of an equilateral triangle has 12 elements. We denote this larger group by D_{3h}, or $\overline{6}2m$ [ITC], or $*223$ [Conway].

Notice that the action of D_{3h} on the plane of the triangle in R^3 is non-effective, since reflection in that plane leaves all its points fixed. This action, described by the homomorphism $D_{3h} \xrightarrow{\rho} \text{Aut } R^2$, has a non-trivial kernel, $\text{Ker}\rho = Z_2$, the group of two elements (the identity and reflection in the plane).

Returning now to the definition of group action, we introduce the following notation. Since the action of a group G on a mathematical structure M is specified by the homomorphism $\rho(g)$ for all $g \in G$, we will write $\rho(g)(m)$ for the transform of any $m \in M$ by $g \in G$, and abbreviate it to $g.m$.[1]

Now we come to a key idea in group action.

Definition: group orbit. The orbit of m (under G) is the set of transforms of m by G; we denote this by $G.m$.

For example (see Figure 2.1), each of the following sets is an orbit of D_3 acting on the two-dimensional plane containing an equilateral triangle:

- three points equally distanced from the center, one on each of the three reflection lines;
- the centroid or, equivalently, the center of mass of the triangle; and
- any set of six distinct points related by the reflections and rotations of the symmetry group D_3.

Figure 2.2 shows orbits of C_{3v}, D_3, and D_{3h} acting on an equilateral triangle in R^3.

Under the action of a finite group, the number of elements in an orbit cannot be larger than the order of the group, and this number always divides the group order. Belonging to an orbit is an equivalence relation on the elements of M and thus M is a disjoint union of its orbits.

For continuous groups an orbit can be a manifold whose dimension cannot exceed the number of continuous parameters of the group. The simplest examples of continuous symmetry groups are the group of rotations of a circle, $SO(2) = C_\infty$, and the circle's complete symmetry group, $O(2) = D_\infty$, which includes reflections. Both C_∞ and D_∞ act effectively on the plane in which the circle lies. In fact their orbits coincide (see Figure 2.3): there is one one-point orbit, the fixed point of the group action, and a continuous family of one-dimensional orbits, each of them a circle.

A second key notion is the stabilizer of an element of M.

Definition: stabilizer. The stabilizer of an element $m \in M$ is the subgroup

$$G_m = \{g \in G,\ g.m = m\}$$

of elements of G which leave m fixed.

If $G_m = G$, then this orbit has a single element and m is said to be a fixed point of M (see Figure 2.1a and Figure 2.3a).

If G is finite, then the number of points in the orbit $G.m$ is $|G|/|G_m|$. Thus if, as in Figure 2.1 b, the stabilizer of a D_3 orbit is a subgroup of order 2, the orbit consists of three points. If $G_m = 1 \equiv e$, the group identity

[1] When G is Abelian and its group law is noted additively, we may use $g + m$ instead of $g.m$ as short for $\rho(g)(m)$, though this use of $+$ is an "abus de langage," since g and m may not be objects of the same type.

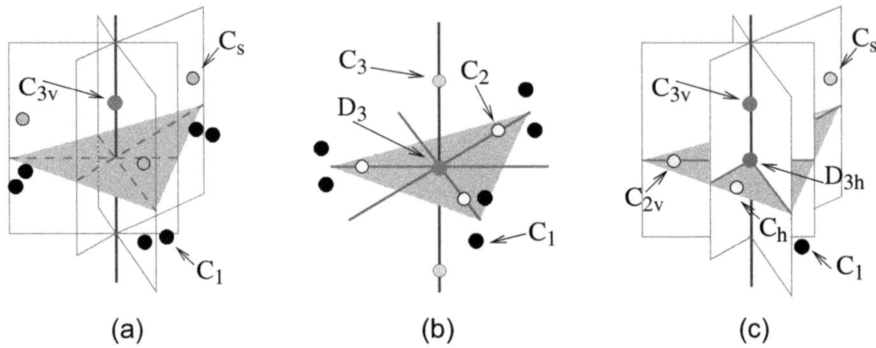

FIG. 2.2 – Generalizing the action of D_3 from R^2 to R^3. (*a*) Action of the group C_{3v}: three orbits with stabilizers C_{3v} , C_s, and C_1 are shown (s stands for reflection in the indicated plane); (*b*) Action of the group D_3: four orbits with stabilizers D_3, C_3, C_2, and C_1 are shown. (*c*) Action of the group D_{3h}: one point from each of six different orbits (D_{3h}, C_{3v}, C_{2v}, C_s, C_h, and C_1) is shown.

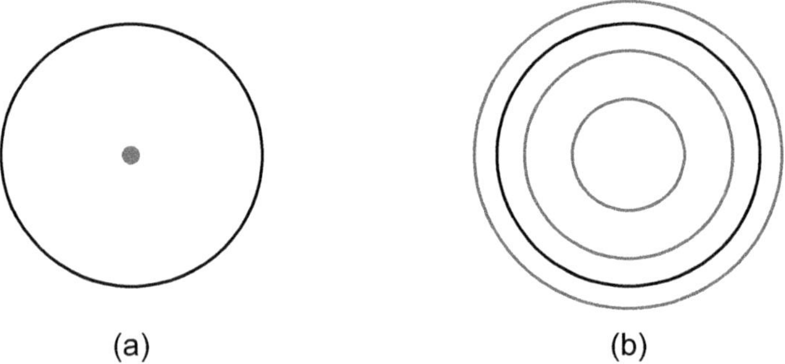

FIG. 2.3 – Orbits of the action of C_∞ and D_∞ on the 2D-plane. (*a*) The fixed point of these group actions on R^2. (*b*) Continuous circular orbits.

(Figure 2.1 c), then the size of the orbit is $|G|$ and the orbit is said to be *principal.*[2]

It is easy to prove that $G_{g.m} = gG_mg^{-1}$, from which it follows that the set of stabilizers of the elements of an orbit is a conjugacy class $[H]_G$ of subgroups of G. For example, the stabilizers of the three vertices of an equilateral triangle are the three reflection subgroups, r_i, of D_3, which are conjugate by rotation. This fact allows us to classify (or to label) orbits by their stabilizers, i.e. by the conjugacy classes of subgroups of group G. We recall that the conjugacy

[2] Orbits with trivial stabilizer 1 are always principal but for continuous group actions principal orbits can have nontrivial stabilizers. In that case principal orbits are defined as orbits forming open dense strata, see below.

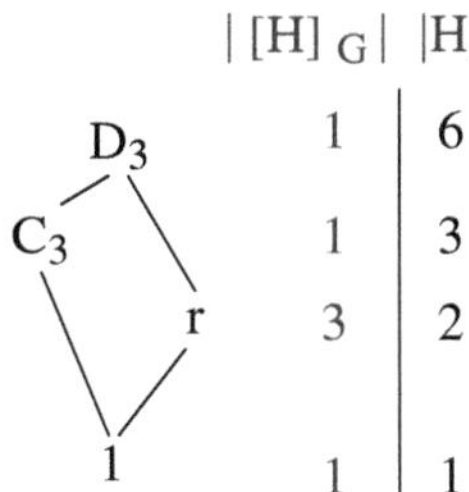

FIG. 2.4 – The lattice of conjugacy classes of subgroups of D_3 group. The table on the right shows, in column 1, the number of elements $|[H]_G|$ in the conjugacy class $[H]_G$ of each type of subgroup. The numbers in the right-hand column are the orders of the subgroup $|H|$.

classes of subgroups of any given group form a partially ordered set: one class is "smaller" than another if it contains a proper subgroup of a group in the other conjugacy class. This partial ordering for D_3 is shown in Figure 2.4.

Orbits with the same conjugacy class of stabilizers are said to be of the same type.

Next, we define the very important notion of stratum.

Definition: stratum. In a group action, a stratum is the union of all points belonging to all orbits of the same type.

By definition, two points belong to the same stratum if, and only if, their stabilizers are conjugate. Consequently we can classify and label the strata of a group action by the conjugacy classes of subgroups of the group.

The three strata of the action of D_3 on R^2 are shown in Figure 2.5. They include the centroid of the triangle (D_3's zero-dimensional stratum), three mirror lines without their intersection point (the one-dimensional stratum), and the complement of these two strata (the two dimensional stratum).

A disc D, minus its center, is one stratum of the action of D_∞ on D; the center is the other.

When they exist, as in the case of the D_3 action on R^2 (Figure 2.1) or the C_∞ action (Figure 2.3) the fixed points form one stratum and the principal orbits form another. Belonging to the same stratum is an equivalence relation for the elements of M or for orbits of a G-action on M. Thus M can be considered as a disjoint union of strata of different dimensions.

We will denote the set of orbits of the action of G on M by $M|G$ and the corresponding set of strata as $M||G$. To belong to the same stratum is an equivalence relation for the elements of M and for elements of the set of orbits, $M|G$. The set of strata $M||G$ is a (rather small in many applications) subset of the set of conjugacy classes of subgroups of G. Thus $M||G$ too has the structure of a partially ordered set $S_i \in M||G$, where by $S_1 < S_2$ we mean that the local symmetry of S_1 is smaller than that of S_2 – i.e. the stabilizers of the points of S_1 are, up to conjugation, subgroups of those of S_2.

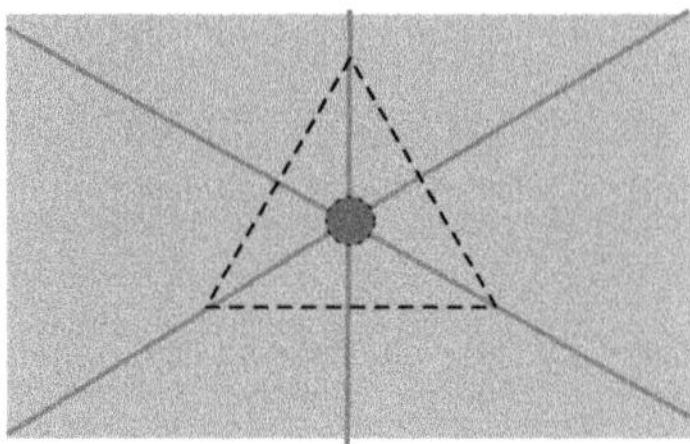

FIG. 2.5 – The strata of D_3 action on R^2. Black point in the center represents the zero-dimensional D_3-stratum. The rays without their common intersection point form the one-dimensional r-stratum. The six two-dimensional regions of the plane form the two-dimensional principal stratum with trivial stabilizer.

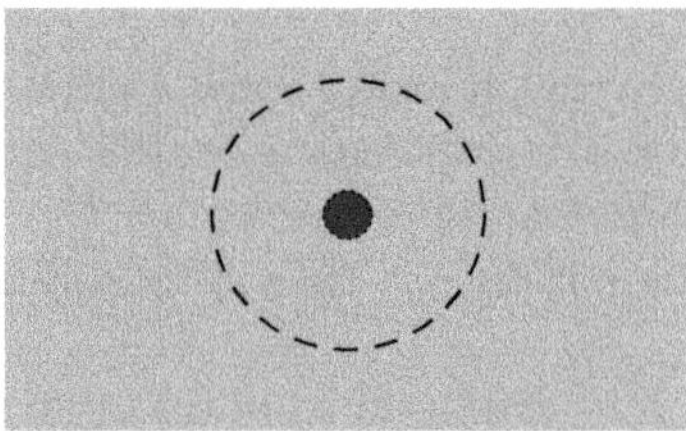

FIG. 2.6 – Strata of the action of C_∞ (or D_∞) on R^2. The black point forms the zero-dimensional stratum. The whole plane without the point is the two-dimensional principal stratum.

Beware: a less symmetric stratum might have a larger dimension than a more symmetric one. The set of strata is partially ordered by local symmetry, not by size.

The example of the action of D_3 on R^2, discussed above, leads to three strata: the zero dimensional stratum D_3, the one-dimensional stratum r, and the two-dimensional principal stratum 1, which is open and dense. Only three conjugacy classes of subgroups of D_3 (see Figure 2.5) appear as local symmetry of strata. The natural partial order between strata is $1 < r < D_3$.

The action of C_∞ (the group of pure rotational symmetries of a circle or of a disk) on R^2 leads to two strata (see Figure 2.6). The zero-dimensional stratum consists of one point, the center. The two-dimensional principal stratum is the whole plane minus that point. Note that the action of D_∞ on R^2 has the same two strata, but their stabilizers now are different.

Finally, we define the notions of orbit space and orbifold.

Definition: orbit space. The set of orbits appearing in an action of G on M is the orbit space $M|G$.

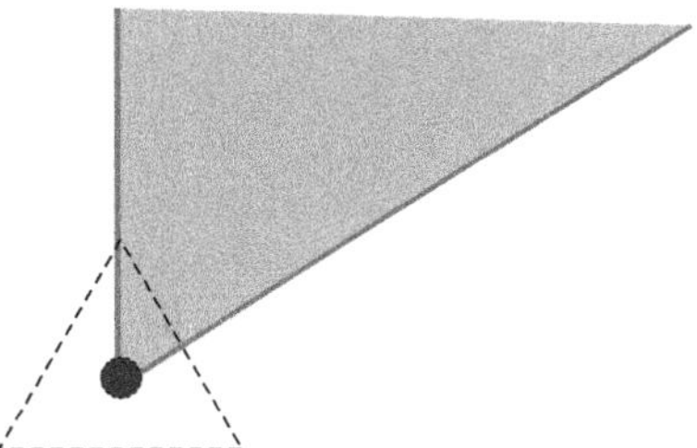

FIG. 2.7 – Orbifold of D_3 action on R^2. The black point represents the D_3-orbit consisting of one point. Two rays form 1D-set of r-orbits consisting each of three points. The shaded region is a two-dimensional set of principal $C_1 \equiv 1$ orbits, consisting each of six points.

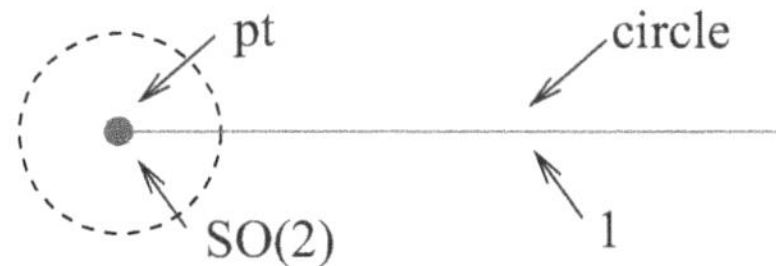

FIG. 2.8 – Orbifold of the action of C_∞ on R^2. Black filled point - the orbit with stabilizer $SO(2)$ consisting of one point. Solid line - the set of 1D-orbits, each orbit being a circle.

If M contains only one orbit, i.e. if any $m \in M$ can be transformed into any other element of M by the group action, the action is said to be *transitive* and M is called a *homogeneous space* (with respect to G and ρ). Examples of homogeneous spaces and their associated groups include

- a circle (not a disk!), $G = D_\infty$;
- R^n, and G the group of translations in R^n.
- a sphere S_n in $(n+1)$-dimensional space and $G = SO(n+1)$.

Definition: orbifold. The orbifold of a group action is a set consisting of one representative point from each of its orbits.

Thus, the space of orbits for the action of D_3 on R^2 can be represented as a sector of the plane (see Figure 2.7). The space of orbits for the action of C_∞ on R^2 can be represented as a one-dimensional ray with a special point at the origin (see Figure 2.8).

Let us consider the space of orbits of a (three-dimensional) D_3 action on a two-dimensional sphere surrounding an equilateral triangle (see Figure 2.9). Assume that its action on R^2 (see Figure 2.2 b) coincides with the action of the 2D-point group D_3.

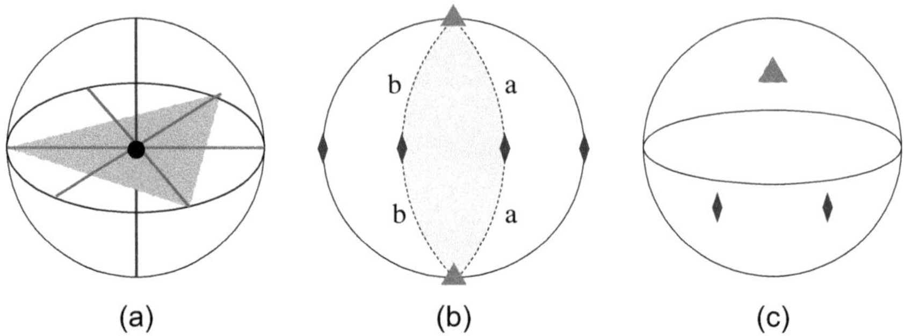

FIG. 2.9 – (*a*) Schematic view of the sphere surrounding an equilateral triangle. (*b*) Action of group D_3 on the sphere represented on orthographic projection. The shaded region represents a fundamental domain. Parts of boundaries indicated by the same letters should be identifies. (*c*) Orbifold for action of D_3 group on the sphere - sphere with three special points, 223.

The action of D_3 on the two-dimensional sphere yields one orbit with stabilizer C_3. This orbit consists of two points (two poles of a sphere lying on the C_3 axis). Another zero-dimensional stratum is formed by two three-point orbits with stabilizer C_2. These points have stabilizers C_2^i, $i = 1, 2, 3$. The three subgroups C_2^i of order two belong to the same conjugacy class C_2, which is used to label these orbits. All other points of the sphere belong to principal orbits with stabilizer 1; each of these orbits consists of six points.

To construct the orbifold (or the space of orbits) we take one representative point from each orbit. From the physical point of view this procedure corresponds to selecting a fundamental domain of the group action (the choice is not unique). The so obtained space of orbits is (from the topological point of view) a two-dimensional sphere with three singular points, corresponding to three isolated orbits. One isolated orbit has stabilizer C_3 and forms itself C_3 stratum. Two other isolated orbits (each consisting of three points) have stabilizer C_2 and form another zero-dimensional stratum. The topology of an orbifold can be quite complex; for a primer on orbifold construction, see [4], [21], and a number of examples in chapter 4.

2.1 The action of a group on itself

Let us consider the set of elements forming group G. Then Aut G is the permutation group of the elements of the set G.

Example 1 G acts on its elements by conjugation. That is, $M = G$, and $\rho(g)(m) = gmg^{-1}$. Then Ker $\rho = C(G)$, the center of G (the subset of elements of G commuting with all $g \in G$). Im ρ is the group of inner automorphisms of G.

The orbit of $x \in G$ is called the *conjugacy class* of x in G; we denote it by $[x]_G$. The fixed points of this action are the elements of the center of G. If G is Abelian, then there is only one stratum, that of fixed points.

We illustrate the action of G on itself by conjugation with the example of D_{3h}, the symmetry group of an equilateral triangle in 3D space (see Figure 2.2, c). Group D_{3h} consists of the 12 elements listed in the first line and the first column of Table 2.1. Column x and line g intersect in the entry gxg^{-1}. All of the entries in each column belong to the same orbit, that is, they form one conjugacy class. The notation for the conjugacy class and the number of elements in it are listed in the two last lines of the table. The elements $x \in G$ invariant under conjugation with one $g \in G$ constitute the stabilizer of g, listed in the last column. In group theoretical terminology the stabilizer of a group element is its *centralizer*.

The action $G \xrightarrow{\rho}$ Aut M defines an action of G on the subsets of M in a natural way. In particular, the action of G on its elements by conjugation induces the action of G on the set of its subgroups. The orbit $G.H$ of a subgroup H is the conjugacy class $[H]_G$ of the subgroups of G conjugate to H. The stabilizer G_H is the *normalizer* of H in G, $N_G(H)$. If H is fixed by this action, it is by definition an *invariant subgroup* of G.

The lattice of subgroups of D_{3h} is shown in Figure 2.10, and the action of D_{3h} on its subgroups is illustrated in Table 2.1. Note that D_{3h} has several subgroups of order two describing reflection in different planes. There are three vertical planes and one horizontal plane (see figure 2.2, c). The three subgroups of reflections in vertical planes form one conjugacy class. The subgroup C_h of reflection in the horizontal plane h is an invariant subgroup. Moreover, C_h is the center of D_{3h}.

We denote the set of subgroups of G by $\{\leq G\}$ and the set of conjugacy classes of subgroups of G by $\{[\leq G]_G\}$. For a large family of groups – including all those we will meet in this monograph – there is a natural partial ordering on $\{[\leq G]_G\}$ by subgroup inclusion up to conjugation. By definition, the set of possible types of G-orbits defines a partial ordering on the stratum space $M\|G$. (As we shall show, the role of this space is essential.) Its elements correspond to the different symmetry types of the elements of M.

For infinite groups, $|\{[\leq G]_G\}|$ is infinite in general[3], but in most problems we shall study, $M\|G$ is finite. In that case, there exist maximal and minimal strata, corresponding to maximal and minimal symmetry.

The set of strata of D_{3h} in R^3 consists of six elements (see figure 2.2, c): D_{3h}, C_{3v}, C_{2v}, C_s, C_h, C_1. They form the partially ordered lattice shown in Figure 2.11. The maximal stratum is the zero dimensional D_{3h} stratum,

[3] This is the case, for example, for U_1, the one dimensional unitary group i.e. the multiplicative group of complex numbers of modulus 1. This group is Abelian and has an infinite number of subgroups $\mathbb{Z}_n$, the cyclic group of n elements. Moreover, since the $\mathbb{Z}_n$ for different n are not isomorphic, every group containing U_1 has an infinite set of conjugacy classes of subgroups. That is also the case of O_n and $GL(n, R)$ for $n > 1$.

TAB. 2.1 – Example of the action of the group D_{3h} on itself. For each $x \in D_{3h}$ the result of the D_{3h} action is given. The stabilizer (or the centralizer) of each element $g \in D_{3h}$ is given in the last column. The last two lines show the classes of conjugate elements to which an element $x \in D_{3h}$ belongs and the number of elements in the corresponding class.

$g \backslash x$	E	C_3	C_3^{-1}	U_2^a	U_2^b	U_2^c	σ_h	S_3	S_3^{-1}	σ_v^a	σ_v^b	σ_v^c	Stabilizer
E	E	C_3	C_3^{-1}	U_2^a	U_2^b	U_2^c	σ_h	S_3	S_3^{-1}	σ_v^a	σ_v^b	σ_v^c	D_{3h}
C_3	E	C_3	C_3^{-1}	U_2^b	U_2^c	U_2^a	σ_h	S_3	S_3^{-1}	σ_v^b	σ_v^c	σ_v^a	C_{3h}
C_3^{-1}	E	C_3	C_3^{-1}	U_2^c	U_2^a	U_2^b	σ_h	S_3	S_3^{-1}	σ_v^c	σ_v^a	σ_v^b	C_{3h}
U_2^a	E	C_3^{-1}	C_3	U_2^a	U_2^c	U_2^b	σ_h	S_3^{-1}	S_3	σ_v^a	σ_v^c	σ_v^b	C_{2v}^a
U_2^b	E	C_3^{-1}	C_3	U_2^c	U_2^b	U_2^a	σ_h	S_3^{-1}	S_3	σ_v^c	σ_v^b	σ_v^a	C_{2v}^b
U_2^c	E	C_3^{-1}	C_3	U_2^b	U_2^a	U_2^c	σ_h	S_3^{-1}	S_3	σ_v^b	σ_v^a	σ_v^c	C_{2v}^c
σ_h	E	C_3	C_3^{-1}	U_2^a	U_2^b	U_2^c	σ_h	S_3	S_3^{-1}	σ_v^a	σ_v^b	σ_v^c	D_{3h}
S_3	E	C_3	C_3^{-1}	U_2^b	U_2^c	U_2^a	σ_h	S_3	S_3^{-1}	σ_v^b	σ_v^c	σ_v^a	C_{3h}
S_3^{-1}	E	C_3	C_3^{-1}	U_2^c	U_2^a	U_2^b	σ_h	S_3	S_3^{-1}	σ_v^c	σ_v^a	σ_v^b	C_{3h}
σ_v^a	E	C_3^{-1}	C_3	U_2^a	U_2^c	U_2^b	σ_h	S_3^{-1}	S_3	σ_v^a	σ_v^c	σ_v^b	C_{2v}^a
σ_v^b	E	C_3^{-1}	C_3	U_2^c	U_2^b	U_2^a	σ_h	S_3^{-1}	S_3	σ_v^c	σ_v^b	σ_v^a	C_{2v}^b
σ_v^c	E	C_3^{-1}	C_3	U_2^b	U_2^a	U_2^c	σ_h	S_3^{-1}	S_3	σ_v^b	σ_v^a	σ_v^c	C_{2v}^c
$[x]_{D_{3h}}$	$[E]$	$[C_3]$			$[U_2]$		$[\sigma_h]$	$[S_3]$			$[\sigma_v]$		
$\|[x]\|$	1	2			3		1	2			3		

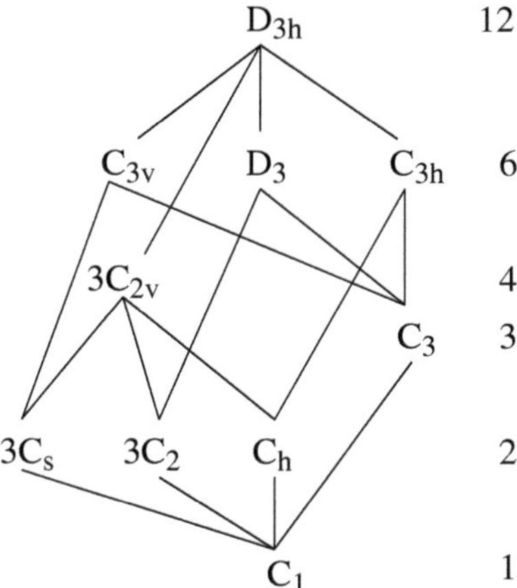

FIG. 2.10 – Lattice of subgroups of D_{3h} group. The order of subgroups is indicated in the right column.

TAB. 2.1 – Action of the group D_{3h} on the set of its subgroups.

H	$G \cdot H$	Normalizer
D_{3h}	D_{3h}	D_{3h}
C_{3v}	C_{3v}	D_{3h}
C_{3h}	C_{3h}	D_{3h}
D_3	D_3	D_{3h}
C_{2v}^a	$\{C_{2v}^a, C_{2v}^b, C_{2v}^c\}$	C_{2v}^a
C_{2v}^b	$\{C_{2v}^a, C_{2v}^b, C_{2v}^c\}$	C_{2v}^b
C_{2v}^c	$\{C_{2v}^a, C_{2v}^b, C_{2v}^c\}$	C_{2v}^c
C_3	C_3	D_{3h}
C_s^a	$\{C_s^a, C_s^b, C_s^c\}$	C_{2v}^a
C_s^b	$\{C_s^a, C_s^b, C_s^c\}$	C_{2v}^b
C_s^c	$\{C_s^a, C_s^b, C_s^c\}$	C_{2v}^c
C_2^a	$\{C_2^a, C_2^b, C_2^c\}$	C_{2v}^a
C_2^b	$\{C_2^a, C_2^b, C_2^c\}$	C_{2v}^b
C_2^c	$\{C_2^a, C_2^b, C_2^c\}$	C_{2v}^c
C_h	C_h	D_{3h}
C_1	C_1	D_{3h}

while the minimal is a generic C_1 three-dimensional stratum with a trivial stabilizer.

Example 2 G acts on itself by left multiplication: $g.m = gm$.

Under this action, G has a single orbit, the entire set G is a single G-orbit. That is, G is a principal orbit. If we restrict this G-action to a proper

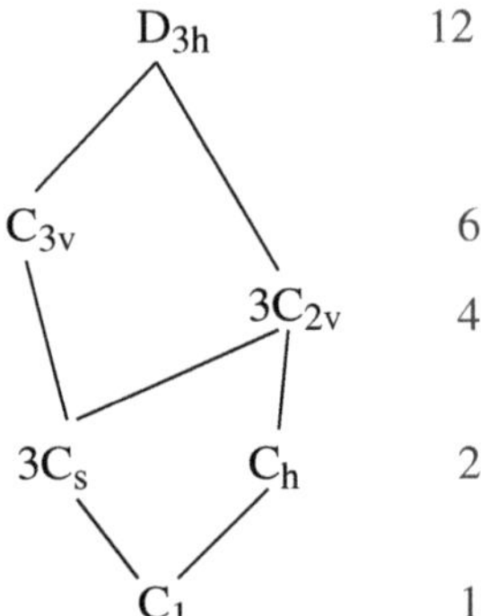

FIG. 2.11 – Lattice of strata of D_{3h} group action on three dimensional space. Numbers before symbols of conjugacy classes of subgroups indicate the number of subgroups in the class. Right column shows the order of the stabilizer written on the left in the diagram.

subgroup H, then the orbit of $x \in G$ is the *right* coset Hx. The set $(G : H)_R$ of *right* cosets Hx is the orbit space $G|H$.

The group action of G on G by right multiplication is defined by $g_r.x = xg^{-1}$. Restricting to $H < G$, the H-orbits are the *left* cosets xH and the orbit space can be identified with $(G : H)_L$, the set of left H-cosets.

2.2 Group action on vector space

Let M be an n-dimensional real vector space V_n and $GL_n(\mathbb{R})$ the real general linear group. Then Aut $M = GL_n(\mathbb{R})$ and the action $GL_n(\mathbb{R}) \xrightarrow{\rho}$ Aut V_n defines a real linear representation of $GL_n(\mathbb{R})$ on V_n.

The elements of V_n are called vectors; we denote them by $\vec{x}, \vec{y}, \ldots$. The action of $GL_n(\mathbb{R})$ on V_n has only two orbits, the origin o, which is fixed, and the rest of the space. We leave it as an exercise to the reader to find the stabilizer of a nonzero vector[4].

Two linear representations ρ and ρ' are said to be *equivalent* if they are conjugate under $GL_n(\mathbb{R})$:

$$\rho \equiv \rho' \ \Leftrightarrow \ \exists \gamma \in GL_n(\mathbb{R}), \ \forall g \in G, \ \rho'(g) = \gamma\rho(g)\gamma^{-1}. \tag{2.1}$$

Moreover, for $\alpha \in GL_n(\mathbb{R})$, the determinant $\det(\alpha)$ defines a homomorphism $GL_n(\mathbb{R}) \xrightarrow{\det} \mathbb{R}$ whose kernel is the *special linear group* $SL_n(\mathbb{R})$, the group of matrices with determinant $+1$.

[4] The answer is given later in this subsection.

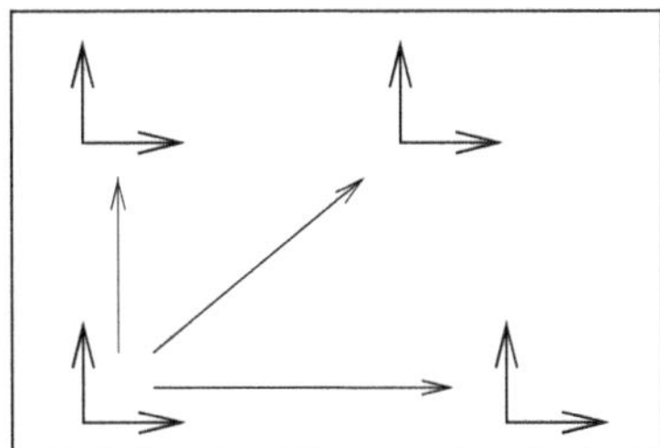

FIG. 2.12 – Action of a group of translations on two-dimensional Euclidean space leads to a parallel displacement of a reference frame. A space with this action is said to be "*homogeneous*".

Example 3 For any two bases $\{\vec{b}_i\}$ and $\{\vec{b}'_i\}$ of V_n, there is a unique $g \in GL_n(\mathbb{R})$ transforming $\{\vec{b}_i\}$ into $\{\vec{b}'_i\}$: in the basis $\{\vec{b}_i\}$, the elements of the jth column of the matrix representing g are the components of the vector $\vec{b}'_j$. Thus the set of bases $\mathcal{B}_n$ is a principal orbit of $GL_n(\mathbb{R})$.

V_n together with the scalar product $(\vec{x}, \vec{y})$ is an orthogonal space that we denote by E_n. Then Aut $E_n = O_n$, the n-dimensional orthogonal group, and ρ defines an orthogonal representation of G. When we are only interested in the Abelian group structure of the elements of V_n or E_n, we use the notation $\mathbb{R}^n$. Figures 2.12 and 2.13 illustrate the action of translations and rotations on two-dimensional space.

Example 4 When $n > 0$, Aut $E_n = O_n$ and there are only two strata: the fixed point $\vec{0}$, and n-dimensional open dense stratum formed by points with stabilizers belonging to the conjugacy class $[O_{n-1}]_{O_n}$. The orbits of this stratum are the spheres (centered at the origin) of vectors of the same norm.

The set of elements of V_n with translation, the "natural" action of $\mathbb{R}^n$, is an affine space that we denote $\mathcal{V}_n$; it is a principal orbit of $\mathbb{R}^n$. We denote its elements, the points, by $x, y, \ldots$.

Let x, y be any pair of points in V_n. The unique translation vector taking x to y will be denoted by $\vec{t} = y - x$ or by $\vec{t} = \overrightarrow{xy}$.

By extension, every algebraic sum of points of the affine space, the sum of whose coefficients is 1, is a well defined point of $\mathcal{V}_n$.

Any $m+1$ points of $\mathcal{V}_n$, $m \leq n$, are said to be *independent* if they span an m-dimensional linear manifold. A *simplex* is the convex hull of $n+1$ independent points in $\mathcal{V}_n$; the independent points are its vertices. Two dimensional simplices are triangles; in three dimensions they are tetrahedra.

The affine space $\mathcal{E}_n$ built from an orthogonal space E_n has a richer structure than $\mathcal{V}_n$, as it inherits a metric from the orthogonal scalar product of E_n: the distance $d(x, y)$ between the pair of points $x, y \in \mathcal{E}_n$ is the positive square root of the scalar product, or norm, $N(y - x)$.

FIG. 2.13 – The action of a group of rotation on two-dimensional Euclidean space around a fixed point shows that the space is isotropic. The dashed and dash-dot frames are the results of action on the initial black frame.

Example 5 Special cases of affine objects include:

- For any $\lambda \in \mathbb{R}$, $\lambda x + (1-\lambda)y$ is the straight line defined by the two distinct points $x, y \in \mathcal{V}_n$;
- The sum $\sum_i \alpha_i x_i$, where $\sum_i \alpha_i = 1$ is the linear manifold defined by the points x_i.
- When $0 \leq \lambda \leq 1$, $\lambda x + (1-\lambda)y$ is the line segment joining x, y; when $\sum_i \alpha_i = 1$ and $0 \leq \alpha_i \leq 1$, $\sum_i \alpha_i x_i$ is the convex hull of the points x_i.

Similarly, any algebraic sum of points, the sum of whose coefficients vanishes, defines a unique translation of $\mathbb{R}^n$; e.g. $a + c - b - d = \overrightarrow{ba} + \overrightarrow{dc} = \overrightarrow{da} + \overrightarrow{bc}$. An arbitrary choice of a point of $\mathcal{V}_n$, called the "origin", reconstructs the structure of a vector space in the affine space.

Example 6
The affine and Euclidean groups

$$\mathrm{Aff}_n = \mathbb{R}^n \rtimes GL_n(\mathbb{R}) = \mathrm{Aut}\ \mathcal{V}_n, \text{ and } \mathrm{Eu}_n = \mathbb{R}^n \rtimes O_n = \mathrm{Aut}\ \mathcal{E}_n. \quad (2.2)$$

are semi-direct products (see appendix A).

To represent the action of Aff_n on $\mathcal{V}_n$ by matrices, we have to choose a basis in the underlying vector space V_n and an origin o in $\mathcal{V}_n$. This yields a system of coordinates: the coordinates of o vanish and those of $o + \overrightarrow{ox}$ (or, more simply, $o + \vec{x}$) are the coordinates of the vector $\vec{x}$. An element of Aff_n can be written in the form $\langle \vec{a} A \rangle$ with $\vec{a} \in V_n$, $A \in GL_n(\mathbb{R})$. Then the group law of Aff_n is

$$\langle \vec{a}, A \rangle \langle \vec{b}, B \rangle = \langle \vec{a} + A\vec{b}, AB \rangle, \quad \langle \vec{a}, A^{-1} \rangle = \langle -A^{-1}\vec{a}, A^{-1} \rangle \quad (2.3)$$

and the action on $\mathcal{V}_n$ is, explicitly,

$$\langle \vec{a}, A \rangle .x = o + \vec{a} + A\vec{x}. \tag{2.4}$$

Note that the map τ given by

$$\tau(\langle \vec{a}, A \rangle) = \begin{pmatrix} 1 & 0 \\ \vec{a} & A \end{pmatrix} \tag{2.5}$$

gives a $(n+1)$-dimensional linear representation of Aff_n.

By definition, the contragredient representation of τ is

$$\hat{\tau}(\langle \vec{a}, A \rangle) = \begin{pmatrix} 1 & -(A^{-1}\vec{a})^\top \\ 0 & (A^{-1})^\top \end{pmatrix}; \qquad \det \hat{\tau}(\langle \vec{a}, A \rangle) = (\det A)^{-1}. \tag{2.6}$$

Since the determinant is invariant under conjugation, τ and $\hat{\tau}$ are inequivalent representations.

Now we can give the answer to the exercise proposed in example 2. Im $\hat{\tau}$ leaves invariant the vectors whose first coordinate is the only nonzero one, and obviously no larger subgroup does. So the stabilizers of Aut V_n for the non-vanishing vectors of V_n form the conjugacy class of Im $\hat{\tau}$.

Example 7 Euclidean geometry.
$\mathcal{E}_n$ is the principal orbit of $\mathbb{R}^n$ or, equivalently, the orbit of $Eu_n : O_n$. Let $\mathcal{E}_n^{\times 2}$ be the set of pairs $x \neq y$ of distinct points of $\mathcal{E}_n$. Its dimension is $2n$. The action of Eu_n on this set contains a unique generic open dense stratum formed by a continuous set of orbits each labeled by a positive real number, the distance $d(x, y)$. Each orbit is a $2n - 1$-dimensional subspace of $\mathcal{E}_n^{\times 2}$. In order to find the stabilizer let m be the midpoint of the segment $\overline{xy}$, and $\mathcal{E}_{n-1}$ the bisector hyperplane of the pair x, y. Figure 2.14 illustrates schematically this construction. It is easy to see from the figure that the stabilizer of this stratum is $[O_{n-1} \times \mathbb{Z}_2]_{Eu_n}$ where O_{n-1} is the stabilizer of m in the Euclidean group of $\mathcal{E}_{n-1}$ and $\mathbb{Z}_2$ is the 2 element group generated by the reflection (in E_n) through the hyperplane $\mathcal{E}_{n-1}$.

Let us now consider the more interesting case of the action of Eu_n on $\mathcal{E}_n^{\times 3}$, the set of triplets x, y, z of distinct points of $\mathcal{E}_n$.

The distances ξ, η, ζ between the 3 pairs of points are a Euclidean invariant, but they are not arbitrary positive numbers. We will choose three invariants λ, μ, ν defined by the conditions

$$\xi = d(y, z) = \frac{1}{2}(\mu + \nu) > 0, \quad \eta = d(z, x) = \frac{1}{2}(\nu + \lambda) > 0,$$

$$\zeta = d(x, y) = \frac{1}{2}(\lambda + \mu) > 0. \tag{2.7}$$

Then

$$\lambda = -\xi + \eta + \zeta \geq 0, \quad \mu = \xi - \eta + \zeta \geq 0, \quad \nu = \xi + \eta - \zeta \geq 0, \tag{2.8}$$

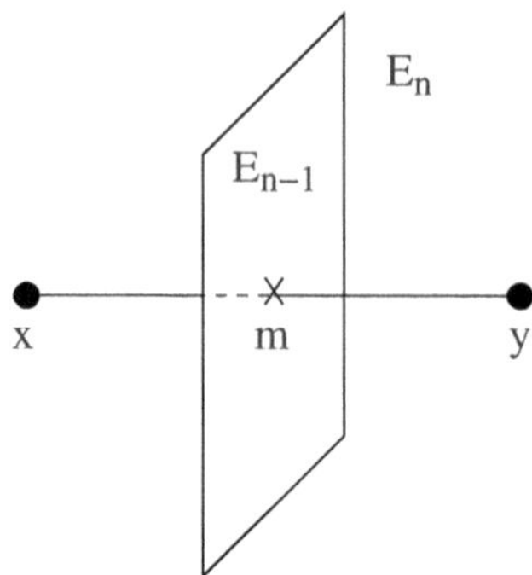

FIG. 2.14 – Construction of orbits of the action of Eu_n on $\mathcal{E}_n^{\times 2}$, the set of pairs of distinct points of $\mathcal{E}_n$.

and – it is easy to prove – no more than one of these 3 invariants λ, μ, ν vanishes. It is sufficient to verify that if any two of invariants become zero, the third is zero as well and the three points coincide. The surface $s(x, y, z)$ of the triangle (x, y, z) satisfies

$$4s(x, y, z)^2 = (\lambda + \mu + \nu)\lambda\mu\nu. \tag{2.9}$$

This implies that if one of the parameters λ, μ, or ν equals zero, the three points belong to a single line.

We have the one-to-one correspondence between orbits and points in the three dimensional space of parameters λ, μ, ν situated in the octant $\lambda \geq 0, \mu \geq 0, \nu \geq 0$ excluding three axes $\lambda = \mu = 0$, $\lambda = \nu = 0$, and $\mu = \nu = 0$.

To find the stabilizers we need first to distinguish two cases and several subcases.

i) None of λ, μ, ν is equal to zero:

- **a)** The 3 invariants have different values. There exists a three-parameter family of orbits corresponding to generic triangles with three different sides.
- **b)** Exactly two of the parameters are equal. This is a two-parameter family of orbits corresponding to isosceles triangles.
- **c)** Three parameters are equal. The triangles are equilateral. In this case there exists a one-parameter family of orbits.

ii) Among the three invariants λ, μ, ν exactly one is zero. Then the 3 points are collinear.

- **a)** The 3 invariants have different values. There is a two-parameter family of orbits corresponding to three points on a line with different distances between them.

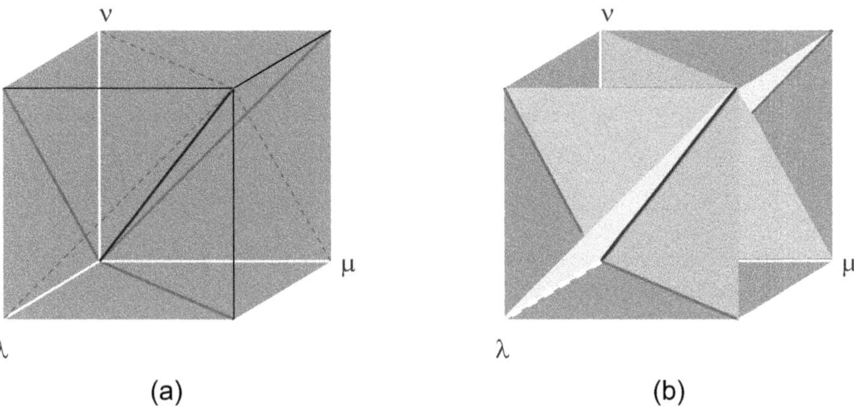

FIG. 2.15 – Orbifold for Eu_n action on $\mathcal{E}_n^{\times 3}$ represented in the space of λ, μ, ν parameters. Two images (a) and (b) are given in order to see better the stratification. The boundary is shown by a shaded area in subfigure (a). It corresponds to orbits of three aligned points with different distances between them. The ray $\lambda = \mu = \nu$ corresponds to one-parameter family of equilateral triangles. Three rays $\lambda = \nu, \lambda = \mu$, $\mu = \nu$ correspond to a one-parameter family of orbits associated with three points on a line with equal distance between them. Three internal differently shaded planes shown in subfigure (b) correspond to a two-parameter family of isosceles triangles.

b) Two invariants are equal and positive, while the third is zero. This means that one point is the midpoint of the segment formed by the other two. There is a one-parameter family of such orbits.

We shall determine the stabilizers of orbits in the 2- and 3-dimensional cases.

Orbits corresponding to a generic triangle have a trivial stabilizer in the 2D-case and a C_h stabilizer in the 3D-case. The symmetry transformation leaving a generic triangle in the 3D-space invariant includes reflection in the plane of a triangle.

Orbits corresponding to isosceles triangles have the stabilizer $\mathbb{Z}_2$ in the two-dimensional case. This group is generated by reflection through the symmetry axis of the triangle. In the three-dimensional case the stabilizer of the isosceles triangles is the C_{2v} group generated by two reflections (in the plane of the triangle and in the plane orthogonal to the triangle and passing through the symmetry axis of the triangle).

For equilateral triangles the stabilizers are respectively (see Figure 2.16) a group of permutation of three objects $S_3 \equiv D_3$ and D_{3h}.

Three points on a line with different distances possess in the 2D-case only one non-trivial symmetry transformation leaving these configuration of points invariant, namely reflection in that line. In the three dimensional case this

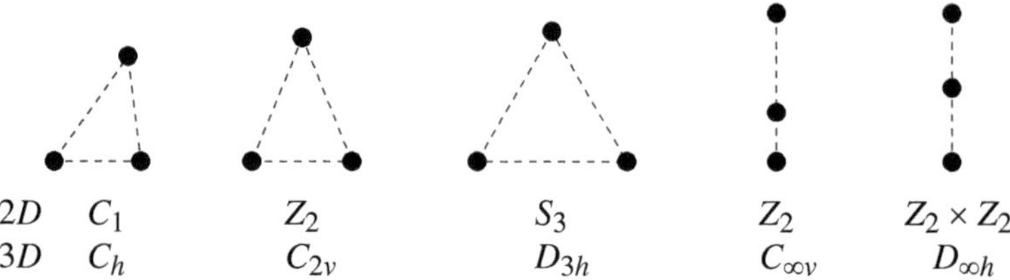

FIG. 2.16 – Point configurations for different orbits of Eu_n action on $\mathcal{E}_n^{\times 3}$ together with their stabilizers for 2D- and 3D-cases. See text for details.

configuration of points is invariant with respect to any rotation around the line and any reflection in planes passing through this line. This group is known as the $O(2)$ group of orthogonal transformations or $C_{\infty v}$ - the three dimensional point symmetry group.

At last, three points equally spaced on the line have in the two-dimensional case the stabilizer $\mathbb{Z}_2 \times \mathbb{Z}_2$ generated by in line reflection and reflection in the orthogonal line. In the three-dimensional case the stabilizer is $D_{\infty h} \equiv O(2) \times \mathbb{Z}_2$.

To summarize, in the case of two-dimensional space we have found 4 strata:

- the minimal one (trivial stabilizer), which corresponds to generic triangles; its dimension is six;
- the unique strata above it (stabilizer $\sim \mathbb{Z}_2$), which contains the orbits of the same type for two different kinds of geometric objects, cases i-b) and ii-a); both components of this stratum have a dimension of five;
- two maximal strata, i-c) (equilateral triangles) and ii-b) (equidistant points on a line) with stabilizers isomorphic to $\mathcal{S}_3$ and $\mathbb{Z}_2^2$ respectively. Both these maximal strata have a dimension of four.

In the $n = 3$ case and even in any $n \geq 3$ space, there are the same five different geometric arrangements of three non-equal points. The difference with the $n = 2$ case consists in the following fact. Now all five arrangements have different stabilizers and consequently there are five strata. In the three dimensional case the stabilizers are

- The three invariants have different values. The stabilizer is $C_s = \mathbb{Z}_2$ - reflection in the plane of triangle. The dimension of the C_s stratum is nine.
- Exactly two of the parameters are equal. The stabilizer is the $C_{2v} = \mathbb{Z}_2 \times \mathbb{Z}_2$ group including C_2 rotation around the bissectrisse (symmetry axis) of the triangle, reflection in plane of the triangle, and reflection in the plane orthogonal to the triangle and including the C_2 axis. The dimension of the C_{2v} stratum is eight.

- Three invariants are equal. The triangle is equilateral. The stabilizer is the D_{3h} group or $\mathcal{S}_3 \times \mathbb{Z}_2$. The dimension of the D_{3h} stratum is seven.
- One of invariant is zero, two other are non-zero and different. The stabilizer is the $O(2) = C_{\infty v}$ group, the continuous group of rotations and reflections around the line going through three points. The dimension of the $C_{\infty v}$ stratum is seven.
- One invariant is zero, two others are equal and non-zero. The stabilizer is $O(2) \times \mathbb{Z}_2 = D_{\infty h}$. The dimension of the $D_{\infty h}$ stratum is six.

Chapter 3

Delone sets and periodic lattices

3.1 Delone sets

We begin our study of lattices in a more general setting.

In the 1930s B.N. Delone (Delaunay) and his colleagues in Moscow began a long-term project of reconstructing mathematical crystallography from the bottom up. The family of point sets we now call Delone sets was their principal tool. The Delone school called them (r, R) systems but after Delone's death in 1980 they were renamed to honor him.

Delone sets are used to model very different phases of matter, from gases to liquids, glasses, quasicrystals and periodic crystals, and the differences are instructive. Delone sets are characterized by two simple but surprisingly powerful postulates inspired by physics: a "hard-core" condition – two atoms cannot overlap; and a "homogeneity" requirement – atoms are distributed more or less homogeneously throughout the medium.

The mathematical setting for Delone sets is a real orthogonal space E_n, by which we mean a vector space V_n endowed with a positive definite scalar product $(\vec{x}, \vec{y})$. We associate to E_n a principal orbit of its translation group; we call this orbit a Euclidean space $\mathcal{E}_n$. (For the definition of "orbit" and related group-theoretic concepts, see Chapter 2.) We choose an origin in $\mathcal{E}_n$ arbitrarily and label it o.

The length of a vector $\overrightarrow{ox} \in E_n$ is the square root of the scalar product $(\vec{x}, \vec{x})$; its squared length is the *norm* $N(\vec{x})$ of $\vec{x}$ (we abbreviate $\overrightarrow{ox}$ to $\vec{x}$). Obviously $N(\vec{x}) < N(\vec{y})$ if and only if $\vec{x}$ is shorter than $\vec{y}$.

The distance between two points x and y of $\mathcal{E}_n$ is the length of the vector $\overrightarrow{x - y} \in E_n$, which is the square root of $N(\overrightarrow{x - y})$.

In this abstract setting the "hard-core" and "homogeneity" conditions translate into axioms: there must be a minimal distance r_0 between any two points of a Delone set Λ, and the radius of a sphere containing *no* points of Λ cannot exceed a fixed positive number R_0.

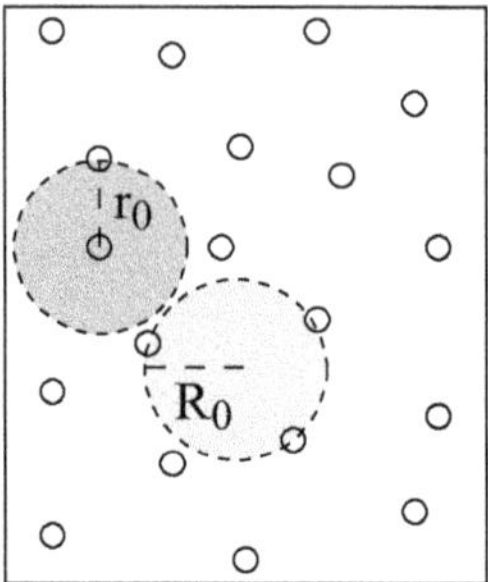

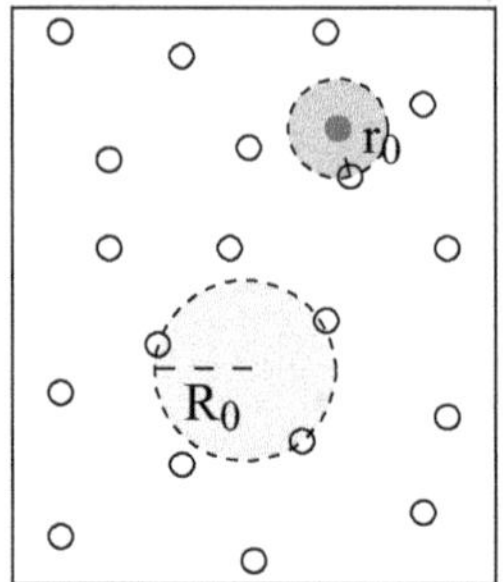

FIG. 3.1 – A portion of a two-dimensional Delone set. The parameter r_0 is the minimal distance between two points of the set; R_0 is the maximum for the radius of any empty hole in the set. Delone set on the right subfigure differs from the Delone set on the left subfigure by adding one extra point (marked by the black dot). This local effect results in drastically decreasing of the r_0 parameter.

That is, we define:

Definition: uniformly discrete A point set $\Lambda \subset \mathcal{E}_n$ is uniformly discrete if there is an $r_0 > 0$ such that every open ball of radius r_0 contains at most one point of Λ.

Definition: relatively dense A point set $\Lambda \subset \mathcal{E}_n$ is said to be relatively dense (in $\mathcal{E}_n$) if there is an $R_0 > 0$ such that every closed ball of radius R_0 contains at least one point of Λ.

With this terminology we say:

Definition: Delone set An n-dimensional Delone set is a point set $\Lambda \subset \mathcal{E}_n$ that is uniformly discrete and relatively dense in $\mathcal{E}_n$.

Note that r_0 can be less than, equal, or greater than R_0. For an example of the case of $r_0 > R_0$, note that the Euclidean plane can be tiled (that is, it can be covered without gaps or overlaps) by congruent equilateral triangles of edge-length 1. Let Λ be the set of vertices of this tiling. Then Λ is a Delone set with parameters $r_0 = 1$ and $R_0 = 1/\sqrt{3}$, the radius of the circle circumscribing any triangle.

For a Delone set of dimension $n > 1$, the minimal ratio R_0/r_0 is the ratio of the radius of the sphere circumscribing a regular n-simplex to the length of an edge; the formula is

$$\frac{R_0}{r_0} = \sqrt{\frac{n}{2(n+1)}}.$$

This formula is easy to prove if we situate the n-dimensional simplex in $(n+1)$-dimensional space. The points $(1, 0, \ldots 0), \ldots (0, \ldots, 0, 1)$ are the vertices of a regular n-dimensional simplex in the hyperplane $x_1 + \cdots + x_{n+1} = 1$; their

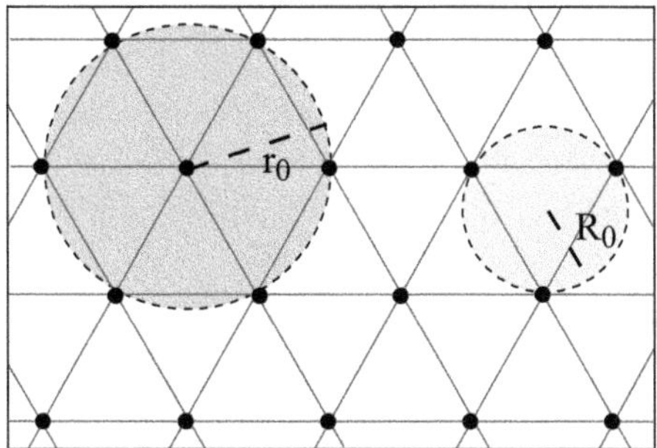

FIG. 3.2 – The vertices of a tiling of the plane by equilateral triangles is a Delone set.

barycenter is $\left(\frac{1}{n+1}, \frac{1}{n+1}, \ldots, \frac{1}{n+1}\right)$. Thus in every dimension n we have

$$1/2 \leq \frac{R_0}{r_0} < \frac{1}{\sqrt{2}},$$

with the ratio approaching the upper bound as $n \to \infty$.

Proposition 1 *A Delone set Λ is countably infinite.*

Proof. Λ is infinite, otherwise all of its points would lie in some half-space, contradicting relative density. Countability follows from uniform discreteness: E^n can be partitioned into a countable number of unit cubes and, since a unit cube can contain only a finite number of balls of radius r_0, there is only a finite number of points of Λ in each cube. □

To study Delone sets we begin, as Delone did, with the "method of the empty sphere." Consider an n-dimensional sphere S in $\mathcal{E}_n$ which contains no points of Λ in its interior. S may, or may not, have points of Λ on its boundary. If these points – that is, the set $S \cap \Lambda$ – lie in an $(n-1)$-dimensional subspace, then as we increase the radius of S it will remain empty. Indeed, by moving the sphere if needed, we can increase its radius until $S \cap \Lambda$ contains $n+1$ linearly independent points. (We say that $n+1$ points of $\mathcal{E}_n$ are linearly independent if they span $\mathcal{E}_n$.)

Definition: hole (of a Delone set) An empty hole or, more simply, a hole (of $\mathcal{E}_n$ with respect to a Delone set Λ) is a sphere S with no points of Λ in its interior and at least $n+1$ linearly independent points of Λ on its boundary.

The maximal radius of an empty hole of $\mathcal{E}_n$ (with respect to Λ) is the parameter R_0 of Λ.

Proposition 2 *Let Λ be a Delone set. We can cover $\mathcal{E}_n$ by closed balls containing $n+1$ independent points of Λ on their boundaries and no points of Λ in their interiors.*

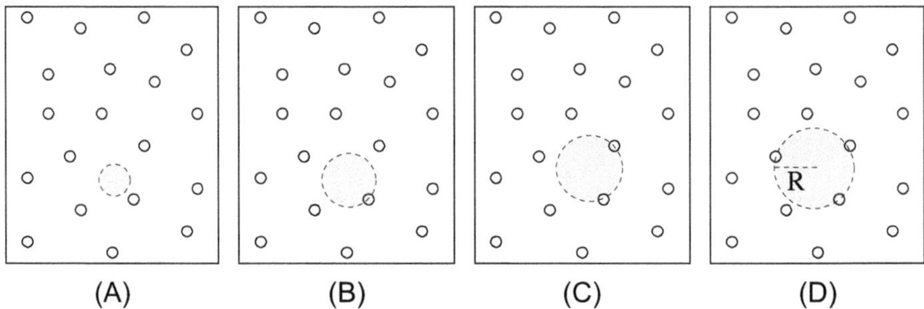

FIG. 3.3 – Consecutive steps in the construction of an empty hole for a two-dimensional Delone set. (A) A sphere S containing no points of the Delone set, Λ. (B) Increase the radius of the sphere untill one point appears on its boundary (i.e., until $S \cap \Lambda$ spans a zero-dimensional subspace). (C) Increase the radius of the sphere keeping one point on its boundary untill the second point appears on its boundary (now $S \cap \Lambda$ consists of two points and spans a one-dimensional subspace). D - Increase the radius of the sphere keeping the two points on it untill the third point appears on its boundary (now $S \cap \Lambda$ consists of three points and spans the plane). The resulting sphere is an empty hole of radius $R \leq R_0$, where R_0 is the parameter of the Delone set.

Proof. The proof uses the method of the empty sphere. Let x be any point of $\mathcal{E}_n$; we will show that it lies in at least one such sphere. Let p be a point of Λ at minimal distance r from x. Then

$$r = |\overrightarrow{px}| \leq R_0.$$

Let $B_x(r)$ be the sphere of radius r with the point x as its center, and suppose that it contains fewer than $n+1$ independent points Λ. Then $B_x(r) \cap \Lambda$ lies in a hyperplane H of dimension $d \leq n-1$. Leaving x fixed, we can expand the sphere along the $(n-d)$-dimensional subspace orthogonal to H until it encounters a point of Λ independent of those in H. We continue this process until the sphere contains $n+1$ independent points. □

The convex hull of the points of Λ on the boundary of a hole H is a polytope L_H, called a *Delone polytope.* (This terminology and notation follows the Russian tradition.)

We will show in Chapter 5 that just as $\mathcal{E}_n$ is covered by the holes of Λ, it is tiled by the Delone polytopes $\{L\}$ of its holes: that is, the Delone polytopes of Λ fit together with no gaps or overlaps.

The Delone polytopes show the empty spaces of Λ in $\mathcal{E}_n$; another construction, called the Voronoï construction, focuses attention instead on the regions "belonging" to the points of Λ.

Definition: Voronoï cell The Voronoï cell $\mathsf{D}(p)$ of $p \in \Lambda$ is the set of points $x \in E^n$ which are at least as close to p as to any other point of Λ.

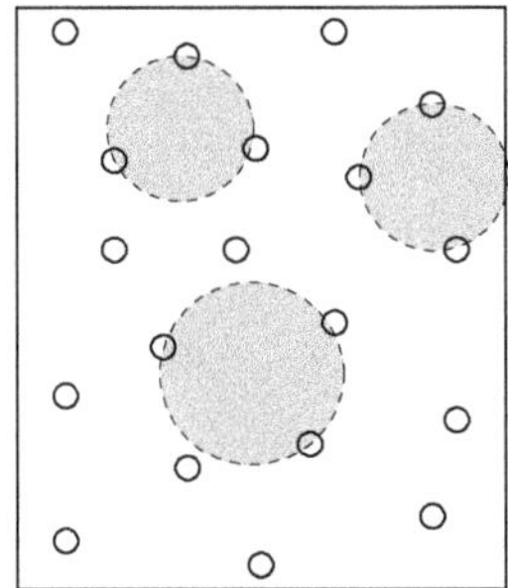
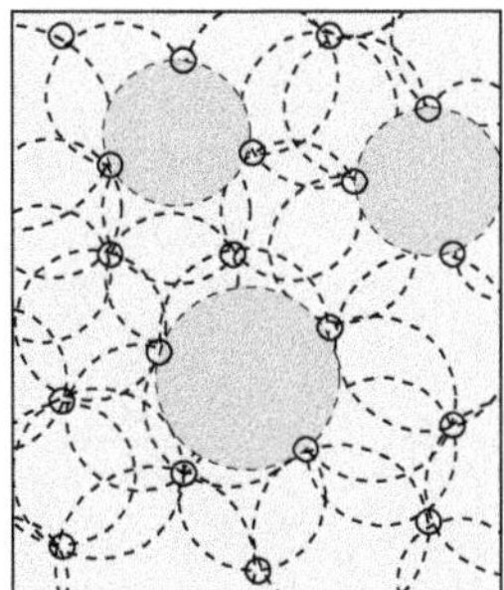

FIG. 3.4 – A covering of a Delone set represented in Figure 3.1 by holes. Left: Arbitrarily chosen three initial holes. Right: Complete set of overlapping holes.

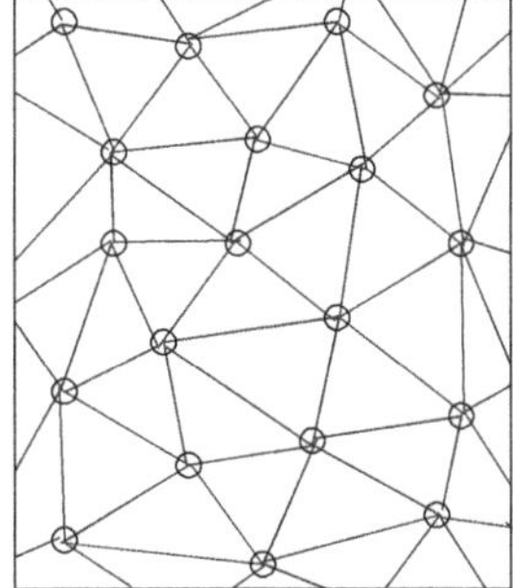

FIG. 3.5 – A Delone set of point with Delone polygons drawn in.

That is,

$$\mathsf{D}(p) = \{x \in E^n | N(x-p) \leq N(x-q), \forall q \in \Lambda\}. \tag{3.1}$$

Voronoï cells – which appear in many contexts and variations – are evidently very old. In 1644, Descartes used what appears to be a variant to describe the structure of the heavens [40] but he did not bother to explain it. The construction first appeared in mathematics in 1850 in the context of the arithmetic theory of quadratic forms; Dirichlet proved that the cells of two dimensional lattices are either rectangles or centrosymmetric hexagons (see [45]). This is why Voronoï cells are also known as Dirichlet domains (as well as by other names, since the construction has been rediscovered many times). We call them Voronoï cells because Voronoï performed the first deep study of their properties for point lattices in an arbitrary dimension n (see [94]), but we denote them by the letter D in honor of Dirichlet's contribution.

To construct the Voronoï cell of a point $p \in \Lambda$, we note that, for any point $q \in \Lambda$, the hyperplane orthogonally bisecting the vector $\overrightarrow{qp}$ divides $\mathcal{E}_n$ into two half spaces, one of them the set of points in $x \in \mathcal{E}_n$ for which

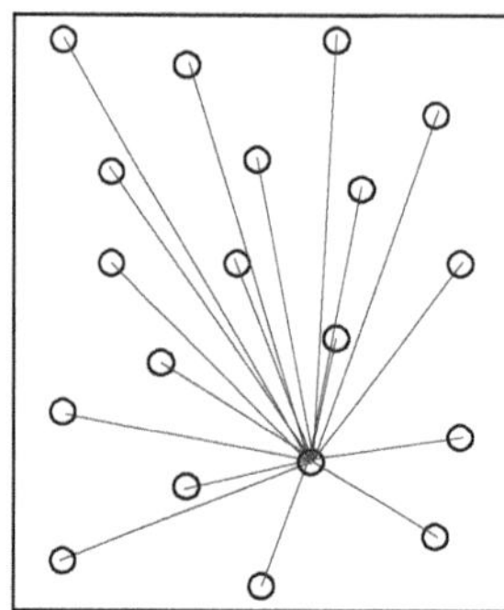 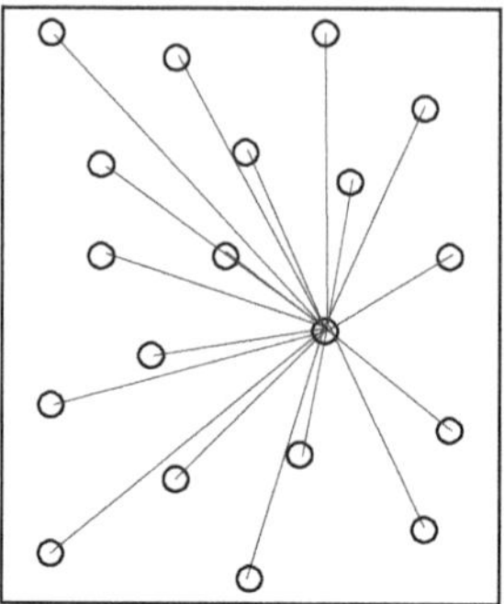

FIG. 3.6 – Two stars of two different points for a Delone set shown in Figure 3.1. Only "arms" between a chosen point and points of the Delone set represented on a fragment are shown. (The star is partial because the figure is finite.)

$N(x-q) < N(x-p)$, and the other the xs for which $N(x-p) < N(x-q)$. Points lying on the bisecting hyperplane are equidistant from p and q.

Next we define:

Definition: global star (of a point of a Delone set) The global star $ST_p(\Lambda)$ of a point p of a Delone set Λ is the configuration of line segments obtained by joining p to all of the other points of Λ.

Since Λ is countable, the star has a countable number of "arms".

To construct $\mathsf{D}(p)$, we orthogonally bisect the arms of $ST_p(\Lambda)$ by $(n-1)$-dimensional hyperplanes. Then $\mathsf{D}(p)$ is the smallest polytope about p bounded by such hyperplanes.

Fortunately it is not necessary to bisect a countable infinity of line segments to construct $\mathsf{D}(p)$:

Theorem 1 *Let Λ be a Delone set with parameters r_0 and R_0. The Voronoï cell of any point $p \in \Lambda$ is contained in the ball $B_p(R_0)$.*

Proof. Assume $\mathsf{D}(p) \not\subset B_p(R_0)$, i.e., that $\exists x \in \mathsf{D}(p)$ such that $|\overrightarrow{xp}| > R_0$. Then $B_x(|\overrightarrow{xp}|) \cap \Lambda = \emptyset$, since x is nearer to p than to any other point of Λ. But this contradicts the assumption that R_0 is the maximum radius of an empty hole. □

This means that $\mathsf{D}(p)$ is completely determined by a *finite* set of vectors issuing from p, all of length $\leq 2R_0$. To say this concisely, we define

Definition: local star (of a point of a Delone set) The r-star $ST_p(\Lambda, r)$ of a point p of a Delone set Λ is the configuration of line segments obtained by joining p to all of the other points of Λ that lie within a sphere about p of radius r: $ST_p(\Lambda, r) = ST_p(\Lambda) \cap B_p(r)$.

Thus

Corollary 1 *$\mathsf{D}(p)$ is completely determined by the $2R_0$-star of p, $ST_p(\Lambda, 2R_0)$.*

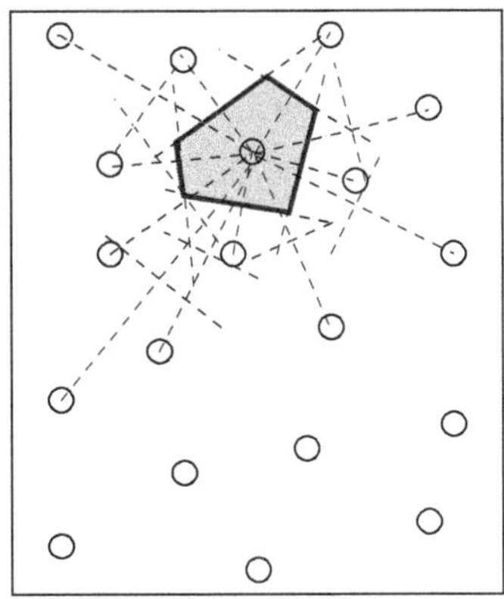

FIG. 3.7 – Constructing the Voronoï cell of a point of the Delone set represented in Figure 3.1.

Proof. By Theorem 1, all the points of Λ contributing faces to $\mathsf{D}(p)$ must lie in $B_p(2R_0)$. □

Voronoï cells play a large role in lattice theory, as we will see in Chapter 5. We note here (but will prove there, proposition 11) that the Voronoï cells of the points of Λ also tile E^n, and this Voronoï tiling is orthogonally dual to its Delone tiling. (That is, each k-dimensional face of one tiling corresponds to an $(n-k)$-dimensional face of the other, and the corresponding faces are orthogonal.) In particular, each edge (1-face) of a Delone tile is orthogonal to a facet ($(n-1)$-face)of a Voronoï cell. We will also see that the vertices of the Voronoï cells of Λ are the centers of its holes.

3.2 Lattices

We denote the number of congruence classes of stars of a Delone set Λ by $|ST(\Lambda)|$. This number is a very rough measure of the randomness of Λ. Thus if we are using Λ to model the set of centers of atoms in a gas or liquid (distributed homogeneously in infinite space), we would expect the number of congruence classes to be countably infinite; that is, $|ST(\Lambda)| = \aleph_0$. If on the other hand $|ST(\Lambda)|$ is finite, then the Delone set is highly ordered. In this case Λ is said to be *multiregular*. Some authors (see [46]) call a multiregular Delone set an *ideal crystal*, because (one can prove that) it is a union of a finite number of orbits of a "crystallographic" group (for more on these groups, see below).

A *regular system of points* is the special case of a multiregular Delone set when all stars are congruent:

Definition: regular system of points The Delone set Λ is said to be a regular system of points when $|ST(\Lambda)| = 1$.

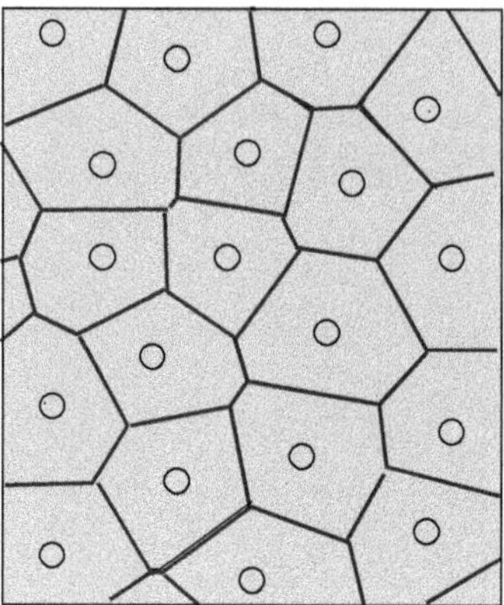

FIG. 3.8 – System of Voronoï cells for the Delone set represented in Figure 3.1.

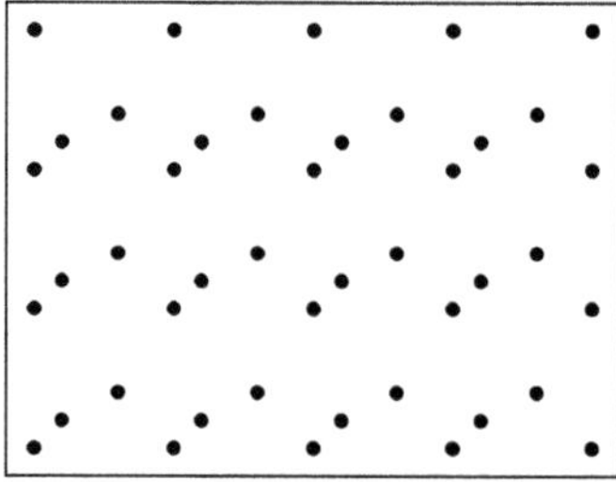

FIG. 3.9 – A multiregular system of points formed by three orbits of the symmetry group; here $|ST(\Lambda)| = 3$.

Delone introduced "r, R systems" in the 1930s to focus crystallographer's attention on local order. In 1976 he and his students Shtogrin, Dolbilin, and Galiulin proved the remarkable fact that global regularity – in the sense of a regular system of points – is a consequence of local regularity: a Delone set is a regular system of points if all its *local* stars of a certain radius are congruent.

Theorem 2 *Let Λ be a Delone set in $\mathcal{E}^n$ with parameters r_0 and R_0. There exists a $C = C(R_0/r_0, n) > 0$ such that if $r > CR_0$ and $|ST(\Lambda, r)| = 1$, then Λ is a regular system of points.*

For a proof see [43] and [46].

In two dimensions, $C = 4$; the exact value of the constant C has not been determined for Delone sets of any higher dimension. There is an analogous result for multiregular Delone sets [46].

The symmetry group of a regular system of points in E^n is still called a *crystallographic group* for historical reasons, though today the definition of "crystal" has been broadened to include non-periodic crystals. In 1910

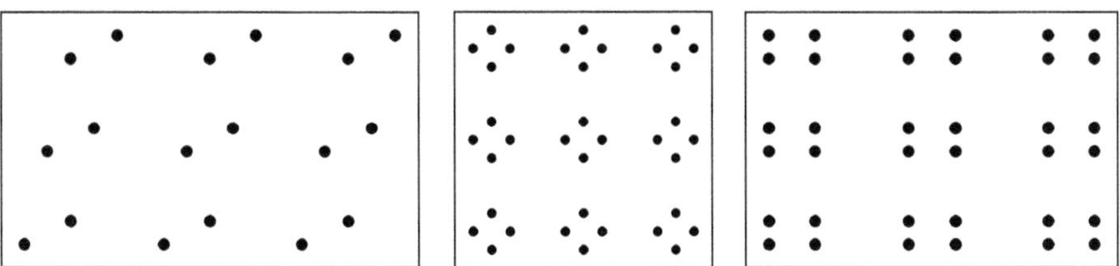

FIG. 3.10 – Three examples of regular systems of points which are more general than a lattice.

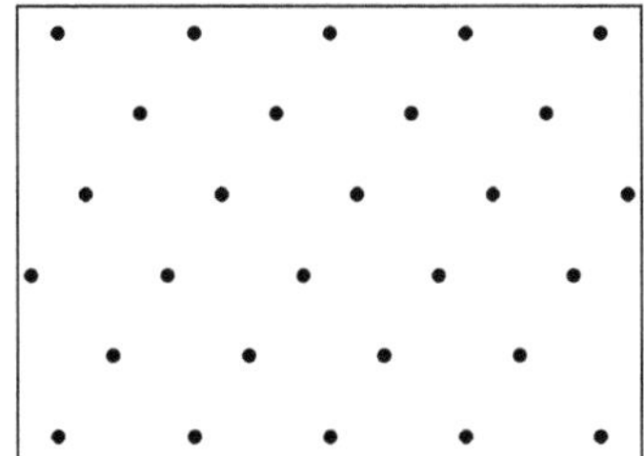

FIG. 3.11 – A point lattice in the plane.

addressing the 18th problem on Hilbert's famous list (1900), Ludwig Bieberbach proved that every crystallographic group $G < E^n$ has an invariant subgroup of translations T of rank n [28]. In slightly different words this means that every group of symmetry operations in E^n which acts transitively on a regular point system X contains n linearly independent translations.

Definition: point lattice A point lattice in $\mathcal{E}^n$ is a regular system of points whose stars are orbits of a rank-n translation group $T \subset E^n$.

Because a point lattice is an orbit of a translation group, the Voronoï cells of its points are congruent polytopes that tile $\mathcal{E}^n$ by translation; the technical term for polytopes with that property is *parallelotope*.

In group theory, the word "lattice" is also used for the translation subgroup T of which the Delone set is an orbit. Thus an n-dimensional lattice is any subgroup of a real vector space V_n that is isomorphic to $\mathbb{Z}^n$.

Considering a lattice L as a Delone set, we have $r_0 = d(L)$, where $d(L)$ is the length of the shortest vector in the lattice.

A point lattice can also be defined as an orbit of a crystallographic group with stabilizer of maximal symmetry. We will discuss lattices from this point of view in Chapter 4.

Definition: basis A basis for an n-dimensional lattice L is any set of n vectors $\{\vec{b}_j\} \subset L$, $1 \leq j \leq n$ such that every vector in L is an integral linear combination of the vectors $\vec{b}_j$.

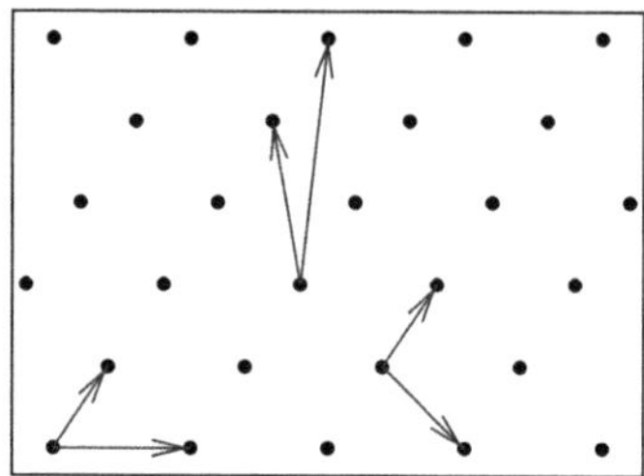

FIG. 3.12 – Three different bases of the point lattice represented in Figure 3.11.

That is, with respect to a given basis $\{\vec{b_j}\}$ the lattice vectors have integral coordinates:

$$\forall \vec{\ell} \in L, \quad \vec{\ell} = \sum_j k_j \vec{b}_j, \quad k_j \in \mathbb{Z}. \tag{3.2}$$

The *determinant* of the vectors of a basis is the oriented volume of the parallelepiped built on it. (We will see in Chapter 5 that this is also the volume of a Voronoï cell of the lattice.)

The basis of a lattice L is not unique: any set of n vectors in L with determinant ± 1 is a basis.

Let $\{\vec{b}'_i\}$ be another basis for L and m_{ij} the coordinates of the vector $\vec{b}'_i$ in the basis $\{\vec{b}_j\}$:

$$\vec{b}'_i = \sum_j m_{ij} \vec{b}_j, \quad m_{ij} \in \mathbb{Z}. \tag{3.3}$$

Since every basis has determinant ± 1, the integers m_{ij} are the elements of a unimodular integral matrix A. Similarly the components of the vectors $\{\vec{b}_j\}$ in the basis $\{\vec{b}'_i\}$ form the matrix A^{-1} which is also integral. Thus $A \in GL_n(\mathbb{Z})$, the group of $n \times n$ integral matrices.

Each matrix $A \in GL_n(\mathbb{Z})$ corresponds to a basis in L and left multiplication by elements of $GL_n(\mathbb{Z})$ maps each basis to the others. Thus

Proposition 3 *The set of bases of a lattice L is an orbit of* $GL_n(\mathbb{Z})$.

For specificity and for computation, it is useful to work with a specific representative of this conjugacy class. As we will see in later chapters, the various methods of classifying lattices are all concerned with this problem.

The elements of V_n/L, the quotient group of the vector space V_n by the lattice, are identified with the cosets $\vec{x}+L$, $\vec{x} \in V_n$. A choice of representatives of each of these cosets constitutes a *fundamental domain* of the translation group T. For example, the interior of the parallelepiped formed by any set of k basis vectors is a fundamental domain for T. Or, given a basis $\{\vec{b}_i\}$, one can choose as fundamental domain

$$\mathcal{P}_b = \{\vec{x} = \sum_i \xi_i \vec{b}_i, \ -1/2 \leq \xi_i < 1/2\}. \tag{3.4}$$

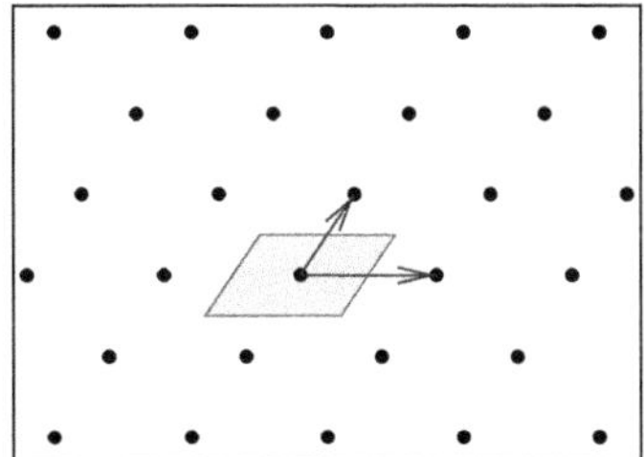

FIG. 3.13 – A fundamental domain for the lattice of Figure 3.11. The choice of fundamental domain follows equation (3.4).

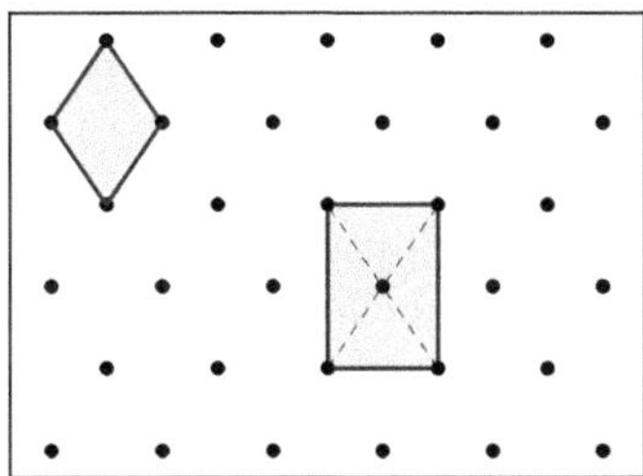

FIG. 3.14 – Primitive and non-primitive cells for the lattice of Figure 3.11. The volume of the non-primitive cell is twice as large as the volume of the primitive cell.

The topological closure $\overline{\mathcal{P}}_b$ of $\mathcal{P}_b$ is obtained by replacing $<$ by $\leq$ in (3.4). $\overline{\mathcal{P}}_b$ is the translate, by the vector $\vec{v} = -(\frac{1}{2}, \ldots, \frac{1}{2})$, of a parallelepiped that in crystallography is called a *primitive unit cell.*

When its opposite faces are identified, $\overline{\mathcal{P}}_b$ becomes a torus; Thurston and Conway made this property the basis of their "orbifold notation", which we describe in Chapter 4.

A primitive unit cell has lattice points only at its vertices. Crystallographers often prefer to work with non-primitive cells (unions of two or more primitive cells) to maximize symmetry, but we mainly use primitive cells in this book.

3.3 Sublattices of L

A sublattice is a lattice L' which is a subset of another lattice L (a subgroup if we are speaking of groups).

If L is one-dimensional, then it has one generator $\vec{a}$, and $L = \{j\vec{a}, j \in \mathbb{Z}\}$. Any sublattice of L has one generator too, say $m\vec{a}$; the sublattice is the set mL, and the quotient $\frac{L}{L'}$ is the cyclic group of order m.

Every lattice L, of any dimensions, has sublattices consisting of the vectors $m\vec{\ell}$, $\vec{\ell} \in L$, $1 < m \in \mathbb{Z}$. The quotient group L/mL has m^n elements and its automorphism group is $GL_n(\mathbb{Z}/m\mathbb{Z})$. As we shall see in Chapter 5, the case $m = 2$ plays a key role in the theory of Voronoï cells.

Most sublattices of L are not of this special type. For example, if the vectors $\vec{b}_1, \ldots, \vec{b}_n$ are a basis for L, then $\vec{b}_1, \ldots, \vec{b}_{n-1}, 2\vec{b}_n$ generate a sublattice of index 2.

For lattices of dimension $n > 1$ we distinguish two types of sublattices: those for which the dimension is also n, and those of lower dimension. Subgroups of the first type are of finite index and the corresponding quotient is a finite group; we consider them first. The ratio $volL/volL'$ is the index of L' in L.

Each sublattice L' of finite index of an n-dimensional lattice L is characterized by an integer matrix A', whose columns are the coordinates of its basis. L' has, of course, a countable infinity of bases and thus is described by a conjugacy class of matrices under the action of $GL_n(\mathbb{Z})$. Again, it is convenient to select a basis; that is, to work with a specific representative of this conjugacy class. The Hermitian normal form serves our purposes here.

A matrix is in *Hermite normal form* if it is upper triangular, all matrix elements are non-negative, and each column has a unique maximum entry, which is on the main diagonal.[1] For example, the matrix

$$\begin{pmatrix} 3 & 1 & 3 \\ 0 & 4 & 5 \\ 0 & 0 & 7 \end{pmatrix}$$

is in Hermitian normal form.

Any integer matrix can be transformed to the Hermitian normal form by left multiplication by a unimodular integer matrix. The form is unique in its conjugacy class. Thus it identifies the sublattice. We call the columns of the Hermitian matrix the sublattice's *Hermitian basis.*

Figure 3.15 illustrates different choices of sublattices of index two and three. Basis vectors for these sublattices corresponding to the Hermitian normal form are respectively:

$$a: \left\{\begin{pmatrix} 1 \\ 0 \end{pmatrix}, \begin{pmatrix} 0 \\ 2 \end{pmatrix}\right\};\ b: \left\{\begin{pmatrix} 2 \\ 0 \end{pmatrix}, \begin{pmatrix} 0 \\ 1 \end{pmatrix}\right\};\ c: \left\{\begin{pmatrix} 1 \\ 0 \end{pmatrix}, \begin{pmatrix} 1 \\ 2 \end{pmatrix}\right\}; \tag{3.5}$$

$$d: \left\{\begin{pmatrix} 1 \\ 0 \end{pmatrix}, \begin{pmatrix} 0 \\ 3 \end{pmatrix}\right\};\ e: \left\{\begin{pmatrix} 3 \\ 0 \end{pmatrix}, \begin{pmatrix} 0 \\ 1 \end{pmatrix}\right\};\ f: \left\{\begin{pmatrix} 1 \\ 0 \end{pmatrix}, \begin{pmatrix} 1 \\ 3 \end{pmatrix}\right\};\ g: \left\{\begin{pmatrix} 1 \\ 0 \end{pmatrix}, \begin{pmatrix} 2 \\ 3 \end{pmatrix}\right\}.$$

What is the number of distinct sublattices of a lattice L of a given index m, and how can we describe their bases explicitly? The answer to both questions is to list the $n \times n$ Hermitian normal forms of determinant m.

Consider, for example, the case where m is a prime p. Since the determinant of a triangular matrix is the product of its diagonal entries, if the index is a prime p, one diagonal entry must be p and the others 1. By the definition

[1] Some definitions specify lower triangular matrices; either can be transformed into the other. For more on this, and how the transformation is effected, see [16].

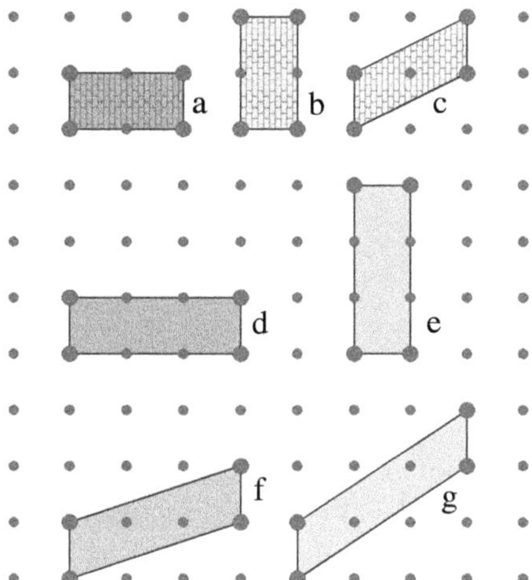

FIG. 3.15 – L, the square lattice in the plane, is shown here with the "Hermitian choice" of basis vectors (3.5) for its three sublattices (a, b, c) of index 2 and four sublattices (d, e, f, g) of index 3.

of Hermitian form, all the entries above each 1 must be 0; thus we need to fill in only the column containing p. Suppose, for example, that $n = 3$ and $k_{22} = 5 = p$. Then the (single) entry above 5 can be any of $0, 1, 2, 3, 4$.

Considering all possible positions for p, we see that the complete number of different $n \times n$ Hermite normal matrices with prime determinant p is

$$1 + p + p^2 + p^3 + \cdots + p^{n-1} = \frac{p^n - 1}{p - 1}. \tag{3.6}$$

We immediately have the useful corollary that the number of sublattices of index 2 for an n-dimensional lattice is $2^n - 1$. Otherwise, for the two-dimensional lattice the number of sublattices of index p, with p being prime, is $p+1$. See figure 3.17 for an explicit example of three sublattices, D_2^r, D_2^-, D_2^+, of index 2 for the D_2^ω lattice.

If the index m is not prime, we first find all factorizations of m into primes, and then calculate the number of different choices for the off-diagonal elements for each diagonal pattern.

Figure 3.15 suggests that "distinct" sublattices may or may not be of the same "type." This raises the question of equivalence of lattices (and sublattices), and questions of symmetry. We turn to them in Chapter 4.

To conclude this subsection, we mention briefly sublattices of L that are not of finite index.

The intersection of L with an arbitrary j-dimensional subspace of V_n spans a vector subspace of dimension $j' \leq j$. In general, $j' < j$; it is useful to give a name to the subspaces V_j such that $j' = j$.

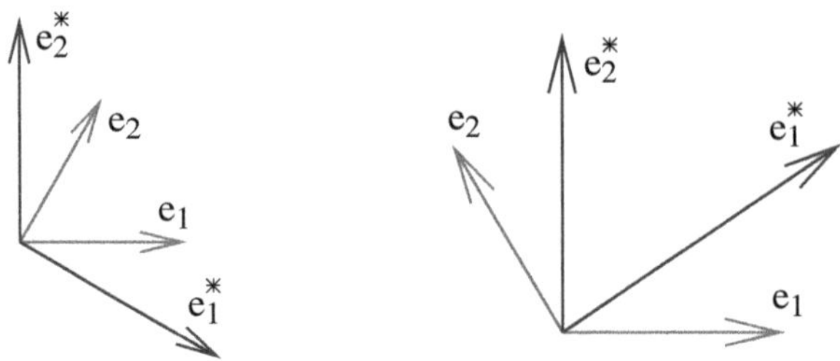

FIG. 3.16 – Examples of construction of dual bases.

Definition: j-plane of a lattice. The vector subspace $V_j < V_n$ is called a j-plane of the lattice L if $L \cap V_j$ is a j-dimensional lattice.

Definition: j-sublattice. An L-subgroup isomorphic to $\mathbb{Z}^j$, $j < n$, is called a j-sublattice if it is the intersection of L by a lattice j-plane.

A good algorithm for studying the sublattices of L and, for any pair of sublattices, their intersection and the sublattice they generate, is given in [39].

3.4 Dual lattices.

The scalar product allows us to define duality between lattices of the same rank in the same vector space. The lattices L and L^* are said to be dual if the scalar product $(\vec{\ell}, \vec{\ell^*})$ of any pair of vectors, one from each lattice, is an integer.

Definition: dual lattice. The dual lattice L^* of the lattice L is defined by $\{\vec{y} \in E_n,\ \forall \vec{\ell} \in L,\ (\vec{y}, \vec{\ell}) \in \mathbb{Z}\}$.

Properties of dual lattices following directly from the definition include:

1. $L^{**} = L$.
2. If $B = \{\vec{b}_j\}$ is a basis of L, then the vectors $\vec{b}_i^*$, $i = 1, \dots, k$ satisfying $(\vec{b}_i^*, \vec{b}_j) = \delta_{ij}$ are a basis B^* for L^*;
3. $B^* = (B^{-1})^\top$;
4. $\mathrm{vol}(L)\mathrm{vol}(L^*) = 1$;
5. If L is a sublattice of L', then the dual of L' is a sublattice of the dual of L: $L'^* < L^*$;
6. The quotient groups L'/L and L^*/L'^* are isomorphic.

An interesting particular case of lattices are the:

Definition: integral lattice. Integral lattices are defined by

$$\forall \vec{\ell}, \vec{\ell}' \in L, \quad (\vec{\ell}, \vec{\ell}') \in \mathbb{Z}, \tag{3.7}$$

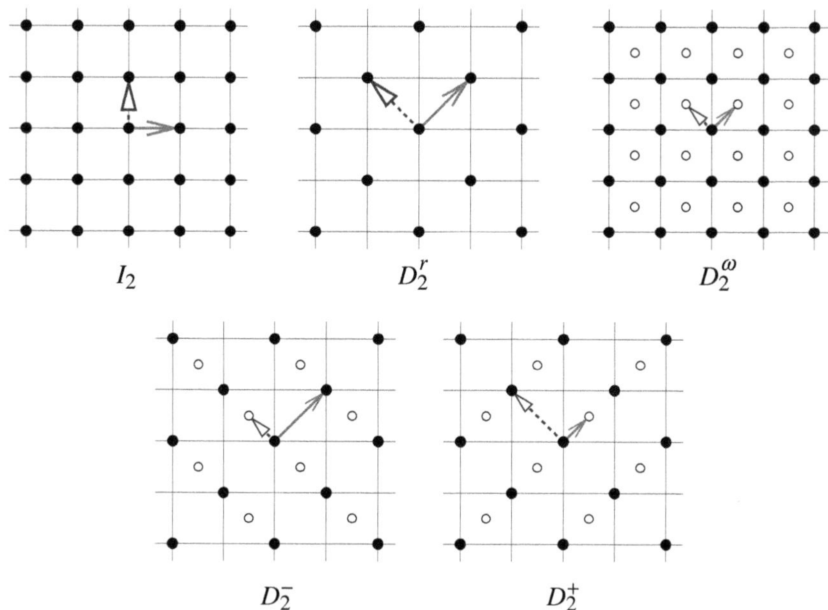

FIG. 3.17 – Examples of two-dimensional lattices. I_2 - simple quadratic lattice. D_2^r - index two sublattice of I_2, defined in (3.10). D_2^ω is the lattice dual to D_2^r. Open circles indicate points added when compared with the I_2 lattice (see eq. (3.11)). $D_2^\pm$ are two intermediate lattices between D_2^r and its dual D_2^ω. D_2^+ is dual to D_2^-. Open circles indicate points added when compared with the D_2^r lattice (see eq. (3.14)). The basis vectors are represented by solid and dash lines in such a way that for dual lattices the scalar product of basis vectors of the same type is equal to 1, and the scalar product of basis vectors of different types is zero.

and the set of integers $(\vec{\ell}, \vec{\ell}')$ is reduced, i.e. they have no common divisor > 1.

From (3.7), a lattice is *integral* if and only if it is contained in its dual. The following relation:

$$L < L' \text{ integral}, \qquad L < L' \leq L'^* < L^*. \tag{3.8}$$

will be very useful. Particular examples of integral lattices are the *self-dual* ones:

Definition: self dual lattice. A lattice L is said to be self-dual if $L = L^*$.

As a consequence of property $\text{vol}(L)\text{vol}(L^*) = 1$, if L is self-dual then $\det(L) = 1$.

Dual lattices play an important role in the physics of x-ray diffraction by crystals; they are the "reciprocal" lattices observed in diffraction diagrams.

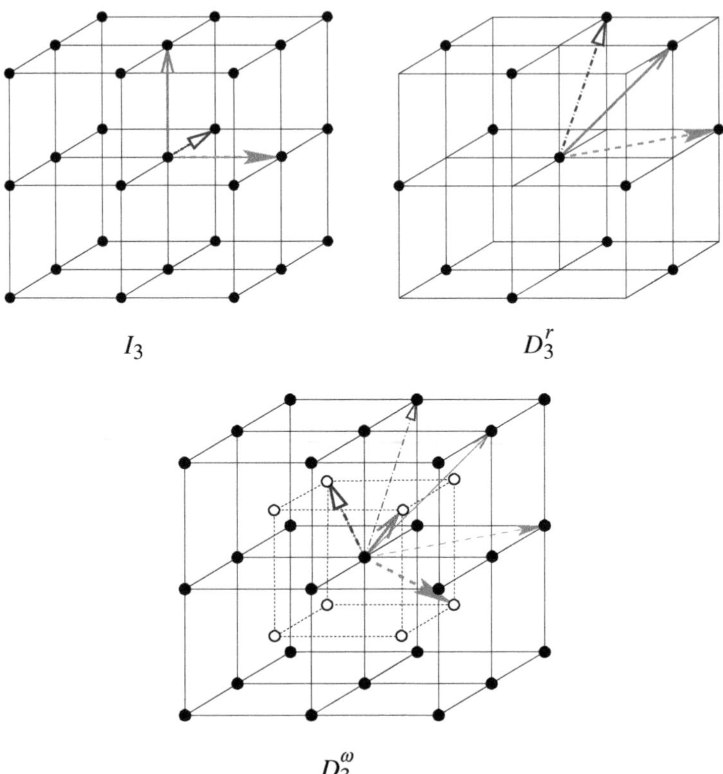

FIG. 3.18 – Examples of three-dimensional lattices. I_3 - simple cubic lattice. D_3^r - index two sublattice of I_3. This sublattice is defined in eq. (3.10). D_3^ω is the lattice dual to D_3^r. Open circles indicate points added when compared with the I_3 lattice (see eq. (3.11)). The basis vectors are represented by solid, dash, and dash-dot lines in such a way that for dual lattices the scalar product of basis vectors of the same type is equal to 1, and the scalar product of basis vectors of different types is zero. Basis vectors of the D_3^r lattice are shown on D_3^ω by thin lines.

Examples of lattices: We give here some examples of dual and self-dual n-dimensional lattices together with their most frequent notations. Figures 3.17, 3.18 illustrate discussed examples in two- and three-dimensional cases.

a) The n-dimensional lattice generated by an orthonormal basis is often denoted by I_n:

$$(\vec{e}_i, \vec{e}_j) = \delta_{ij}, \quad I_n = \sum_i \lambda_i \vec{e}_i, \ \lambda_i \in \mathbb{Z}; \quad \mathrm{vol}(I_n) = 1. \tag{3.9}$$

In crystallography it is called the cubic P lattice. It is self-dual.

b) A sublattice of index 2 of I_n is

$$D_n^r = \left\{ \sum_i \lambda_i \vec{e}_i, \ \sum_i \lambda_i \in 2\mathbb{Z} \right\}; \quad I_n/D_n^r = \mathbb{Z}_2; \quad \mathrm{vol}(D_n^r) = 2. \tag{3.10}$$

Note that D_n^r is an even integral lattice.

c) The dual lattice of D_n^r is usually denoted by D_n^w. With the use of (3.8) we find:

$$D_n^r < I_n < D_n^w := (D_n^r)^* = I_n \cup (\vec{w}_n + I_n), \tag{3.11}$$

with

$$\vec{w}_n = \frac{1}{2} \sum_i \vec{e}_i \quad \mathrm{vol}(D_n^w) = \frac{1}{2}.$$

With the remark that $2w_n \in D_n^r$ when n is even and $2w_n \notin D_n^r$ when n is odd, one easily proves:

$$D_n^w / D_n^r = \begin{cases} \mathbb{Z}_2^2 & \text{when } n \text{ is even,} \\ \mathbb{Z}_4 & \text{when } n \text{ is odd.} \end{cases} \tag{3.12}$$

So when n is even, there must be three intermediate lattices (corresponding to the three subgroups of index 2 of $\mathbb{Z}_2^2$) between D_n^r and its dual. To construct them we define:

$$n \text{ even,} \quad \vec{w}_n^{\pm} = \frac{1}{2} \left(\pm \vec{e}_n + \sum_{i=1}^{n-1} \vec{e}_i \right); \quad N(\vec{w}^{\pm}) = \frac{n}{4}, \quad (\vec{w}_n^+, \vec{w}_n^-) = \frac{n-2}{4}. \tag{3.13}$$

We have seen that I_n is one of these intermediate lattices. The two others are:

$$n \text{ even}: \ D_n^{\pm} = D_n^r \cup (\vec{w}_n^{\pm} + D_n^r); \quad \det(D_n^{\pm}) = 1. \tag{3.14}$$

With the remark that $\forall \vec{\ell} \in D_n^r$, $(\vec{\ell}, \vec{w}^{\pm}) \in \mathbb{Z}$ and equations ((3.13) and (3.14)) one obtains

Proposition 4 *For $n \equiv 0$ mod 4, $D_n^{\pm}$ are self-dual lattices. For $n \equiv 2$ mod 4, they are dual of each other: $D_n^- = (D_n^+)^*$.*

We note (see chapter 7 for more details) that the two lattices $D_{4m}^{\pm}$ are identical and that $D_4^+ = I_4$ and D_8^+ is the remarkable lattice E_8.

Chapter 4

Lattice symmetry

4.1 Introduction

In this chapter we study periodic lattices from the point of view of their symmetry. That is, we describe the different classes of transformations leaving lattices invariant. Depending on the class of allowed transformations the symmetry of lattices will be different and thus symmetry classification can be more or less detailed. For physical applications we choose the classification best suited to the problem.

A related important notion is the equivalence of lattices. We need to specify which two lattices could be considered equivalent and which should be treated as different, and this varies with the type of classification.

For a simple example, let us consider three dimensional physical space as a realization of an abstract Euclidean space $\mathcal{E}_n$ with a chosen basis defining a frame $\mathcal{F}$. This allows us to associate with each point P of the three dimensional space three real numbers x, y, z, the coordinates of the point P in the frame $\mathcal{F}$.

Since $\mathcal{E}_n$ is homogeneous and isotropic, two lattices related by an arbitrary translation or rotation should be considered equivalent (or, simply, to be the same intrinsic lattice). Sometimes simultaneous scaling of the coordinates also can be treated as "uninteresting" and the lattice can be supposed to be normalized, that is, the volume of its primitive parallelepiped (primitive cell) can be chosen to be equal to one.

Obviously the same lattice can be constructed in different frames and the corresponding transformation between different frames can also be treated as a symmetry transformation of the lattice.

4.2 Point symmetry of lattices

Let us start by looking for the groups of orthogonal transformations leaving one lattice point invariant; these are called point groups. (The point symmetry groups of lattices are also called the holohedries in crystallography.)

FIG. 4.1 – The only point group for a one-dimensional lattice is the group of order two.

Looking at the actions of O_n on n-dimensional lattices that fix at least one point of the lattice, and finding their stabilizers we establish the classification of lattices by their point symmetry.

If one such transformation exists, any power of this transformation is a symmetry transformation as well. If there exist only a finite number of different powers, these transformations form a cyclic subgroup of the symmetry group of a lattice. Finding cyclic groups compatible with the existence of the lattice is a first step in the description of lattice symmetry. The restrictions imposed by the lattice have, historically, been called "crystallographic restrictions", though this terminology is out of date after the discovery of aperiodic crystals (quasicrystals).

In this section we find the cyclic groups compatible with one-, two-, and three-dimensional lattices. Generalizations to the arbitrary n-dimensional case will follow in section 4.6.

One-dimensional lattices. All one-dimensional lattices have the same point group, the group $C_2 = 2$ of order two consisting of the identity transformation and reflection (inversion) in one point (see Figure 4.1).

Two-dimensional lattices. Two-dimensional lattices can have as symmetry elements only rotation axes of order 2, 3, 4, 6 and reflection. This restriction is rather obvious (see figure 4.2). Let o be a center of k-fold rotation of the lattice and op be the shortest translation for the lattice. Then p is also a center of k-fold rotation. Let the rotation through $2\pi/k$ about o transform p into p', and let the same kind of rotation about p (realized in the opposite direction) transform o into p''. If $k = 6$ the points p' and p'' coincide. In all other cases we must have $p'p'' \geq op$, since a lattice is a Delone set. This is possible only if $k \leq 4$. Thus, the only possible rotational symmetries for two-dimensional lattices are $k = 2, 3, 4, 6$.

The point group of a lattice in any dimension has the subgroup of order two generated by reflection in a fixed point. This restricts the possibilities for two-dimensional lattices to four point groups. We give here both the Schoenflies and ITC notations[1]: $C_2 = 2$ (oblique), $D_2 = 2mm$ (rectangular), $D_4 = 4mm$ (square), $D_6 = 6mm$ (hexagonal). The associated polygons are shown in figure 4.3 together with their symmetry elements.

Three-dimensional lattices. The crystallographic restrictions for three-dimensional lattices are exactly the same as for two-dimensional: only reflections and rotations of order 2, 3, 4, and 6 are allowed. We accept this fact, for now, without proof; in section 4.6.3 we will explain that more generally the

[1] See Appendix C for discussion of different notations.

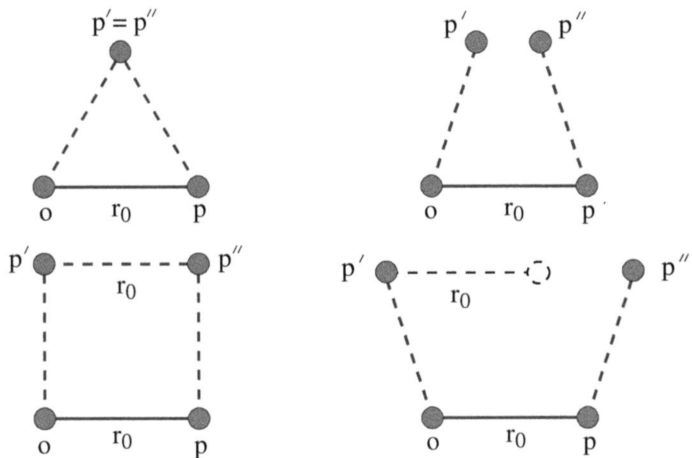

FIG. 4.2 – Crystallographic restrictions for two-dimensional lattices.

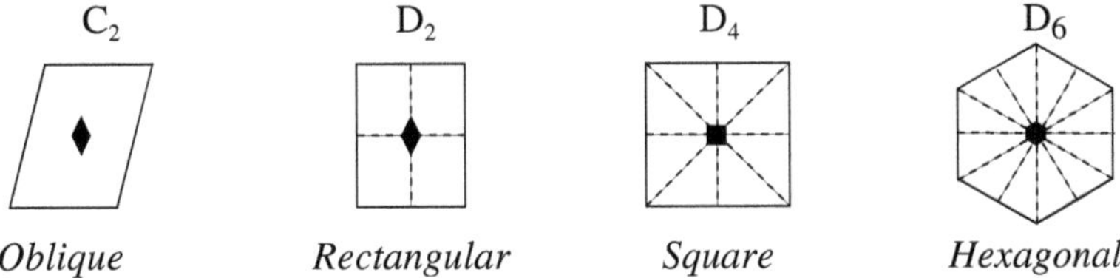

FIG. 4.3 – Four point groups for two dimensional lattices. Black rhombus - rotation axis of order two; black square - rotation axis of order four; black hexagon - rotation axis of order six. Dashed lines - reflection lines.

crystallographic restrictions for lattices of dimensions $2k$ and $2k+1$ coincide, for any positive integer k.

Every lattice in any dimension is invariant with respect to reflection in a fixed point. In three-dimensional space, inversion is an "improper" orthogonal transformation ("improper" means its determinant is -1). Consequently, the point groups of three-dimensional lattices have subgroups of index two consisting of proper orthogonal transformations (pure rotations). Thus point groups are characterized by their rotation subgroups; indeed their axes of order two suffice. Any rotation of higher order for three-dimensional lattices is generated by axes of order two (see, for example [42], Ch.1, sect. 5).

There are three possibilities:

- The point group has no axes of order two.
- The point group has only one axis of order two.
- The point group has several axes of order two.

If the point group has more than one axis of order two, the crystallographic restriction implies that the angles between the axes must be $\pi/6, \pi/4, \pi/3$ or $\pi/2$. These four possibilities yield five different sub-cases.

- $\pi/6$: system of axes of a hexagonal prism;
- $\pi/4$: system of axes of a quadratic prism;
- $\pi/3$: system of axes of a rhombohedron;
- $\pi/3$: system of axes of a cube.
- $\pi/2$: system of axes of an orthogonal parallelepiped.

The specific arrangements of these two-fold axes are shown in Figure 4.4, where the cases with several order-two axes are labeled H, Q, R, C, and O respectively and shown together with case M (one order-two axis) and case T (no two-fold axes).

Adding inversion we get the complete set of generators for the seven lattice point groups listed in Table 4.1.

4.3 Bravais classes

In the last section, we classified lattices by their point groups. But this classification is not fine enough for applications in crystallography and physics. Figure 4.5 shows a pair of two-dimensional lattices that are evidently "different" – the primitive cell of one is a rectangle, while the primitive cell of the other is a rhombus. Yet they have the same point group, $2mm = D_2$.

How can we characterize this difference mathematically? Let us use bases shown in figure 4.5. The matrices σ_3 and σ_1 that describe reflections across the vertical mirror lines in these two lattices are, left to right:[2]

$$\sigma_3 = \begin{pmatrix} 1 & 0 \\ 0 & -1 \end{pmatrix}; \qquad \sigma_1 = \begin{pmatrix} 0 & 1 \\ 1 & 0 \end{pmatrix}. \tag{4.1}$$

In fact σ_3 and σ_1, though they describe the "same" reflection, are not interchangeable, in the sense that neither matrix can be obtained from the other

[2] This notation introduced by Pauli is usual in physics. In 1925, Pauli wrote the first paper in quantum mechanics computing the spectrum of the hydrogen atom in a vacuum and in a constant magnetic or electric field including the spin effects.

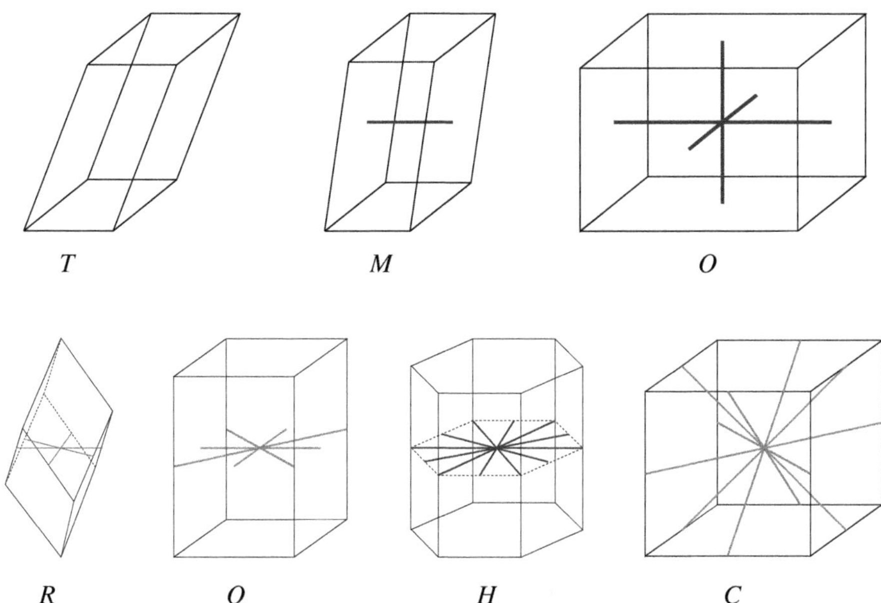

FIG. 4.4 – The seven three dimensional point groups for lattices represented through arrangements of their order-two symmetry axes. T - triclinic crystallographic system has no two-fold axes. M - monoclinic crystallographic system has one two-fold axis. O - orthorhombic system has three mutually orthogonal order two axes. R - rhombohedral (or trigonal) system has three two-fold axes belonging to plane with $\pi/3$ angle between them. Q - Tetragonal system has four two-fold axes belonging to the plane with $\pi/4$ angle between them. H - hexagonal system has six two-fold axes belonging to the plane with $\pi/6$ angle between them. C - cubic system has six two-fold axes of a cube with $\pi/3$ or $\pi/2$ angles between them.

TAB. 4.1 – The seven three dimensional point groups for lattices and the associated names of Bravais crystallographic systems.

Bravais CS	Triclinic	Monoclinic	Orthorhombic	Tetragonal	Rhombohedral	Hexagonal	Cubic
Abbreviation	T	M	O	Q	R	H	C
Schoenflies	C_i	C_{2h}	D_{2h}	D_{4h}	D_{3d}	D_{6h}	O_h
ITC	$\bar{1}$	$2/m$	mmm	$4/mmm$	$\bar{3}m$	$6/mmm$	$m\bar{3}m$

by a change of lattice basis. That is, though these matrices are conjugate in the general linear group $GL_2(\mathbb{R})$, *they are not conjugate in* $GL_2(\mathbb{Z})$. To convince yourself, let $A = \begin{pmatrix} a & b \\ c & d \end{pmatrix}$ be any matrix in $GL_2(\mathbb{Z})$; that is,

FIG. 4.5 – These lattices have the same point group – the points of each are stabilized by a pair of orthogonal mirror lines – yet they are "different."

a, b, c, d are integers and $ad - bc = 1$. Then $A^{-1} = \begin{pmatrix} d & -b \\ -c & a \end{pmatrix}$ and an easy computation shows that there is no choice of *integer* entries for A for which $A\,\sigma_3\,A^{-1} = \sigma_1$.

These two lattices are said to be different *Bravais types*. Since the other three two-dimensional point groups do not subdivide in this way, there are five Bravais lattices in two dimensions.

Bravais himself classified lattices by choosing minimal possible cells (preferably rectangular) which keep the point symmetry of the lattice. Lattices having the same point symmetry group but associated with different cells are referenced now as belonging to different Bravais classes.

In more formal mathematical terms

the conjugacy class $[P_L^z]_{GL_n(\mathbb{Z})}$ *defines the Bravais class of* L*;*

the conjugacy class $[P_L]_{GL_n(\mathbb{R})}$ *defines the crystallographic system of* L*.*

Correspondingly, we call P_L^z the *Bravais group* of L, while P_L is the point group of L.

It is clear that several Bravais classes $[P_L^z]_{GL_n(\mathbb{Z})}$ can correspond to the same conjugacy class $[P_L]_{GL_n(\mathbb{R})}$ defining the point symmetry group of the lattice up to conjugation within the $GL_n(\mathbb{R})$ group because $GL_n(\mathbb{Z})$ is obviously a subgroup of $GL_n(\mathbb{R})$.

To see more examples of lattices with a given point symmetry group and a different number of associated Bravais classes, let us now consider the n-dimensional case.

In every dimension the generic lattices form only one Bravais class: by definition the point group includes only $\{I_n, -I_n\}$.

The situation changes, however, if we consider lattices with just one additional reflection symmetry. In every dimension $n \geq 2$, although reflections through a hyperplane are all conjugate in $GL_n(\mathbb{R})$, this is not true in $GL_n(\mathbb{Z})$. To see this, from the matrices σ_1 and σ_3 above, we build two reflection matrices $M_i = \sigma_i \oplus I_{n-2}$. They cannot be conjugate in $GL_n(\mathbb{Z})$; if they were, this would also be true of the two matrices $I_n + M_3$ and $I_n + M_1$. That is not possible: indeed, the greatest common divisor (gcd) of the elements of these matrices is 2 for the former and 1 for the latter; but conjugation by an element of $GL_n(\mathbb{Z})$ cannot change the gcd of the elements of a matrix.

So there are at least two conjugacy classes of reflections in $GL_n(\mathbb{Z})$; in fact there are only two. Here we give a direct proof for $n = 2$. A reflection in $GL_2(\mathbb{Z})$ has trace 0 and determinant -1; so its general form is

$$S = \begin{pmatrix} a & b \\ c & -a \end{pmatrix}, \quad a, b, c \in \mathbb{Z}, \quad a^2 + bc = 1. \tag{4.2}$$

If $S \neq \pm\sigma_3$, it is not diagonal. We may not be able to diagonalize it by conjugation in $GL_2(\mathbb{Z})$, but we can make it upper triangular. Indeed, corresponding to the eigenvalue 1, it has, up to a sign, a unique integral, visible eigenvector $\vec{v} = \binom{\alpha}{\beta}$ with $\alpha = b/k$, $\beta = (1-a)/k$ where $k = \gcd(b, 1-a)$. Then we can choose a pair α', β' of relatively prime integers such that $\alpha\beta' - \beta\alpha' = 1$, to complete a conjugating matrix:

$$\begin{pmatrix} \beta' & -\alpha' \\ -\beta & \alpha \end{pmatrix} \begin{pmatrix} a & b \\ c & -a \end{pmatrix} \begin{pmatrix} \alpha & \alpha' \\ \beta & \beta' \end{pmatrix} = \begin{pmatrix} 1 & x \\ 0 & -1 \end{pmatrix} = T, \tag{4.3}$$

where $x = 2a\alpha'\beta' + b\beta'^2 - c\alpha'^2$. Depending on whether x is even ($x = 2y$), or odd ($x = 2y+1$), the matrix T can be conjugate to σ_3 or σ_1 by the matrices $\begin{pmatrix} 1 & y \\ 0 & 1 \end{pmatrix}$ and $\begin{pmatrix} 1 & y \\ 1 & 1+y \end{pmatrix}$ respectively. Thus there are exactly two conjugacy classes of reflections in $GL_2(\mathbb{Z})$.

Although explicit expression for x given above depends on a, b, c and α', β', it is possible to give more direct and more simple formulae expressing parity of x or equivalently class mod 2 in terms of matrix elements of matrix S only.

Proposition. $x \bmod 2 = (b + c + bc) \bmod 2$.

We leave derivation of this expression for the interested reader.

The two classes of reflections corresponding to the classes $0, 1$ of $x (\mathrm{mod} 2)$ are labeled in [14] by *pm* and *cm* respectively. These symmetry groups cannot be symmetry group of a lattice because lattice symmetry always includes point reflection. These groups are subgroups of a larger lattice symmetry group and naturally subgroups of $GL_n(\mathbb{Z})$. More generally, classes of conjugated subgroups of $GL_n(\mathbb{Z})$ are named *"arithmetic classes"*.

A reflection from either class and $-I_2$ generates a point group isomorphic to $\mathbb{Z}_2^2$; they define two Bravais classes, *pmm* and *cmm*. Since the two matrices of (4.1) have the same determinant and the same trace, they have the same characteristic polynomial, so they are conjugate in $GL_2(\mathbb{R})$. This conjugacy class describes the $2D$-crystallographic system called *rectangular* or orthorhombic.

Generalizing the n-dimensional lattices of the orthorhombic crystal system leads to lattices with point symmetry described by the group of $n \times n$ diagonal matrices with diagonal elements ± 1. The conjugacy class in $GL_n(\mathbb{R})$ of this group is named $A_1^n = A_1 \times A_1 \cdots \times A_1$ in the spirit of notations used for groups generated by reflections and Coxeter groups [5, 7].[3] The number of

[3] $n = 3$: ITC=mmm, SCH=D_{2h}. For $n = 2$, ITC=$2mm$, SCH=C_{2v}.

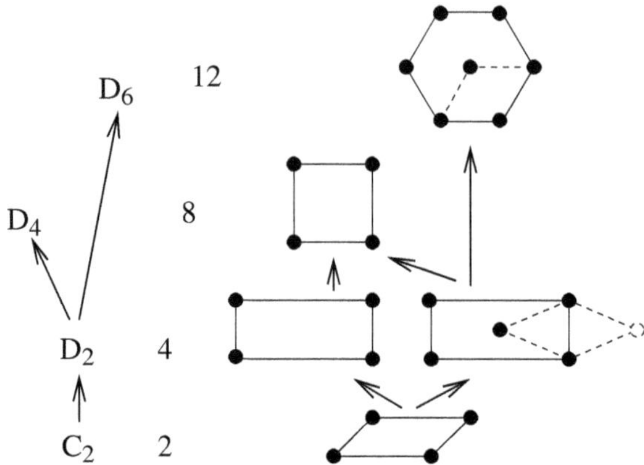

FIG. 4.6 – 2D Bravais crystallographic systems (left) and corresponding Bravais classes of lattices (right).

corresponding Bravais classes for $n = 2, 3, 4$ is 2, 4, and 8. With increasing n, the number of Bravais classes grows exponentially.

Another example of a family of lattice point symmetry groups and corresponding Bravais classes defined for arbitrary n is given by the symmetry group of the cube (or the cross-polytope) in dimension n. Three Bravais classes correspond to this conjugacy class in $GL_n(\mathbb{R})$ for every n except $n = 1, 2, 4$; there is one Bravais class for $n = 1, 2$ and two for $n = 4$. Following crystallographic convention, for $n = 3$ one calls the three Bravais classes Cubic P (or simple), Cubic F (or face centered), Cubic I (or body centered).

4.4 Correspondence between Bravais classes and lattice point symmetry groups

In any given dimension n all lattice point symmetry groups form a partially ordered set of subgroups of $O(n)$ or $GL_n(\mathbb{R})$. Considered up to conjugation in $GL_n(\mathbb{R})$ they characterize crystallographic systems.

In a similar way, the Bravais classes of lattices (as subgroups of $GL_n(\mathbb{Z})$) form themselves a partially ordered set of subgroups of $GL_n(\mathbb{Z})$. There exists correspondence between these two partially ordered sets which maps all isomorphic Bravais groups on corresponding crystallographic system.

Figures 4.6 and 4.7 show this correspondence for two-dimensional lattices, where only one among four existing crystallographic systems has more than one Bravais class.

The same correspondence for three-dimensional lattices is given in Figures 4.8 and 4.9.

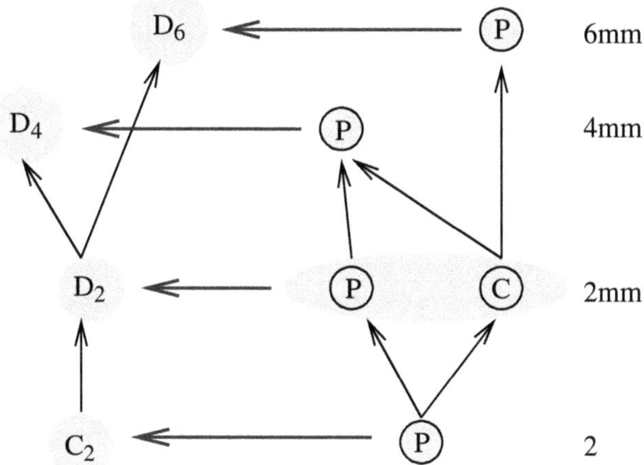

FIG. 4.7 – Surjective map $\{BC\}_2 \rightarrow \{BCS\}_2$ from the partially ordered set of Bravais classes (right) to a partially ordered set of Bravais crystallographic systems (left) for two-dimensional lattices.

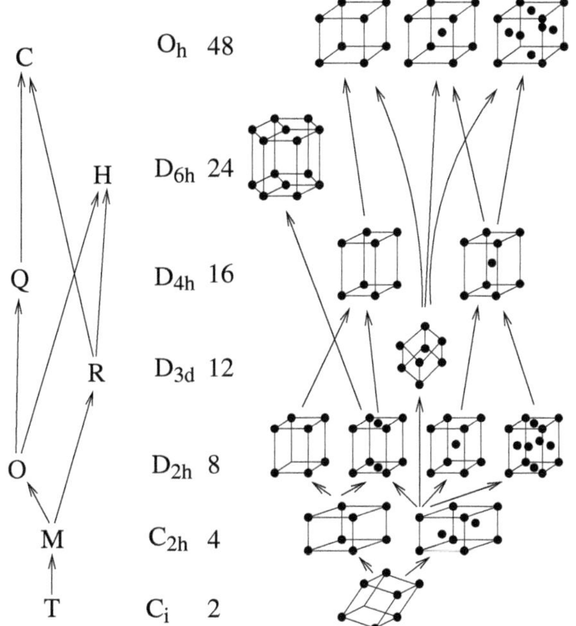

FIG. 4.8 – 3D Bravais crystallographic systems (left) and corresponding Bravais classes of lattices (right).

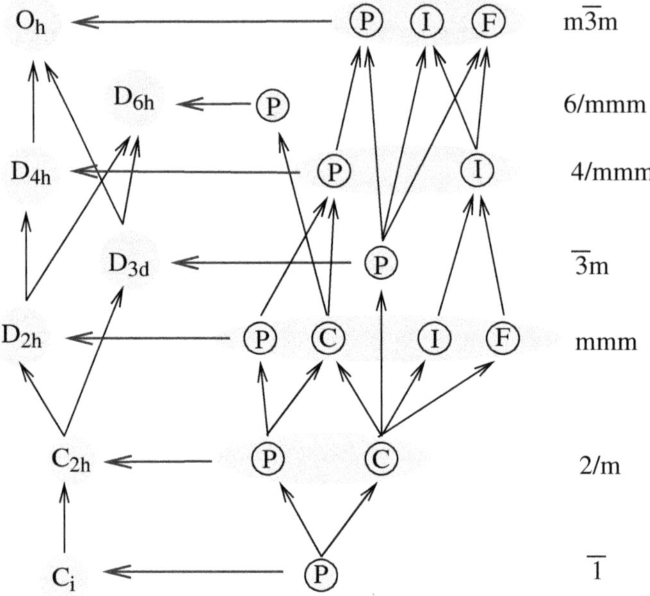

FIG. 4.9 – Surjective map $\{BC\}_3 \to \{BCS\}_3$ from the partially ordered set of Bravais classes (right) to partially ordered set of Bravais crystallographic systems (left) for three-dimensional lattices.

4.5 Symmetry, stratification, and fundamental domains

The symmetry group of a lattice acts on the ambient Euclidean space and thus we can classify all points of space into orbits according to their stabilizers. Orbits of the same type (i.e., those whose stabilizers are conjugate within the symmetry group) form strata. Selecting one point from each orbit, we get a fundamental domain of the lattice. This suffices to describe any local properties of the physical system because any properties at other points can be obtained by applying symmetry operations to points of the fundamental domain. Moreover, the topological properties of the fundamental domain, i.e. topological properties of the space of orbits, correspond to important global topological properties of physical systems.

In this subsection we describe the strata, fundamental domains, and orbifolds for two- and three-dimensional lattices.

We will use orbifold notion interchangeably with the space of orbits when we want to introduce or to stress the topological representation of the fundamental domain of the lattice taking into account the symmetry group.

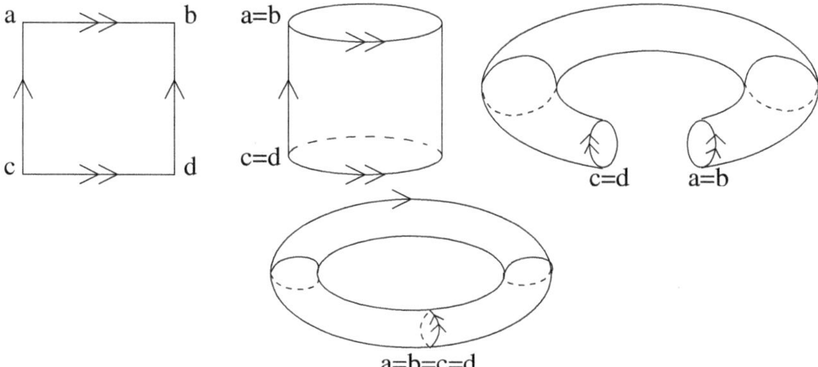

FIG. 4.10 – Torus construction from a plane rectangle by identifying points on its boundary.

We analyze the action of the symmetry group of the lattice on the ambient space, find its orbits, and represent each orbit as a single point. Since each orbit is characterized by its stabilizer and orbits with the same stabilizer form strata, the orbit space is represented as stratified topological space.

If there are no additional symmetry transformations except the translation symmetry defining the lattice, the space of orbits (or orbifold) for a n-dimensional lattice is a n-dimensional torus, obtained by taking the fundamental cell of the lattice and identifying those points on its boundary which belong to the same orbit of the translation group action.

The two-dimensional case can be easily visualized with a little imagination. To pass from the fundamental cell to the orbifold (see Figure 4.10) we can first take a rectangle made of paper and identify respective points on one pair of opposite sides. This gives us a cylinder. Now we need to identify points on the other two sides of the rectangle (which have become circles). Replacing the paper cylinder by an elastic cylindrical tube, we see how this identification leads to a torus.

Adding symmetry transformations of the lattice is equivalent to introducing group action on the torus.

Let now construct the system of strata, fundamental cells, and orbifolds for the five different Bravais symmetry groups of two-dimensional lattices.

The Bravais group $p2$ has four C_2 orbits within a fundamental cell, forming four different zero-dimensional strata as shown in Figure 4.11. In ITC these four C_2 orbits are called Wyckoff positions with site symmetry $2 \equiv C_2$. It is important to note that although these four orbits have the same stabilizer as an abstract group, these four C_2 subgroups are not conjugate and belong to different strata. This can be easily seen because there is no symmetry operation which transform one orbit into another. All other internal points of the fundamental cell belong to generic orbits with trivial stabilizer.

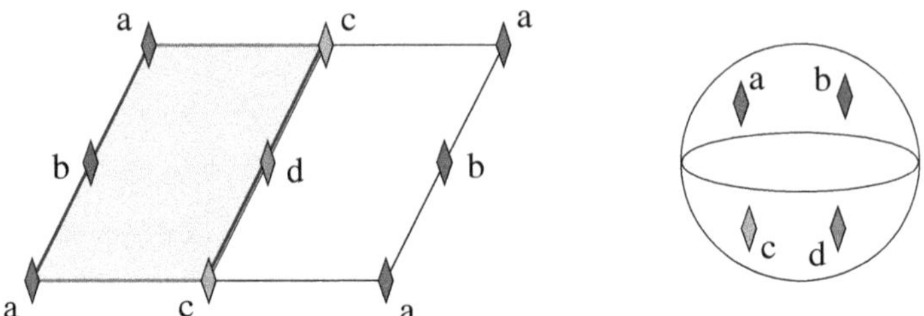

FIG. 4.11 – Group action on the fundamental cell for $p2$ and the corresponding orbifold.

Each generic orbit is formed by two points per primitive cell, related by C_2 symmetry. The pair of points forming the generic orbit transform one into another by C_2 symmetry transformation. To pass from the primitive crystallographic cell representation to the orbifold we need to keep only one representative point from each orbit. For example, we can keep the points in the shaded part of the unit cell and identify points on the boundary of this part which belong to the same orbit. This means that intervals of the boundary labeled by the same letters should be identified. Identifying first two *ab* intervals and next two *cd* intervals we get a topological disk whose boundary consists of two intervals *ac* to be identified as well. This final identification leads to a topological two-dimensional sphere with four special points. For an orbifold which is a topological sphere, the Conway-Thurston notation indicates only the singular points. Thus, the notation for the $p2$ orbifold is 2222.

To see the correspondence with the torus, we note that selecting one point from each generic orbit on the torus is equivalent to taking one half of the torus, which is a cylinder with two boundary circles *aba* and *cdc*. Identifying the two *ab* half circles and two *cd* half circles leads to a topological sphere with four marked points a, b, c, d.

The next two Bravais groups are $p2mm$ and $c2mm$. The action of the symmetry group on the crystallographic cell for these two groups is shown in Figures 4.12 and 4.13.

It is easy to see that the space of orbits for the $p2mm$ group has a boundary formed by symmetry reflection lines. Thus the space of orbits for $p2mm$ is a topological disk. There are four singular points on the disk boundary corresponding to C_2 orbits; they belong to different strata (their stabilizers are not conjugate in $GL_2(\mathbb{Z})$). The boundary of the disk in its turn consists of four intervals again belonging to four different strata.

The orbifold notation indicates the presence of a boundary by a $*$, followed by stabilizers of singular points on the boundary. Thus, the orbifold notation for $p2mm$ is $*2222$.

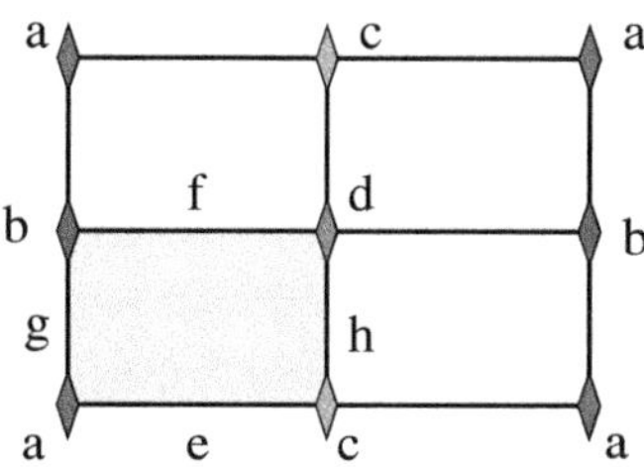

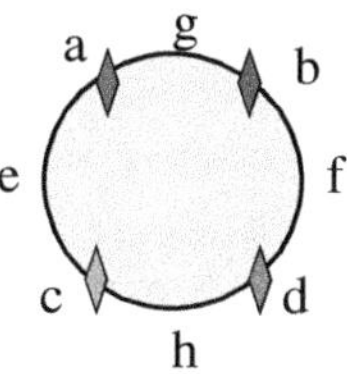

FIG. 4.12 – Group action on the primitive crystallographic cell for *p2mm* and the corresponding orbifold.

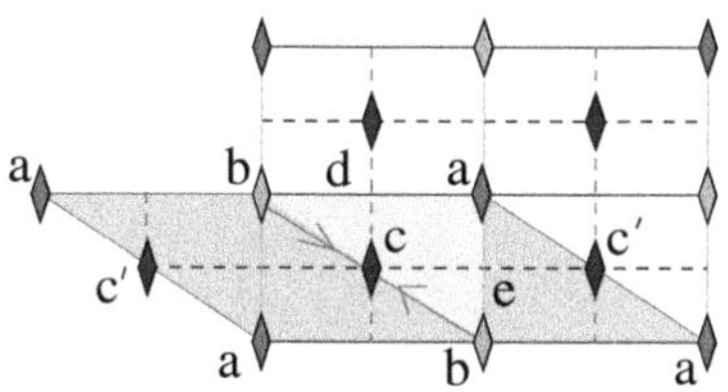

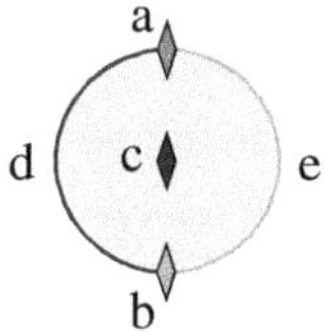

FIG. 4.13 – Group action on the crystallographic cell for *c2mm* and the corresponding orbifold.

To construct the orbifold for the Bravais group *c2mm* we note that the ITC uses a double cell, rather than a primitive cell of this lattice. Figure 4.13 shows the traditional ITC choice of a fundamental (double) cell for the *c2mm* group together with one possible choice of a primitive cell (grey shading). To take one representative point from each orbit of the symmetry group action on the primitive fundamental cell means to take the triangular region shown by light shading together with its two mirror boundaries marked *ab* but belonging to two different strata and to identify two subintervals *bc* on the third boundary. The resulting orbifold is a topological disk with two singular points on its boundary corresponding to two non conjugated stabilizers and one singular point inside. The two intervals of the boundary again correspond to two different stabilizers. The orbifold notation for the space of orbits is 2∗22.

The action of the Bravais group *p4mm* on a primitive fundamental cell of a two-dimensional lattice is shown in figure 4.14, where the primitive cell is drawn. The shade region together with its boundary contains one representative point from each group orbit. Points with different stabilizers belong to different strata and different strata are marked by different letters. The space of orbits is a topological disk with three isolated orbits on its boundary. Two C_{4v} orbits belong to different strata which are not conjugate, the third C_{2v} orbit forms also its own isolated stratum. Three intervals on

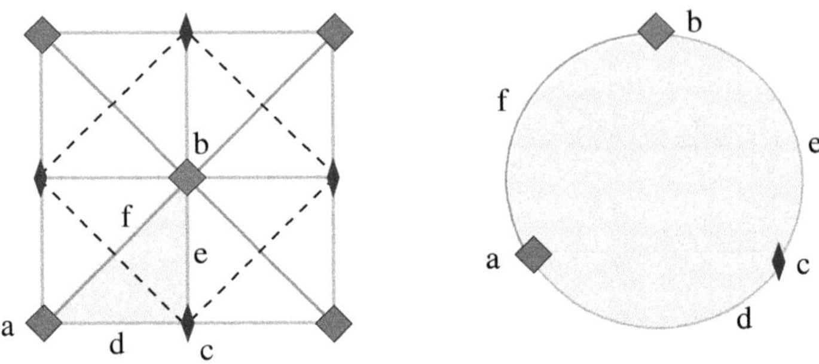

FIG. 4.14 – Orbifold for *p4mm*.

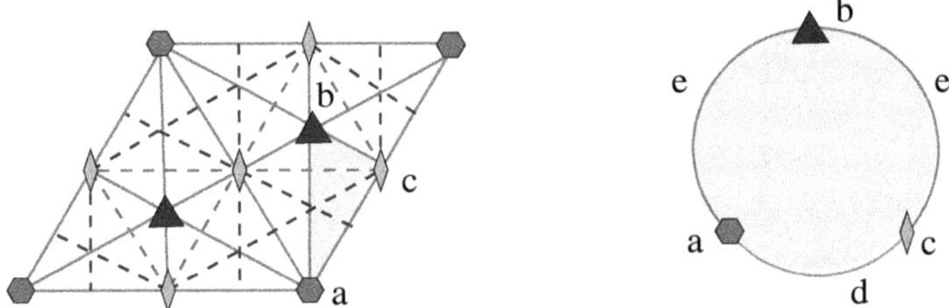

FIG. 4.15 – Orbifold for *p6mm*.

the disk boundary form three different strata. The orbifold notation of *p4mm* orbifold is $*442$.

The action of the Bravais group $P6mm$ on a primitive fundamental cell is shown in figure 4.15. The shade region together with its boundary contains one representative point from each group orbit. Points with different stabilizers belong to different strata and different strata are marked by different letters. The space of orbits is a topological disk with three isolated orbits on its boundary whose stabilizers are C_{6v}, C_{3v}, and C_{2v}. The three intervals on the boundary of the disk are formed by orbits belonging to two different strata. The orbifold notation is $*632$.

The construction of orbifolds for the symmetry groups of three-dimensional lattices is naturally a more complicated task. We need to split this problem into several subproblems.

One sub-problem is to describe local neighborhoods for representative points of different strata of the orbifolds. For this purpose it is sufficient to find spaces of orbits for the local action of the three-dimensional point symmetry groups. Moreover, as soon as we suppose that one point is fixed, we can restrict our analysis from three-dimensional space to a surface of a two-dimensional sphere surrounding that point. (The action of a group fixing one chosen point is independent of the radius.) Naturally, we analyze only point groups compatible with three-dimensional lattices.

Three dimensional group actions can be divided into two cases, called reducible and irreducible. In 1776, Euler proved that every rotation in $\mathbb{R}^3$ is a rotation about an axis which maps all planes orthogonal to that axis into themselves. The so called reducible point groups are those which leave one rotation axis invariant; the irreducible groups are those with no invariant axis. Passing from group action on the space to orbifold, the invariant axis remains an invariant axis of the orbifold. In other words, a fibration of the space becomes a fibration of the orbifold. Each fiber becomes either a circle or an interval.

If you imagine looking along the invariant direction of a fibered symmetry group you will see one of the Euclidean plane groups. Thus orbifolds for reducible symmetry groups can be constructed starting from two-dimensional orbifolds.

Orbifolds for irreducible three dimensional groups must be studied each in turn. Since only the cubic point group, $O_h = Pm\bar{3}m$ is irreducible, only the three corresponding Bravais classes, $Pm\bar{3}m$, $Im\bar{3}m$, and $Fm\bar{3}m$ are irreducible.

Let us consider first spherical orbifolds for point group actions on a two-dimensional sphere. We restrict ourselves to point groups which appear as symmetry groups (holohedries) of three-dimensional lattices, C_i, C_{2h}, D_{2h}, D_{3d}, D_{4h}, D_{6h}, and O_h.

4.5.1 Spherical orbifolds for 3D-point symmetry groups

The lowest symmetry for holohedry of 3D-lattices is the C_i group. The action of the C_i group on a two-dimensional sphere in three-dimensional space leads to only one type of orbit. Each orbit is formed by two opposite points on the sphere (see Figure 4.16). This means that the set of orbits can be equivalently interpreted as a set of straight lines passing through origin in three-dimensional space. (This is real projective space.) Alternatively, the set of orbits can be considered as a set of points on the half-sphere with additional identification of opposite points on the boundary circle.

The action of the three-dimensional C_{2h} point symmetry group on three-dimensional space is shown in Figure 4.17, a. There are four strata formed by orbits of different type. Restriction of this action on a two-dimensional sphere leads to

i) zero-dimensional C_2 stratum which includes two opposite points at the intersection of the C_2 axis and the sphere,

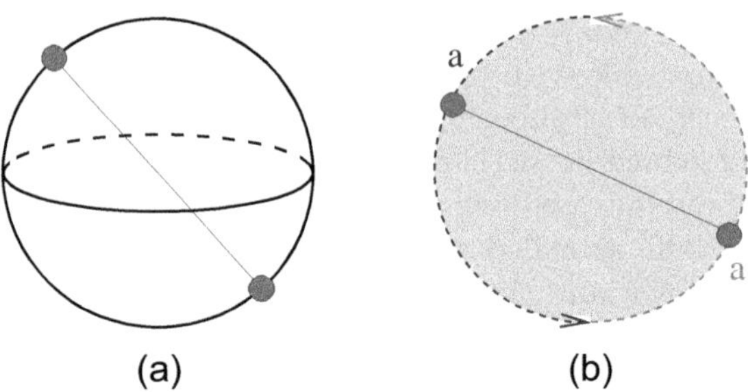

FIG. 4.16 – Construction of the orbifold for the 3D-point group C_i acting on two dimensional sphere. (*a*) Action of the group C_i consists in forming two-point orbits. Each orbit includes two diametrically opposite points on the sphere. (*b*) To represent the space of orbits it is sufficient to take demi-sphere and to identify the diametrically opposite points on the boundary circle. The resulting orbifold is real projective space RP_2.

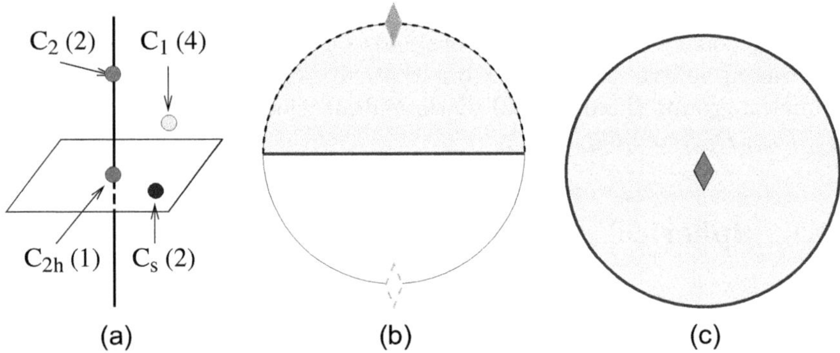

FIG. 4.17 – Construction of the orbifold for the 3D-point group C_{2h} acting on a two dimensional sphere. (*a*) Action of the group C_{2h} on the 3D-space. Stabilizers of points and the number of points in the corresponding orbit are indicated. (*b*) Schematic view of the action of the C_{2h} group on a two-dimensional sphere. The rhombus indicates points belonging to the C_2 orbit. Thick solid line corresponds to the reflection plane. All other points of the sphere belong to the generic C_1 orbits. The fundamental domain of the C_{2h} group action fills half of the upper demi-sphere. Its projection is shown as a shaded region together with its boundary. Two dashed intervals of the boundary of the fundamental domain should be identified. (*c*) Representation of the orbifold $2*$ for the action of the C_{2h} group on the sphere as a disk with one special point (C_2 orbit) inside.

ii) one dimensional stratum formed by points at the intersection of the symmetry plane and the sphere,

iii) two-dimensional generic stratum.

Keeping one point from each orbit leads, for example, to taking half of the upper demi-sphere and to identify points on the meridianal section which belong to the same orbit under the C_2 symmetry operation (see Figure 4.17, b). The resulting space of orbits is a topological disk with one singular point inside (see Figure 4.17, c).

The construction of orbifolds for D_{2h}, D_{4h}, and D_{6h} groups can be studied through analysis of the whole D_{nh} family of groups.

The group D_{nh} has order $4n$. It can be obtained by adding to the C_{nv} group the symmetry reflection plane orthogonal to the C_n symmetry axis. The system of conjugacy classes for the D_{nh} group is quite different for even $n = 2p$ and for odd $n = 2p + 1$. Thus we treat these two cases separately.

Group D_{nh} with $n = 2p \geq 2$ has $2(p + 3)$ conjugacy classes. In particular, there are two different classes of order two rotation axes C_2 and C_2', and of vertical symmetry reflection planes σ_v, σ_d. The third class of symmetry reflection planes includes one element - reflection in the horizontal symmetry plane. There are seven different strata for the action of D_{nh} with $n = 2p \geq 2$ on the sphere. There are three zero dimensional strata with stabilizers C_{nv}, C_{2v} and C_{2v}'. There are three one-dimensional strata with stabilizers C_s, C_s', C_s''. At last, a generic stratum has orbits with trivial stabilizer C_1. There is only one orbit with stabilizer C_{nv} consisting of two points (poles of the sphere). There is one n-point orbit with stabilizer C_{2v} and one n-point orbit with stabilizer C_{2v}'. Each orbit with stabilizer C_s, or C_s' or C_s'' consists of $2n$ points. Each generic orbit consists of $4n$ points. To form the space of orbits we can take the part of the sphere bound by three symmetry planes. As a result the orbifold is a topological disk with a boundary which has three singular orbits C_{nv}, C_{2v}, and C_{2v}'. The regular points on the boundary belong to three different strata C_s, C_s' and C_s''. Three singular orbits on the boundary also belong to three different strata. The notation of the orbifold is $*n22$.

The group D_{nh} with $n = (2p + 1) \geq 3$ has $2(p + 2)$ conjugacy classes. In particular, all vertical symmetry reflection planes belong to the same conjugacy class. All C_2 rotations also belong to the same conjugacy class. These facts modify the stratification of the D_{nh} with $n = (2p + 1) \geq 3$ as compared to D_{nh} with the $n = (2p) \geq 2$ case. There are only two zero dimensional strata with stabilizers C_{nv} and C_{2v}. Stratum C_{nv} includes one two-point orbit. Stratum C_{2v} includes two $2n$-point orbits. There are two one-dimensional strata with stabilizers C_s and C_s'. C_s orbits are formed by points lying on all vertical symmetry planes. C_s' orbits are formed by points belonging to the horizontal symmetry plane. The space of the orbits takes the form of a topological disk with three singular orbits on the boundary. One of these singular orbits is C_{nv}, two others are of the C_{2v} type. Three intervals of

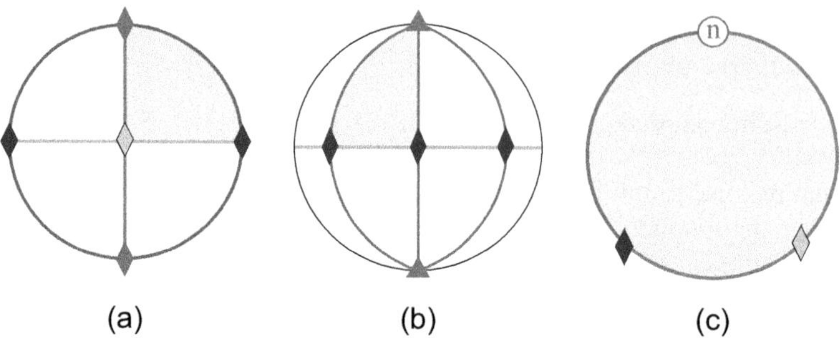

FIG. 4.18 – (*a*) Action of the group D_{2h} on the sphere. (*b*) Action of the group D_{3h} on the sphere. Shade regions represent fundamental domains. (*c*) The orbifold for action of the D_{nh} group on the sphere: $*n22$. There are seven strata for D_{nh} groups with even n; there are five strata for D_{nh} groups with odd n. See text for details.

regular points on the boundary of the orbifold are filled by two types of orbits. There are C_v orbits between C_{nv} and C_{2v}, whereas there are C_h orbits between two C_{2v} singular orbits on the boundary. The notation of the orbifold of the D_{nh} action on the sphere, namely $*n22$ is the same for even n and odd n.

Although we need only a D_{3d} point group for description of holohedries of three-dimensional lattices we can easily consider orbifolds for all D_{nd} point groups simultaneously.

The group D_{nd} with $n \geq 2$ has order $4n$. It can be obtained from D_n by adding n symmetry planes which include a C_n axis and are situated between neighboring C_2 axes. For the D_{nd} group in both cases of even or odd n all symmetry planes belong to the same conjugacy class. All C_2 axes also belong to the same conjugacy class. This means that we can describe strata of the D_{nd} action on a sphere simultaneously for all $n \geq 2$. There are four strata: two zero dimensional, one one-dimensional, and one two-dimensional generic stratum. One zero dimensional stratum is formed by one two-point orbit with stabilizer C_{nv} (two poles of the sphere). Another zero dimensional stratum consists also of one orbit which has $2n$ points. The stabilizer of this $2n$-point orbit is C_2. A one-dimensional stratum is formed by $2n$-point orbits with stabilizer C_s. These points belong to the symmetry planes. Finally the generic stratum is formed by 4n-point orbits. In order to form the space of orbits and to take one representative point from each orbit it is sufficient to take a $2\pi/n$ sector of the north half-sphere together with the boundary formed by the intersection of the sphere with the equatorial plane. Moreover, two halves of the equatorial arc should be identified according to the action of the C_2 rotation. This identification shows that the space of orbits is a topological

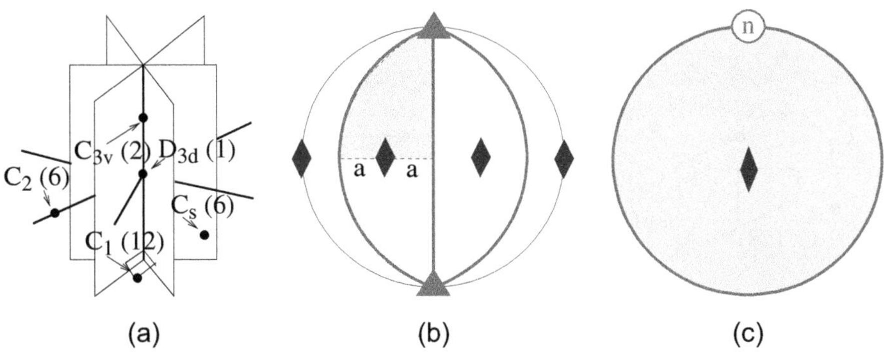

FIG. 4.19 – Construction of the orbifold for the 3D-point group D_{nd} acting on a two dimensional sphere. (*a*) Action of the group D_{3d} on the 3D-space. Stabilizers of points and the number of points in the corresponding orbit are indicated. (*b*) Schematic view of the action of the D_{3d} group on a two-dimensional sphere. The rhombus indicates points belonging to the C_2 orbit. The filled triangles correspond to points belonging to the C_{3v} orbit. The thick solid lines indicate C_s orbits. All other points of the sphere belong to generic C_1 orbits. The fundamental domain of the D_{3d} group action is shown as a shaded region. Respective points on two parts of the boundary marked by letter a should be identified. (*c*) Representation of the orbifold $2{*}n$ of the group D_{nd} action on the sphere as a disc with one special point on the boundary (C_{nv} orbit) and one special point inside (C_2 orbit).

disc with a boundary. There is one singular C_{nv} point on the boundary and one singular point C_2 inside the disk. The notation of the orbifold is $2{*}n$.

The point symmetry group O_h is a full symmetry group of a cube. There are 48 symmetry operations in the O_h group. The presence of two different conjugacy classes of C_2 rotations for the group O implies the existence of two different conjugacy classes of reflection planes for the O_h group. One conjugacy class of reflection planes consists of three planes (orthogonal to the C_4 axes and named often "*horizontal*"). Another conjugacy class consists of six reflection planes. These planes are orthogonal to the C_2 axes of a cube going through the middle of the edges. They are named often "*diagonal*". Taking these facts into account, the action of the group O_h on the sphere yields three zero-dimensional strata, two one-dimensional strata, and one generic two-dimensional stratum. The C_{4v} zero-dimensional stratum consists of one six-point orbit. The C_{3v} zero-dimensional stratum consists of one eight-point orbit. The C_{2v} zero-dimensional stratum consists of one twelve-point orbit. The C_s and C_s' one-dimensional strata are formed by 24-point orbits situated on "horizontal" and "diagonal" planes associated with two different conjugacy classes of elements of the O_h group. These two different strata are marked by solid and dash lines in Figure 4.20, *b*. A generic stratum is formed by 48-point orbits

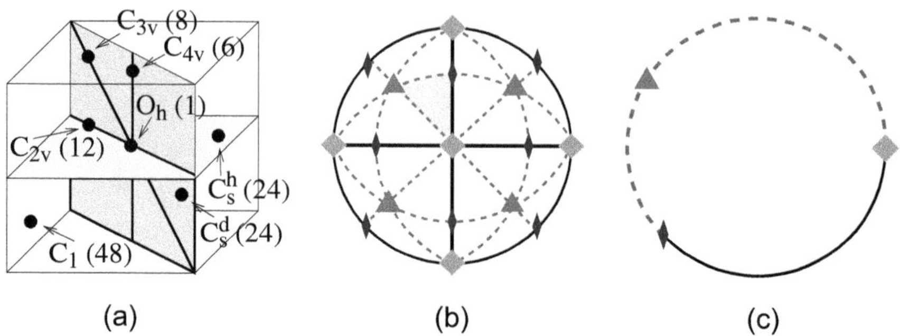

FIG. 4.20 – Construction of the orbifold for the 3D-point group O_h acting on a two dimensional sphere. (*a*) Action of the group O_h on the 3D-space. Stabilizers of points and the number of points in the corresponding orbit are indicated. Only one C_4 axis, one C_3 axis, one C_2 axis, and one symmetry plane from each of the two classes of conjugated elements are shown. (*b*) Schematic view of the action of the O_h group on a two-dimensional sphere. A rhombus indicates points belonging to the C_{2v} orbit. Triangles correspond to points belonging to C_{3v} orbits. Squares show points belonging to the C_{4v} orbit. Thick solid lines correspond to reflection planes forming a conjugacy class of three reflection planes which do not contain C_3 axes. Thick dash lines correspond to six reflection planes containing C_3 axes and forming one conjugacy class of so called "diagonal planes". All other points of the sphere belong to generic C_1 orbits. The fundamental domain of the O_h group action is shown as a shaded region together with its boundary. (*c*) Representation of the orbifold $*432$ as a disk with three special points on its boundary.

with trivial stabilizer. The fundamental domain of the sphere including one point from each orbit can be chosen as the shaded region (Figure 4.20, *b*) with the boundary. This means that the orbifold is a topological disc with three special points (C_{4v} orbit, C_{3v} orbit, C_{2v} orbit). The fact that the boundary is formed by two different strata is ignored in the Conway orbifold notation, $*432$.

4.5.2 Stratification, fundamental domains and orbifolds for three-dimensional Bravais groups

Because the number of three-dimensional Bravais groups is relatively large (14 groups), we treat here only two examples, $P4/mmm$ and $I4/mmm$. Irreducible three-dimensional Bravais groups are illustrated in appendix E.

4.5.3 Fundamental domains for $P4/mmm$ and $I4/mmm$

We have chosen $P4/mmm$ and $I4/mmm$ Bravais lattices to illustrate the stratification and fundamental domain construction because these two Bravais groups belong to the same point symmetry group D_{4h} and being relatively simple they allow us to demonstrate dependence of stratification and topology of orbifolds on Bravais groups within the same holohedry.

The zero-, one-, and two-dimensional strata of $P4/mmm$ are shown in Figure 4.21.

There are four zero dimensional strata with stabilizer $D_{4h} = 4/mmm$, marked on the first sub-figure of Figure 4.21 by small Latin letters a, b, c, d in accordance with ITC notation for Wyckoff positions. These four stabilizers are different non-conjugate subgroups of the Bravais group. Two more zero dimensional strata, e, f have stabilizer D_{2h} and they are also non-conjugate subgroups of $P4/mmm$. There is one point per cell for strata with stabilizer D_{4h} and there are two points per cell for strata with stabilizer D_{2h}. Note that there are eight points of type a which are shown in sub-figure 4.21 because each point a belongs to eight cells and only one point a should be chosen as a representative of its stratum when constructing the fundamental domain and orbifold. In a similar way, there is a quadruplet of points b (each point belongs to four cells) and doublet of points c (each point c belongs to two cells). In contrast, there is only one point d in Figure 4.21 because this point lays inside the primitive cell. Four points of type e are shown in Figure 4.21. This corresponds to two points of type e par primitive cell because each point belongs to two cells. There are eight points of type f with stabilizer D_{2h} because each point f belongs to four cells and this gives exactly two points per cell.

There are two one dimensional strata g, h with stabilizer C_{4h}. Each of these strata has two points per cell in each orbit. Each stratum consists of two intervals per cell situated symmetrically with respect to the middle symmetry plane. Each interval includes one point from every orbit. One pair of intervals is shown for the h stratum because this stratum belongs to the interior of the primitive cell. Four pairs of intervals are shown for the g stratum because now each interval (being an edge of a primitive cell) belongs to four cells.

There are seven different non-conjugate strata i, j, k, l, m, n, o with stabilizer C_{2v}. Each of these strata has four points per primitive cell in each orbit and consists of four intervals per primitive cell. Again each interval includes one point from each orbit with stabilizer C_{2v}.

There are five two-dimensional strata p, q, r, s, t. All have stabilizer C_s, but they are non-conjugate within the lattice symmetry group. Each stratum has eight points per primitive cell in each orbit. Each two-dimensional stratum consists of eight open domains per cell. For p and q strata these domains are triangles. For r, s, t strata the corresponding domains are rectangles. Each such domain includes one point from every orbit belonging to the stratum.

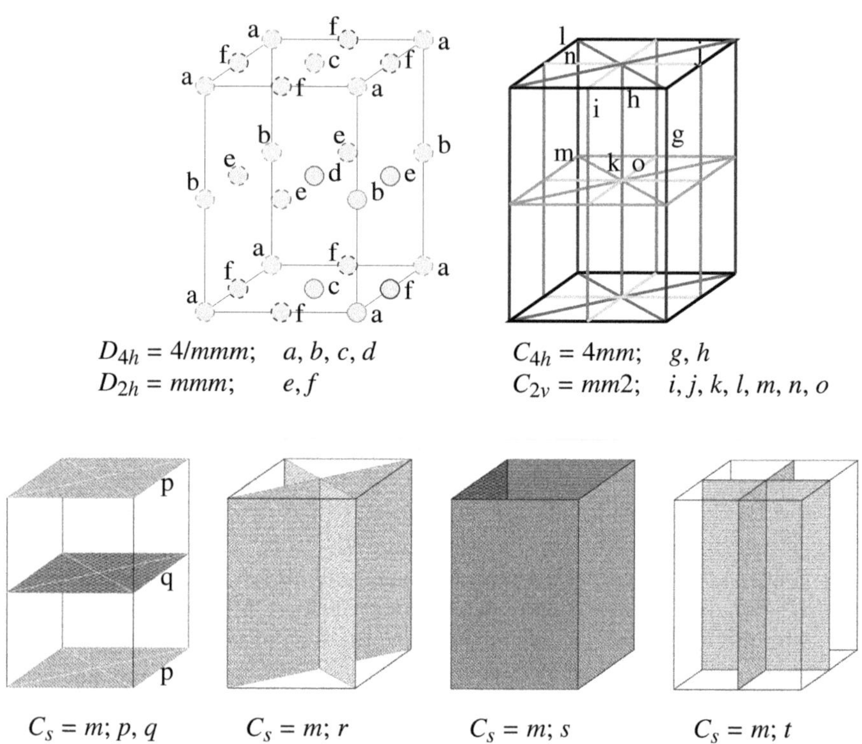

FIG. 4.21 – Different strata for $P4/mmm$.

The detailed description of strata given above and their geometrical representation in Figure 4.21 allows us to single out the fundamental domain for the action of $P4/mmm$ on the three-dimensional space. This fundamental domain including one point from each orbit is shown in Figure 4.22. It is the triangular prism whose internal points are representative of generic orbits with the trivial stabilizer $C_1 = 1$ (stratum u in ITC notation for Wyckoff positions). The boundary of the prism consists of six zero-dimensional strata (vertices of the prism), nine one-dimensional strata (edges of the prism) and five two-dimensional strata faces of the prism). From the topological point of view the fundamental domain (or the space of orbits, or orbifold) is a three-dimensional disk.

Let us now study $I4/mmm$. We can choose a double cell which shows explicitly the D_{4h} point symmetry. The stratification of the double cell by Bravais group action is shown in Figure 4.23 for zero- and one-dimensional strata and in Figure 4.24 for two-dimensional strata. It is instructive to compare the stratification of the ambient space by the $I4/mmm$ group with that by the $P4/mmm$ group studied earlier.

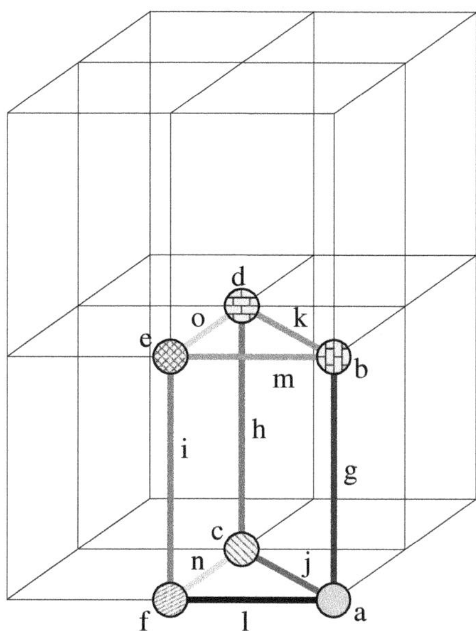

FIG. 4.22 – Representation of the fundamental domain for the $P4/mmm$ three-dimensional Bravais group. Five faces of the prism are formed by five different two-dimensional strata (see Figure 4.21). All internal points belong to generic stratum with the trivial stabilizer C_1.

There are two zero-dimensional strata for $I4/mmm$ action with D_{4h} stabilizer (a, b ITC notation). Each of these strata consists of two points per cell (this reflects the fact that the cell is a double one). Stratum a of $I4/mmm$ includes two strata a and d of $P4/mmm$. Stratum b of $I4/mmm$ includes strata b and c of $P4/mmm$.

There is one stratum c of $I4/mmm$ action with stabilizer D_{2h}. It consists of four points per (double) cell and includes strata e and f of $P4/mmm$ action.

The zero-dimensional stratum d of $I4/mmm$ action has stabilizer D_{2d}. It consists of four points per cell. The zero-dimensional stratum f of $I4/mmm$ action has stabilizer C_{2h}. It consists of eight points per cell. There are no analogs of zero-dimensional strata d and f of the $I4/mmm$ group in the action of the $P4/mmm$ group.

Action of $I4/mmm$ yields formation of six one-dimensional strata. Stratum e with stabilizer C_{4v} includes four points in each orbit per cell and unifies strata g and h of $P4/mmm$ action.

Strata g, h, i, j of $I4/mmm$ action have stabilizer C_{2v} and consequently have eight points of each orbit per (double) cell. Stratum g of $I4/mmm$ action coincides with the stratum i of $P4/mmm$ action. But each eight-point orbit of type g of $I4/mmm$ includes points from two orbits of type i of $P4/mmm$

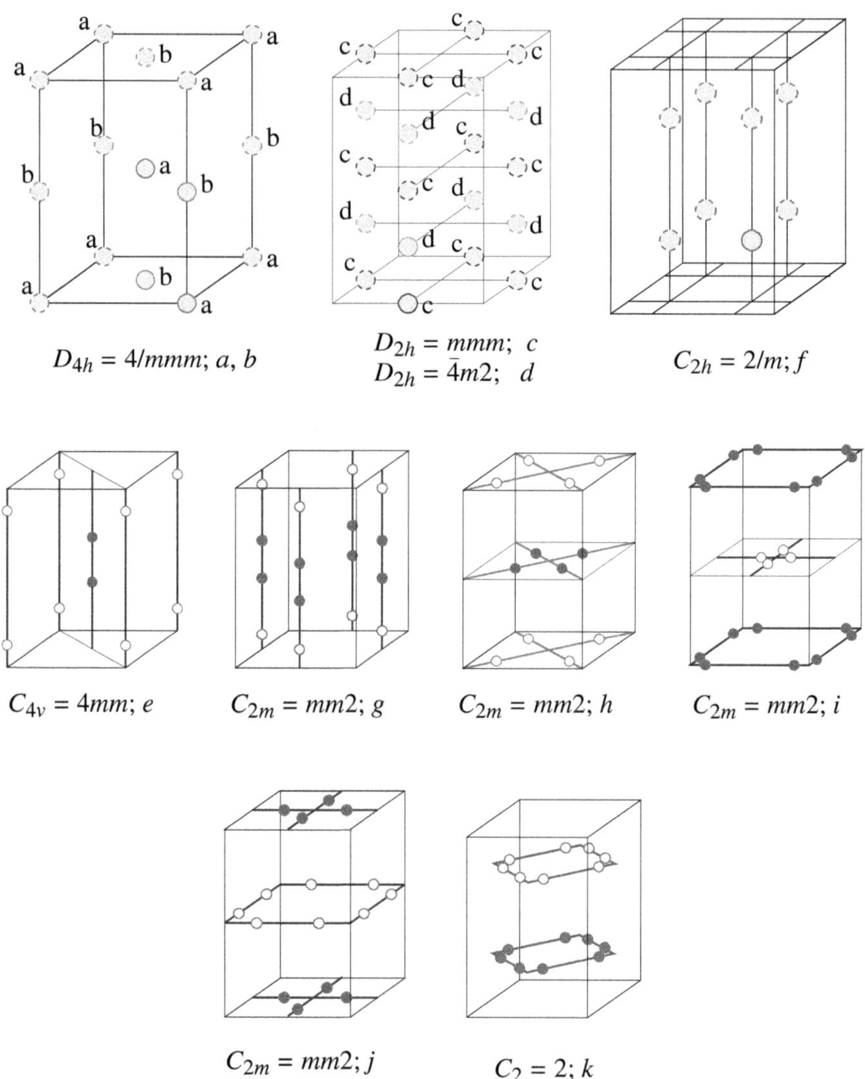

FIG. 4.23 – Different zero- and one-dimensional strata for the $I4/mmm$ three-dimensional Bravais group. For one dimensional strata one orbit is shown by a set of open and filled dots. Filled and open dots distinguish subsets of points related by $GL(2, Z)$ transformation. The point symmetry group transformations relate points of the same type only (transform open points among themselves and filled points among themselves).

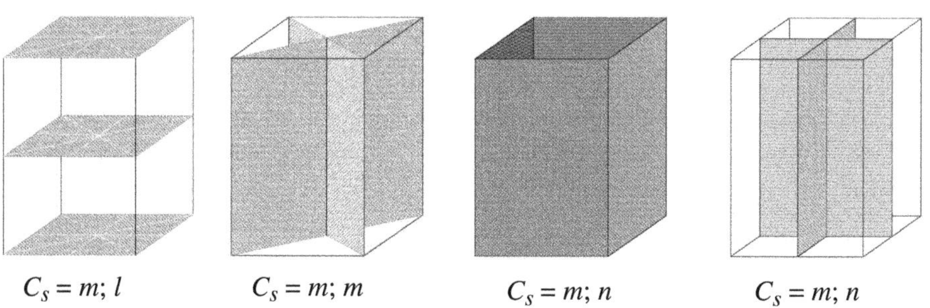

FIG. 4.24 – Different two-dimensional strata for the $I4/mmm$ three-dimensional Bravais group.

action. In a similar way stratum h of $I4/mmm$ includes two strata j and k of the $P4/mmm$ action; the stratum i of $I4/mmm$ includes l and o strata of $P4/mmm$, and stratum j of $I4/mmm$ includes strata m and n of $P4/mmm$.

One dimensional stratum k of $I4/mmm$ has stabilizer C_2. It possesses 16 points per (double) cell in each orbit. There is no analog for this one-dimensional stratum for $P4/mmm$.

There are three two-dimensional strata l, m, n of $I4/mmm$ action. Each stratum consists of 16 points per cell in each orbit. Stratum l consists of 16 open disconnected components (interiors of 16 triangles). Each such triangle includes one point from each orbit. One such triangle should be included in the fundamental domain and in the orbifold.

Stratum m consists of eight open domains (interiors of rectangles - representing 1/4 part of a diagonal section of the prism). Each such rectangle includes two points from each orbit. Consequently the fundamental domain should include 1/8 part of a diagonal section.

Stratum n consists of 16 open rectangles (each rectangle is a quarter of a side of the prism). One such rectangle includes one point from each orbit of the stratum n.

Stratum l of $I4/mmm$ includes strata p and q of $P4/mmm$; stratum m of $I4/mmm$ coincides with stratum r of $P4/mmm$; stratum n of $I4/mmm$ includes strata s and t of $P4/mmm$.

To construct the fundamental domain for $I4/mmm$ action we need to keep one point from each orbit. This should be done with care to exclude appearance of several points from one orbit. We comment now on the construction illustrated in Figure 4.25. We keep five points a, b, c, d, f representing each of five zero-dimensional strata. A one dimensional stratum e is shown in Figure 4.25 consisting of two edges of a prism (without points a and b). These two edges consist of points belonging to different orbits except two upper end points which form one orbit because of the C_2 symmetry transformation present at point f. Four other one-dimensional strata g, h, i, j represented in

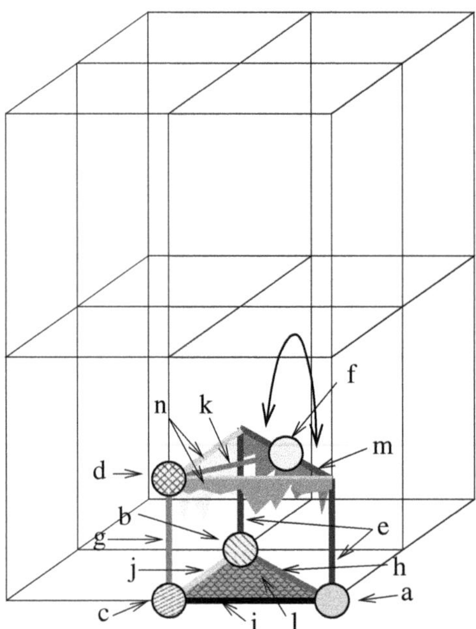

FIG. 4.25 – Representation of the fundamental domain for the $I4/mmm$ three-dimensional Bravais group. Pairs of points on the upper base of the prism situated symmetrically with respect to stratum k should be identified (equivalently, only one point from each pair should be used to represent the fundamental domain).

Figure 4.25 as edges of the prism without end points include each one point from each orbit. The one-dimensional stratum k is represented as a median of the upper face of a prism without end points and also includes one point from each orbit. Thus, they should be included in the fundamental domain and in the orbifold.

The two dimensional stratum l includes all internal points of the triangular base of the prism. All these points belong to different orbits and should be included in the fundamental domain and orbifold.

The two dimensional stratum m fills one rectangular face of the prism but the pairs of points belonging to the upper edge of this face and located symmetrically with respect to point f form one orbit. We need to take only one point from each pairs, or (saying in other way) to identify respective pairs of points on the upper edge. Two other rectangular sides of the prism are filled by points belonging to stratum n. Again all points of these two sides belong to different orbits except for points lying on the two upper edges. These two edges should be identified because C_2 symmetry transformation produced by the stabilizer of k stratum transforms one edge into another.

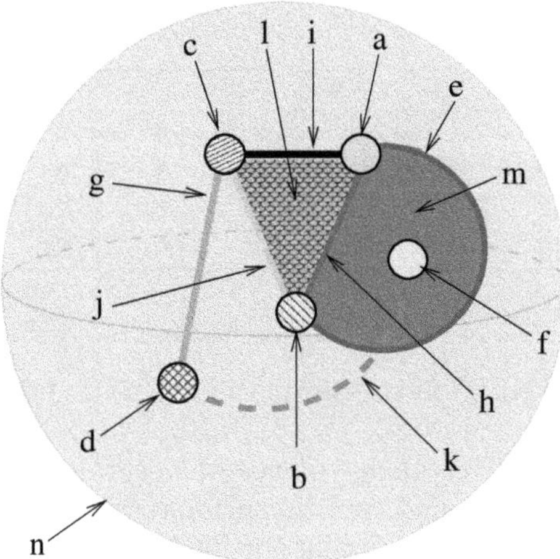

FIG. 4.26 – Schematic representation of the orbifold for the $I4/mmm$ three-dimensional Bravais group as a topological three-dimensional disk. Interior points belong to generic three-dimensional stratum o and to one-dimensional stratum k. All other zero-dimensional (a, b, c, d, f), one-dimensional (e, g, h, i, j, k), and two-dimensional (l, m, n) strata belong to the boundary surface of the disk.

All internal points of the prism represent generic C_1 orbits. The upper face of the prism is formed also by generic C_1 points but points of this face located symmetrically with respect to stratum k belong to the same orbit and should be identified.

It is possible to imagine the topology of the resulting space of orbits by joining two halves of the upper edge of the m face together with two halves of the upper side simultaneously joining two upper edges of the prism. The so obtained object can be described topologically as a three-dimensional disk with five special points on its boundary representing five different zero-dimensional strata. The zero-dimensional strata d and f are connected by one-dimensional stratum k situated inside the disk. All other one-dimensional and two dimensional strata are located on the disk boundary. Their relative positions are shown schematically in Figure 4.26.

4.6 Point symmetry of higher dimensional lattices

In order to describe point symmetry groups for n-dimensional lattices, it is necessary to take into account first of all the crystallographic restrictions

TAB. 4.2 – Several first values of the Euler function.

$\varphi(n)$	+0	+1	+2	+3	+4	+5	+6	+7	+8	+9
0+		1	1	2	2	4	2	6	4	6
10+	4	10	4	12	6	8	8	16	6	18
20+	8	12	10	22	8	20	12	18	12	28
30+	8	30	16	20	16	24	12	36	18	24

on possible types of rotation transformation. The useful observations for this analysis are:

Every rotation in E^n can be represented through rotations in a set of mutually orthogonal one- and two-dimensional subspaces.

Every rotation symmetry of a lattice has a representation in E^n through a unimodular $n \times n$ matrix with integer coefficients.

There is a natural bijection map between the conjugacy classes of the finite subgroups of $GL_n(\mathbb{R})$ and that of O_n.

4.6.1 Detour on Euler function

Definition. Euler function $\varphi(n)$ is an arithmetic function which gives for a positive integer n the number of integers k in the range $1 \leq l \leq n$ for which $\gcd(n,k) = 1$, where "gcd" means greatest common divisor.

The Euler function is multiplicative. This means that if $\gcd(m,n) = 1$, then $\varphi(mn) = \varphi(m)\varphi(n)$.

If p is prime, then evidently $\varphi(p) = p - 1$. For $\varphi(p^2)$ we immediately get $\varphi(p^2) = p^2 - p$ and more generally for any integer $k > 1$ $\varphi(p^k) = p^k - p^{k-1} = p^k(1 - \frac{1}{p})$.

As soon as for any $n > 1$ we have a unique expression $n = p_1^{k_1} \cdot p_2^{k_2} \cdots p_r^{k_r}$ in terms of the prime integers $p_1, p_2, \dots, p_r$ with $k_i \geq 1$, applying repeatedly the multiplicative property of φ and the formula for $\varphi(p^k)$ we get the Euler product formula for $\varphi(n)$:

$$\begin{aligned} \varphi(p_1^{k_1} \cdots p_r^{k_r}) &= \varphi(p_1^{k_1}) \cdots \varphi(p_r^{k_r}) = p_1^{k_1} \cdots p_r^{k_r} \left(1 - \frac{1}{p_1}\right) \cdots \left(1 - \frac{1}{p_r}\right) \\ &= n \left(1 - \frac{1}{p_1}\right) \cdots \left(1 - \frac{1}{p_r}\right). \end{aligned} \tag{4.4}$$

Several initial values of the Euler function are given in Table 4.2

In order to see why the Euler function is relevant to the construction of possible cyclic symmetry groups of n-dimensional lattices let us first remember the relation between the Euler function, roots of unity, cyclotomic polynomials, and companion matrices.

4.6.2 Roots of unity, cyclotomic polynomials, and companion matrices

The roots of $x^m - 1$ are called m-th roots of unity. They are

$$\{e^{2k\pi i/m} = \cos(2k\pi/m) + i\sin(2k\pi/m): \quad k = 1, 2, \ldots, m\}. \tag{4.5}$$

In the complex plane, the roots of unity are placed regularly on the unit circle starting at 1. They form a cyclic group of order m under operation of multiplication of complex numbers. Generators of this group are called primitive m-th roots of unity. Obviously, the root $e^{2k\pi i/m}$ is primitive if $\gcd(k, m) = 1$, where gcd stands for greatest common divisor. Alternatively, we can say that k and m should be relatively prime. Consequently, the number of different primitive m-th roots of unity is given by $\varphi(m)$, Euler's totient function.

The d-th cyclotomic polynomial $\Phi_d(z)$ is defined by

$$\Phi_d(z) = \prod_{\omega}(z - \omega) \tag{4.6}$$

where ω ranges over all primitive d-th roots of unity. By construction the degree of $\Phi_d(z)$ is the values of the Euler function $\phi(d)$. The cyclotomic polynomials $\Phi_d(z)$ have integer coefficients[4].

For d prime the cyclotomic polynomial has degree $d - 1$ and the explicit form

$$\Phi_d(\mathbb{Z}) = \sum_{i=0}^{d-1} z^i \quad \text{for } d \text{ prime}. \tag{4.7}$$

For several other low d values the cyclotomic polynomials are

$$\Phi_1(\mathbb{Z}) = z - 1; \quad \Phi_4(\mathbb{Z}) = z^2 + 1; \quad \Phi_6(\mathbb{Z}) = z^2 - z + 1; \tag{4.8}$$

$$\Phi_8(\mathbb{Z}) = z^4 + 1; \quad \Phi_9(\mathbb{Z}) = z^6 + z^3 + 1; \tag{4.9}$$

$$\Phi_{10}(\mathbb{Z}) = z^4 - z^3 + z^2 - z + 1; \quad \Phi_{12}(\mathbb{Z}) = z^4 - z^2 + 1. \tag{4.10}$$

Using cyclotomic polynomials one can for a given integer m construct a matrix of order m and of dimension $\phi(m) \times \phi(m)$. For this it is sufficient to take a "companion" matrix whose characteristic polynomial is $\Phi_m(z)$. Generically, for a polynomial $p(z) = z^k + b_1 z^{k-1} + \cdots + b_{k-1}z + b_k$ the companion matrix C, i.e. matrix with characteristic polynomial being $p(z)$, has the following

[4] When $k < 105$, all coefficient values are $1, 0, -1$ but for $k = 105$ (this is the smallest integer product of three distinct odd primes), some ± 2 appear. The absolute value of the coefficients of the cyclotomic polynomials are unbounded when $k \to \infty$.

form

$$C = \begin{pmatrix} 0 & 1 & 0 & . & . & . & 0 \\ 0 & 0 & 1 & . & . & . & 0 \\ 0 & 0 & 0 & . & . & . & 0 \\ . & . & . & . & & & . \\ . & . & . & & . & & . \\ . & . & . & & & . & . \\ -b_k & -b_{k-1} & -b_{k-2} & . & . & . & -b_1 \end{pmatrix}. \tag{4.11}$$

In particular in (4.11) we have

$$b_k = (-1)^k \det(C); \quad b_1 = -\mathrm{Tr}(C). \tag{4.12}$$

For a cyclic group $\mathbb{Z}_m$ its regular representation is generated by the "companion matrix" M whose characteristic polynomial is $P_M = z^m - 1$. This polynomial can be expressed as a product

$$Z^m - 1 = \prod_{d|m} \Phi_d(z), \tag{4.13}$$

of cyclotomic polynomials over all divisors d of m. For example, $z^4 - 1 = \Phi_4\Phi_2\Phi_1 = (z^2+1)(z+1)(z-1)$.

Since the coefficients of $\Phi_d(\mathbb{Z})$ are integers the corresponding companion matrix A of $\Phi_d(\mathbb{Z})$ is an integer matrix. Also, since $\Phi_d(\mathbb{Z})$ is an irreducible factor of $\mathbb{Z}^m - 1$, $A^m = I$, because the matrix satisfies its own characteristic equation, and this is true for no lower power of A. This means that the matrix A has order m.

4.6.3 Crystallographic restrictions on cyclic subgroups of lattice symmetry

We formulate now a theorem giving possible orders of elements of finite symmetry groups of a lattice. The formulation of the theorem below and its proof follows [64].

Theorem 3 *Let $m = p_1^{e_1} p_2^{e_2} \cdots p_l^{e_l}$ with $p_1 < p_2 < \cdots < p_l$. Then $GL_n(\mathbb{Q})$, and hence $GL_n(\mathbb{Z})$, has an element of order m if and only if*

$$\sum_{i=1}^{l} (p_i - 1)p_i^{e_i - 1} - 1 \leq n \quad \text{for} \quad p_1^{e_1} = 2, \tag{4.14}$$

or

$$\sum_{i=1}^{l} (p_i - 1)p_i^{e_i - 1} \leq n \quad \text{otherwise}. \tag{4.15}$$

Proof. Let $W : \mathbb{Z} \to \mathbb{Z}$ be defined for $m = p_1^{e_1} p_2^{e_2} \cdots p_l^{e_l}$ by $W(m) = \sum_{i=1}^{l} (p_i - 1) p_i^{e_i - 1} - 1$ when $p_1^{e_1} = 2$ and $W(m) = \sum_{i=1}^{l} (p_i - 1) p_i^{e_i - 1}$ otherwise. Then theorem 3 can be reformulated as: $GL_n(\mathbb{Q})$ has an element of order m if and only if $W(m) \leq n$. Suppose that m is a positive integer with $W(m) \leq n$; we produce an element of $GL_n(\mathbb{Q})$ of order m. First suppose $m = p_1^{e_1} p_2^{e_2} \cdots p_l^{e_l}$ with $p_1^{e_1} \neq 2$. For each i we can construct matrix A_i of dimension $(p_i - 1) p_i^{e_i - 1} \times (p_i - 1) p_i^{e_i - 1}$, i.e of dimension $\varphi(p_i) \times \varphi(p_i)$, and of order $p_i^{e_i}$. Then we can construct matrix B

$$B = A_1 \oplus A_2 \oplus \cdots \oplus A_l = \begin{bmatrix} A_1 & 0 & \cdot & \cdot & \cdot & 0 \\ 0 & A_2 & \cdot & \cdot & \cdot & 0 \\ \cdot & \cdot & \cdot & & & \cdot \\ \cdot & \cdot & & \cdot & & \cdot \\ \cdot & \cdot & & & \cdot & \cdot \\ 0 & 0 & \cdot & \cdot & \cdot & A_l \end{bmatrix} \tag{4.16}$$

which has order m. If $W(m) = n$, then $A = B$ is the desired matrix. If $W(m) < n$, then $A = B \oplus I_s$ is the desired matrix, where $s = n - W(m)$. Now suppose $p_1^{e_1} = 2$. Then $W(m/2) = \sum_{i=2}^{t} (p_i - 1) p_i^{e_i - 1} \leq n$, and applying the previous construction, $GL_n(\mathbb{Q})$ has an element A of order $m/2$. Since $m/2$ is odd, the matrix $(-A)$ has order m. For the proof of the inverse statement which is more technical we refer to [64]. □

Note, that both sums (4.14, 4.15) introduced in theorem 3 are always even. This leads to the following interesting Corollary

Corollary 2 *$GL_{2k}(\mathbb{Q})$ has an element of order m if and only if $GL_{2k+1}(\mathbb{Q})$ does.*

We have already seen that both two-dimensional and three dimensional lattices have elements of order $2, 3, 4$, and 6. Similarly, both four-dimensional and five-dimensional lattices have elements of orders $2, 3, 4, 5, 6, 8, 10$, and 12.

Table 4.3 (taken from [26]) gives the orders of cyclic groups which are allowed symmetries of n-dimensional lattice but do not appear for smaller dimensions due to crystallographic restrictions.

4.6.4 Geometric elements

Now we can construct the geometric elements for n-dimensional lattices. Following Hermann [60], we denote simply by (k) the companion matrix with characteristic polynomial Φ_k. For example:

$$(1) = I = -(2), \quad (3) = \begin{pmatrix} 0 & 1 \\ -1 & -1 \end{pmatrix} = -(6), \quad (4) = \begin{pmatrix} 0 & 1 \\ -1 & 0 \end{pmatrix}$$

$$(5) = -(10) = \begin{pmatrix} 0 & 1 & 0 & 0 \\ 0 & 0 & 1 & 0 \\ 0 & 0 & 0 & 1 \\ -1 & -1 & -1 & -1 \end{pmatrix}, (8) = \begin{pmatrix} 0 & 1 & 0 & 0 \\ 0 & 0 & 1 & 0 \\ 0 & 0 & 0 & 1 \\ -1 & 0 & 0 & 0 \end{pmatrix}$$

TAB. 4.3 – The orders of cyclic groups which are allowed symmetries of an n-dimensional lattice but do not appear for smaller dimensions due to crystallographic restrictions [26].

n	orders of cyclic groups appearing for this n
1	1,2
2	3,4,6
4	5,8,10,12
6	7,9,14,15,18,20,24,30
8	16,21,28,36,40,42,60
10	11,22,35,45,48,56,70,72,84,90,120
12	13,26,33,44,63,66,80,105,126,140,168,180,210
14	39,52,55,78,88,110,112,132,144,240,252,280,360,420
16	17,32,34,65,77,99,104,130,154,156,165,198,220,264,315,330,336,504,630,840
18	19,27,38,51,54,68,91,96,102,117,176,182,195,231,234,260,308,312,390,396, 440,462,560,660,720,1260
20	25,50,57,76,85,108,114,136,160,170,204,208,273,364,385,468,495,520,528, 546,616,770,780,792,924,990,1008,1320,1680
22	23,46,75,95,100,119,135,143,150,152,153,190,216,224,228,238,255,270,286, 288,306,340,408,455,480,510,585,624,693,728,880,910,936,1092,1155,1170, 1386,1540,1560,1848,1980
24	69,92,133,138,171,189,200,266,272,285,300,342,357,378,380,429,456,476, 540,570,572,612,672,680,714,819,858,1020,1040,1232,1365,1584,1638,1820

$$(12) = \begin{pmatrix} 0 & 1 & 0 & 0 \\ 0 & 0 & 1 & 0 \\ 0 & 0 & 0 & 1 \\ -1 & 0 & 1 & 0 \end{pmatrix}. \tag{4.17}$$

The $\mathbb{Q}$-irreducible representations of $\mathbb{Z}_m$ are generated by the matrices (d), $d|m$ with d dividing m; the only faithful one is that generated by (m). But can one obtain faithful reducible representations? Indeed any faithful n-dimensional integral representation of $\mathbb{Z}_m$ is generated by the matrix $A_m = \oplus_i c_i(k_i)$ (the c_i's are the multiplicities of the matrices (k_i)) where the set of different integers k_i satisfies the two conditions: $\sum_i c_i\varphi(k_i) = n$, $\mathrm{lcm}_i(k_i) = m$.

That establishes the "crystallographic condition"; we have proven more since we know how to build all $\mathbb{Q}$-inequivalent integral representations of the cyclic subgroups of $GL_n(\mathbb{Z})$. Their generators are all the possible matrices A_m; they form a set of representatives of the conjugacy classes of the elements of finite order of the group $GL_n(\mathbb{Q})$; they are called the *geometric elements of dimension n* by the crystallographers. We shall use the Hermann notation[5]

[5] In [60] Hermann did not impose to the matrices (k_i) to be distinct, so it did not exhibit the c_i's in the notation.

TAB. 4.4 – Orders of transitive cyclic subgroups of point symmetry groups of n-dimensional lattices. Only orders of transitive groups which do not appear in lower dimensions are indicated for lattices of dimension $n \leq 16$.

n	allowed orders
1	1,2
2	3, 4, 6
4	5, 8, 10, 12
6	7, 9, 14, 18
8	15, 16, 20, 24, 30
10	11, 22
12	13, 21, 26, 28, 36
14	-
16	17, 32, 34, 40, 48, 60

TAB. 4.5 – Number γ_n of geometric elements in dimension n.

n	1	2	3	4	5	6	7	8	9
γ_n	2	6	10	24	38	78	118	224	330

$(\prod_i k_i^{c_i})$ for the matrix $A_m = \oplus_i c_i(k_i)$. We summarize these results by the theorem.

Theorem 4 *The geometric classes of the cyclic point groups in dimension n can be labeled by the Hermann symbols: $(\prod_i k_i^{c_i})$ with $\sum_i c_i \varphi(k_i) = n$. The order of the corresponding cyclic group is $m = \mathrm{lcm}_i(k_i)$, the least common multiple of the k_i's.*

The last equality was introduced in Hermann's paper [60] in the English abstract (the paper is in German). For the dimension n, he called the cyclic groups $\mathbb{Z}_m$ with $\varphi(m) = n$ transitive and called intransitive the cyclic groups $\mathbb{Z}_m$ which are reducible on $\mathbb{Q}$.

A list of orders of transitive cyclic groups which do not appear in smallest dimensions is given in Table 4.4. It follows directly from inversion of the Euler function (see Table 4.2).

We denote by γ_n the number of geometric elements of dimension n. In his paper Hermann gave the value of γ_n for $n \leq 6$ and $n = 8$. Some values of γ_n are listed in Table 4.5:

We illustrate the construction of all geometric elements on the example of an 8-dimensional lattice. We look for different possible splitting of dimension $n = 8$ into $\mathbb{Z}$-irreducible blocks of dimension 8, 6, 4, 2, 1, and count the number of different cyclic groups with a prescribed block structure.

i) First, there are five transitive groups represented by eight-dimensional irreducible (over $\mathbb{Z}$) integer matrices of orders: 15, 16, 20, 24, and 30.

This follows from the inversion table of the Euler totient function (see Table 4.4).

ii) Next, let us consider geometric elements having a 6-dimensional irreducible (over $\mathbb{Z}$) block and two one-dimensional or one two-dimensional block. There are four different choices for a six-dimensional block - elements of order 7, 9, 14, and 18 - as follows from Table 4.4. There are three possibilities for two one-dimensional blocks, $(1^2), (1.2)$, and (2^2) and three possibilities for two-dimensional irreducible blocks, $(3), (4)$, and (6). Combining these six possibilities with four choices for 6-dimensional irreducible blocks we have 24 different elements.

iii) The next possibility is: two irreducible (over $\mathbb{Z}$) blocks of dimension 4. There are 10 different cases: $(5^2), (8^2), (10^2), (12^2), (5.8)$, $(5.10), (5.12), (8.10), (8.12)$, and (10.12).

iv) One four-dimensional irreducible block can be combined with a four-dimensional block formed in its turn from one and two-dimensional blocks. For a four-dimensional block with structure 1111 there are five elements (1^4), $(1^3.2)$, $(1^2.2^2)$, (1.2^3), and (2^4). A four-dimensional block with structure 211 gives nine elements $(m.1^2), (m.2.1)$, and $(m.2^2)$ with $m = 3, 4, 6$. The block structure 22 corresponds to six elements (3^2), (4^2), (6^2), (3.4), (3.6), and (4.6). After combination with four possibilities for a 4-dimensional irreducible block we get 80 geometric elements.

v) Finally we need to analyze eight-dimensional blocks having at most two-dimensional irreducible sub-blocks. There are 15 elements with block structure 2222. There are 30 elements with block structure 22211, 30 elements with the block structure 221111; 21 elements with block structure 2111111, and nine elements with only one-dimensional blocks 11111111.

The total number of geometric elements for an eight-dimensional lattice is equal to 224.

Since $\varphi(m)$ is even when $m > 2$, one obtains these classes for the odd dimension $2n+1$ from those of the even dimension $2n$ by adding one of the two one dimensional matrices (1) or (-1) of order 1, 2, respectively. To compute the values of the table, or more, we define the following expressions:
ν_{2m} = the number of integers k satisfying the equation $\varphi(k) = 2m$ and $\nu_{2m}^{(k)} = \binom{\nu_{2m}-1+k}{k}$ (remark that $\nu_{2m}^{(1)} = \nu_{2m}$). We define also:

$$\mu_0 = 1, \quad \mu_2 = \nu_2, \quad \mu_4 = \nu_4 + \nu_2^{(2)}, \quad \mu_6 = \nu_6 + \nu_4\nu_2 + \nu_2^{(3)},$$
$$\mu_8 = \nu_8 + \nu_6\nu_2 + \nu_4^{(2)} + \nu_4\nu_2^{(2)} + \nu_2^{(4)}, \cdots \tag{4.18}$$

then

$$\gamma_{2n+1} - \gamma_{2n} = \tau_{2n} = \gamma_{2n} - \gamma_{2n-1}, \text{ where } \tau_{2n} = \sum_{m=0}^{n} \mu_{2m}. \tag{4.19}$$

Let us denote by $\rho(m)$, the smallest n such that $GL_n(\mathbb{Q})$ contains a cyclic group of order m. As a corollary of Theorem 4, when m is a power of a prime number, $\rho(m) = \varphi(m)$; but when m is divisible by different primes one has always the inequality[6] $\rho(m) < \varphi(m)$ as was noted first in [60]. Indeed the value of $\rho(m)$ is for all cases:

$$m = 2^k \prod_i p_i^{k_i},$$

$$\rho(m) = (k \geq 2)\varphi(2^k) + \sum_i \varphi(p_i^{k_i}) = (k \geq 2)2^{k-1} + \sum_i (p_i - 1)p_i^{k_i - 1}, \quad (4.20)$$

where p_i are odd primes and $(k \geq 2)$ is an example of a Boolean function; its value is 1 or 0 depending whether the relation between the brackets is true or false. Because $\rho(m)$ is the sum of the φ's of the essential factors of m while $\varphi(m)$ is their product, the more factors has m the smaller is $\rho(m)$ compare to $\varphi(m)$; examples $\rho(210) = 12$, $\varphi(210) = 48$, $\rho(2310) = 22$, $\varphi(2310) = 480$. Notice also that (4.20) shows that the *same orders of cyclic groups appear in* $GL_{2n}(\mathbb{Q})$ *and* $GL_{2n+1}(\mathbb{Q})$.

Table 4.6 gives the list of the geometric elements for dimension 2, 3, and 4; they define the cyclic point groups in these dimensions. Table 4.7 gives for even dimensions up to 10, the geometric elements whose order does not appear in smaller dimensions.

[6] Although many papers, books and dictionaries of mathematics (at the entry "crystallography") state the contrary. This error was already pointed to in [61].

TAB. 4.6 – List of geometric elements in dimension 2, 3, 4. For each dimension, we give first the Hermann notation, then the notation in [14] for $n = 2$, 3; for $n = 3$ we give also the Schoenflies notation in [14] for the generated cyclic group; then the values of the order of the elements, and the values of the independent coefficients of the characteristic polynomial (defined in (4.12)); notation: t=trace, d=determinant.

n=1	(1)	(2)								
n=2	(1^2)	(1.2)	(2^2)	(3)	(4)	(6)				
ITC	1	m	2	3	4	6				
order	1	2	2	3	4	6				
$d = b_2$	1	-1	1	1	1	1				
$t = -b_1$	2	0	-2	-1	0	1				
n=3	(1^3)	$(2^2.1)$	(3.1)	(4.1)	(6.1)	(2^3)	(2.1^2)	(6.2)	(4.2)	(3.2)
ITC	1	2	3	4	6	$\bar{1}$	m	$\bar{3}$	$\bar{4}$	$\bar{6}$
SCH	1	C_2	C_3	C_4	C_6	C_i	C_s	C_{3i}	S_4	C_{3h}
order	1	2	3	4	6	2	2	6	4	6
$d = -b_3$	1	1	1	1	1	-1	-1	-1	-1	-1
$t = -b_1$	3	-1	0	1	2	-3	1	-2	-1	0
n=4	(1^4)	$(1^3.2)$	$(1^2.2^2)$	(1.2^3)	(2^4)	(3.1^2)	(3^2)	(4.1^2)	(4.2.1)	(4.2^2)
order	1	2	2	2	2	3	3	4	4	4
$d = b_4$	1	-1	1	-1	1	1	1	1	-1	1
$t = -b_1$	4	2	0	-2	-4	1	-2	2	0	-2
b_2	6	0	-2	0	6	0	3	2	0	2
n=4	(4^2)	(3.2.1)	(3.2^2)	(6.1^2)	(6.2.1)	(6.2^2)	(6.3)	(6^2)	(5)	(8)
order	4	6	6	6	6	6	6	6	5	8
$d = b_4$	1	-1	1	1	-1	1	1	1	1	1
$t = -b_1$	0	-1	-3	1	0	-1	0	2	-1	0
b_2	2	0	0	4	0	0	1	3	1	0
n=4	(10)	(4.3)	(6.4)	(12)						
order	10	12	12	12						
$d = b_4$	1	1	1	1						
$t = -b_1$	1	-1	1	0						
b_2	1	2	2	-1						

TAB. 4.7 – For each dimension given in the left part, the orders of cyclic groups of $GL_{2n}(\mathbb{Q})$, $1 \leq n \leq 5$ which do not appear in a smaller dimension; in the right part, the Hermann notation of a generator for each representation of this dimension is given for the cyclic group with the new order.

1	1,2,	(1); (2);
2	3,4,6,	(3);(4);(6);
4	5,8,10,12	(5);(8);(10); (12),(3.4),(4.6);
6	7,9,14,15,18,20,	(7);(9);(14);(3.5);(18);(4.5),(4.10);
	24,30,	(3.8),(6.8); (3.10),(6.5),(6.10);
8	16,21,28,36,	(16);(3.7);(4.7),(4.14);(4.9),(4.18);
	40,42,	(5.8),(8.10);(6.7),(6.14);
	60,	(3.4.5),(3.4.10),(4.6.5),(4.6.10),(5.12),(10.12);
10	11,22,35,45,48,56,	(11);(22);(5,7); (5.9);((3.16),(6.16);(8.7),(8.14);
	70,72,	(5.14),(10.14);(8.9),(8.18);
	84,	(3.4.7),(3.4.14),(4.6.7),(4.6.14),(12.7),(12.14);
	90,120	(10.9),(10.18);(3.5.8),(3.10.8),(6.5.8),(6.10.8);

Chapter 5

Lattices and their Voronoï and Delone cells

In this section we study lattices from the point of view of their tilings by polytopes.

5.1 Tilings by polytopes: some basic concepts

Definition: polytope A polytope P is a compact body with a nonempty interior whose boundary ∂P is the union of a finite number of facets, where each facet is the $(n-1)$-dimensional intersection of P with a hyperplane.

Two-dimensional polytopes are called polygons; three-dimensional polytopes are called polyhedra.

Definition: k-face (of a polytope) For $k = 0, \ldots, n-2$, a k-dimensional face (or k-face, for short) of a polytope is an intersection of at least $(n-k)$ facets that is not contained in the interior of a j-face for any $j > k$.

Thus a 0-face of a polytope is a point that lies in the intersection of at least n facets but not in the interior of any 1-face, 2-face, etc. As a customary, we use the terms vertex and edge, respectively for the 0-dimensional and 1-dimensional faces of tiles, and facets for faces of dimension $n-1$.

In the tilings we will study, the tiles will be convex polytopes in E^n. Remember that the polytope P is convex if P contains the line segments joining any two points in P or on its boundary.

Definition: tiling A tiling $\mathcal{T}$ of E^n is a partition of E^n into a countable number of closed cells with non-overlapping interiors:

$$\mathcal{T} = \{T_1, T_2, \ldots\}, \quad \bigcup T_i = E^n, \quad \text{int } T_i \cap \text{int } T_j = \emptyset \text{ if } i \neq j. \tag{5.1}$$

The words *tiling* and *tessellation* are used interchangeably; similarly, *tiles* are often called cells.

Definition: prototile set A prototile set $\mathcal{P}$ for a tiling $\mathcal{T}$ is a set of polytopes such that every tile of $\mathcal{T}$ is an isometric copy of an element of $\mathcal{P}$.

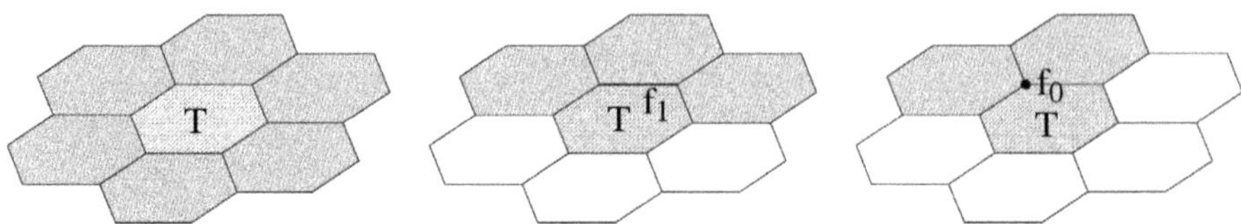

FIG. 5.1 – Two-dimensional tiling with a single prototile T. Left: Corona of a tile T. Middle: Corona of a 1-face f_1 (facet) of a tiling. Right: Corona of a 0-face f_0 (vertex) of a tiling.

When the prototile set contains a single tile T, the tiling is said to be *monohedral*. A prototile set does not, in general, characterize a tiling completely. Indeed a single prototile may admit different tilings. There are uncountably many Penrose tilings of the plane with the same prototile set of two rhombs.

Definition: convex, facet-to-facet, locally finite (tilings) A convex tiling is one whose tiles are convex. A tiling is said to be facet-to-facet if the intersection of the interior of any two facets is either empty or coincides with both facet interiors. A tiling is said to be locally finite if every ball in E^n of finite radius meets only finitely many tiles.

We state without proof the important fact [59]:

Proposition 5 *(Gruber and Ryshkov) A locally finite convex tiling in E^n is facet-to-facet if and only if it is k-face-to-k-face ($k = 0, 1, \ldots, n-2$).*

Definition: corona of a k-face. Let $\mathbf{f_k}$ be a k-face of a tiling $\mathcal{T}$, where $0 \leq k \leq n$. The (first) corona of $\mathbf{f_k}$ is the union of $\mathbf{f_k}$ and the tiles that meet it, i.e., the tiles whose intersection with $\mathbf{f_k}$ is nonempty. When $k = 0$, the corona is called a vertex corona. When $k = n$ (i.e. when $\mathbf{f_k}$ is a tile T) the corona is called the corona of T.

Figure 5.1 shows different corona for an example of a two-dimensional tiling.

Definition: parallelotope A convex prototile P of a monohedral tiling in which the tiles are translates of P is called a parallelotope. Every convex parallelotope admits a facet-to-facet tiling; this is a corollary of the Venkov-McMullen's theorem [92, 67] characterizing convex parallelotopes in arbitrary dimension. To formulate this theorem, we need the concept of a *belt*:

Definition: belt A belt of a parallelotope P is a complete set of parallel $(n-2)$-faces of P.

Note that when $n = 3$, the $(n-2)$-faces of P are edges. Figure 5.2 shows the two belts of a hexagonal prism.

Theorem 5 *(Venkov, McMullen) A convex polytope P is a parallelotope if and only if it satisfies the following three conditions:*

1. *P is centrosymmetric;*
2. *all facets of P are centrosymmetric;*
3. *all belts of P have length four or six.*

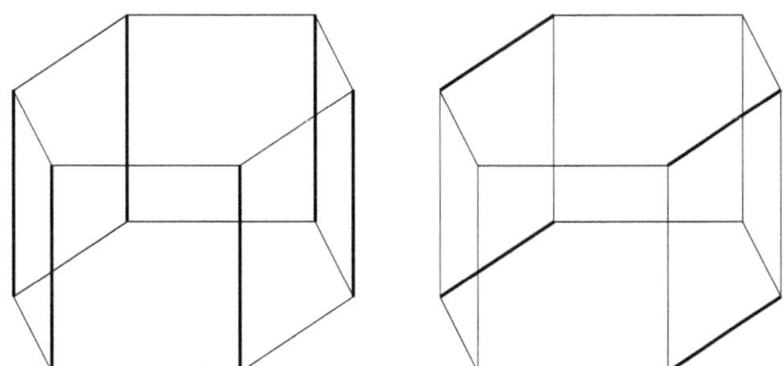

FIG. 5.2 – Different belts for a hexagonal prism. Left: Belt formed by six edges, i.e. by six $(d-2)$-faces. Right: One of three belts formed by four edges.

Corollary 3 *Of the five Platonic (regular) solids in E^3, only the cube is a parallelotope.*

It follows immediately that all other Platonic solids have triangular or pentagonal facets which are not centrosymmetric. (See figure 5.10.)

Central symmetry of faces implies also that within a belt the number of $(n-2)$-faces equals the number of facets.

5.1.1 Two- and three-dimensional parallelotopes

Two-dimensional parallelotopes are called parallelogons; in three dimensions they are parallelohedra. Since a monohedral tiling of the plane by convex polygons can have at most six edges, parallelogons are either parallelograms or centrosymmetric hexagons. To characterize their combinatorial type it is sufficient to use single labels indicating the number of edges (1-faces) or number of vertices (0-faces) which coincide. In order to use the same notation for two-, three-, and arbitrary d-dimensional parallelohedra we prefer to use symbols $\mathbf{N}_{(d-1)}.\mathbf{N}_0$ indicating both, the number of facets, i.e. $(d-1)$-faces, and the number of 0-faces.

Two combinatorial types of two-dimensional parallelogons are therefore **4.4** and **6.6**. They were described by Dirichlet in 1850 [45]. For $n=3$ Fedorov found five combinatorial types of parallelohedra in 1885 [12].[1] We label combinatorial types of three-dimensional parallelohedra by $\mathbf{N}_2.\mathbf{N}_0$ showing number of 2-faces and of 0-faces of a parallelohedron. The five combinatorial types of three-parallelohedra are: the cube **6.8**, the hexagonal prism **8.12**, the rhombic dodecahedron **12.14**, the elongated dodecahedron **12.18**, and the truncated octahedron **14.24**. They are shown in Figures 5.4-5.8.

These five combinatorial types of parallelohedra can be related by the operation consisting in shrinking one of the belts. Such operation is very

[1] In 1929 Delone found 51 combinatorial type for $n=4$; this was corrected to 52 by Shtogrin in 1972 [41, 87]. The number, 103769, of combinatorial types in five dimensions was determined by Engel [51].

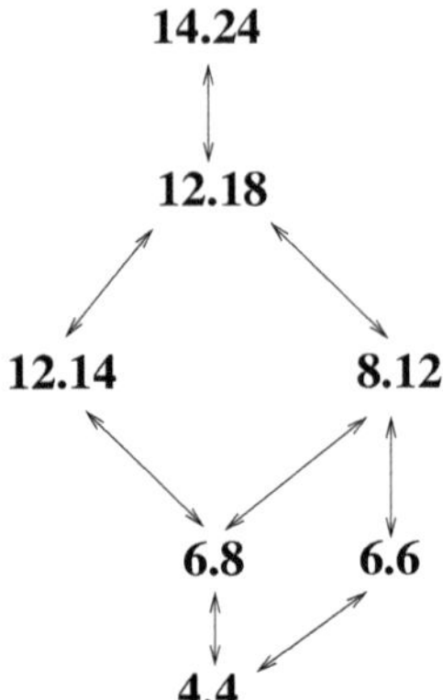

FIG. 5.3 – Zone contraction family of three- and two-dimensional parallelohedra. For three-dimensional polytopes the zone contraction can be equivalently described as belt shrinking.

important for a general classification of parallelohedra in arbitrary dimension. But instead of belts (set of parallel $(n-2)$-faces) one needs to consider zones (the set of all edges (1-faces) parallel to a given vector). Obviously, for three-dimensional parallelohedra zones are equivalent to belts. Nevertheless, to be consistent with more general treatment we prefer to name the operation of shrinking of belts for three-dimensional parallelohedra the zone contraction operation. The zone contraction family of three-dimensional parallelohedra is represented in Figure 5.3. It includes the zone contraction operation which reduces three-dimensional polytopes to two-dimensional ones and also the zone contraction between two-dimensional polytopes. Concrete geometrical visualization of a zone contraction for all possible pairs of three-dimensional Voronoï parallelohedra is shown in Figures 5.4-5.8. Contractions for three dimensional parallelohedra are complemented in Figure 5.3 by zone contraction operations transforming three-dimensional cells into two-dimensional: These are **8.12** $\rightarrow$ **6.6** and **6.8** $\rightarrow$ **4.4**. Also there is one zone contraction between two-dimensional cells: **6.6** $\rightarrow$ **4.4**. Note, that with each zone contraction operation we can associate inverse operation which is named zone extension.

5.2 Voronoï cells and Delone polytopes

We return to Delone sets Λ and to Voronoï cells and Delone polytopes introduced briefly in Chapter 3.

First we note that the Voronoï cells of the points of Λ tile E^n; that is, they fill E^n without gaps or overlapping interiors. We denote the tiling by $\mathcal{T}_\Lambda$. This follows from the fact that every point of E^n is closer to a unique point

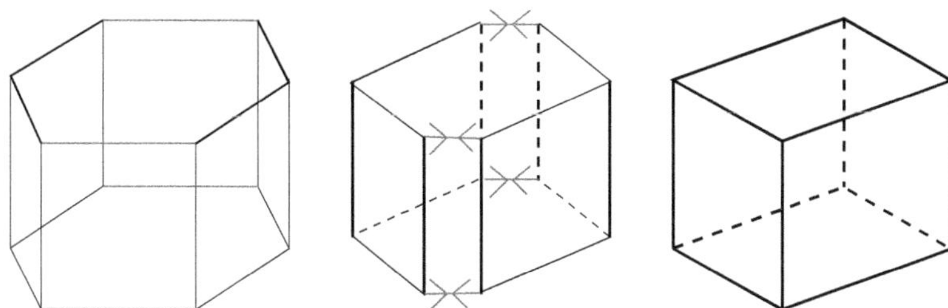

FIG. 5.4 – Contraction of the **8.12** cell (hexagonal prism) into the **6.8** cell (cube). Four edges shrink to zero, two quadrilateral facets disappear, two hexagonal facets transform into quadrilateral ones. There are three 4-belts to shrink.

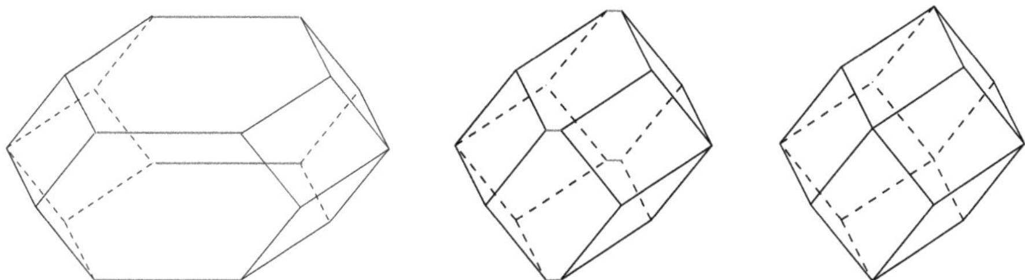

FIG. 5.5 – Contraction of the **12.18** cell (elongated dodecahedron) into the **12.14** cell (rhombic dodecahedron). Four edges shrink to zero and four hexagonal facets transform into quadrilateral ones. There is only one 4-belt to shrink.

of Λ, or is equidistant from two or more of them. The tiling $\mathcal{T}_\Lambda$ is locally finite and facet-to-facet.

Theorem 6 *The vertices of the Voronoï cells of* Λ *are the centers of its holes.*

Proof. A vertex v of a Voronoï cell $D(p)$ is the intersection of at least $n+1$ hyperplanes bisecting the vectors from p to other points $q_1, \ldots, q_k$ of Λ, where $k \geq n$. Consequently, the distances $r_1 = \ldots = r_k = r$ between p and q_i, $i = 1, \ldots, k$ are all the same and v is the center of a ball of radius r. By construction, there is no other point of Λ in this ball. □

Figure 5.9 illustrates construction of the Voronoï cell for a Delone set. The construction consists of two steps:

i) construct the $2R_0$ star for a chosen point p,
ii) construct the orthogonal bisectors of the arms of the star.

Then the Voronoï cell is the intersection of the half-spaces containing p determined by these bisectors.

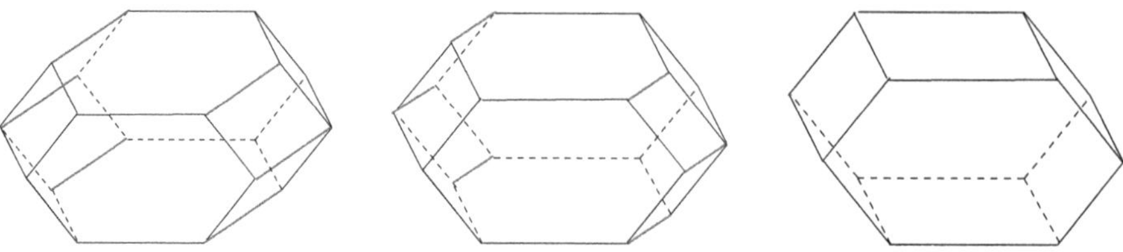

FIG. 5.6 – Contraction of the **12.18** cell (elongated dodecahedron) into the **8.12** cell (hexagonal prism). Six edges shrink to zero and four quadrilateral faces disappear. There are four 6-belts to shrink.

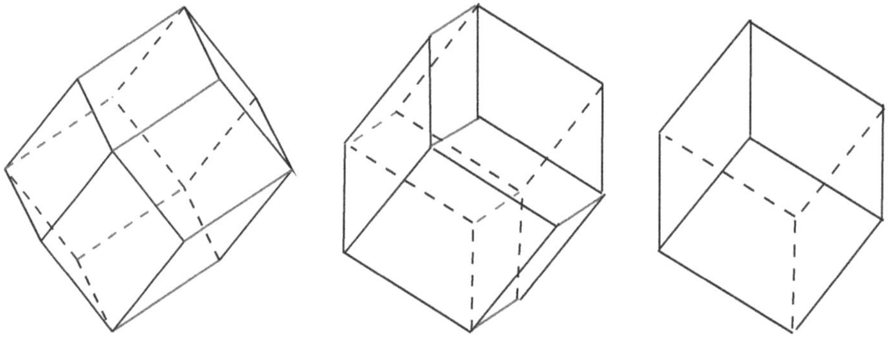

FIG. 5.7 – Contraction of the **12.14** cell (rhombic dodecahedron) into the **6.8** cell (cube). Six edges shrink to zero and six quadrilateral facets disappear. There are four 6-belts to shrink.

Definition: corona vector A vector $\vec{f} \in \Lambda$ is said to be a *corona vector* of the Voronoï cell $D(o)$ if it joins o to the center of a Voronoï cell in the corona of $D(o)$.

We denote the set of corona vectors of $D(o)$ by $\mathcal{C}_o$.

Definition: facet vector A facet vector $\vec{f} \in \Lambda$ is a corona vector joining o to a Voronoï cell with which it shares a facet.

Alternatively we can say that a vector $\vec{f} \in \Lambda$ is a facet vector of the Voronoï cell $D(o)$ if a facet $\mathbf{f}$ of $D(o)$ is contained in its orthogonal bisector. We denote the set of facet vectors of $D(o)$ by $\mathcal{F}_o$.

The equation of the bisecting hyperplane is $(\vec{f}, \vec{x}) = \frac{1}{2}N(\vec{f})$. Thus the Voronoï cell of the point o is the set

$$D(o) = \{x \in E^n | (\vec{x}, \vec{f}) \leq \frac{1}{2}N(\vec{f}), \ \ \forall \vec{f} \in \mathcal{F}\}. \tag{5.2}$$

When $x \in \partial D(o)$, equality must hold in (5.2) for at least one $\vec{f} \in \mathcal{F}$. The definition of facet vector does not imply that the midpoint $\frac{1}{2}\vec{f} \in \mathbf{f}$; it may lie outside of $D(o)$. But $\frac{1}{2}\vec{f} \in \mathbf{f}$ when Λ is a lattice.

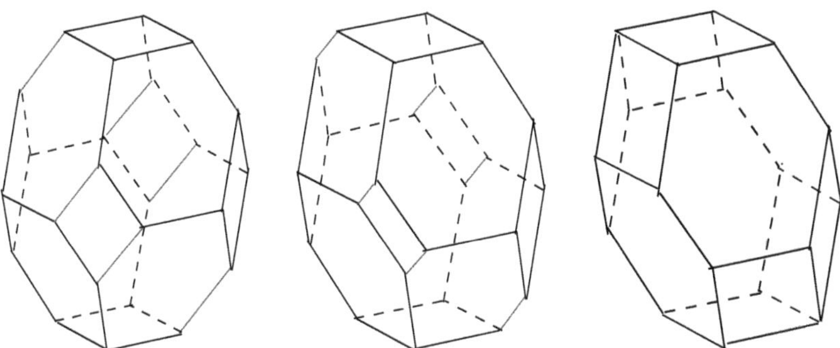

FIG. 5.8 – Contraction of the **14.24** cell (truncated octahedron) into the **12.18** cell (elongated dodecahedron). Six edges shrink to zero and two quadrilateral facets disappear. Four hexagonal facets transform into quadrilateral facets. There are six 6-belts to shrink.

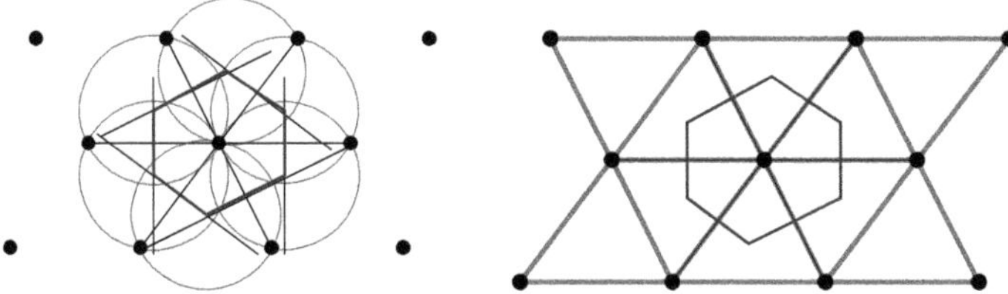

FIG. 5.9 – Construction of the Voronoï and Delone cells for a two-dimensional Delone set.

Proposition 6 *Let* $\mathbf{f}_k$ *be a* k*-face of* $D(o)$, $0 \leq k \leq n-1$, *and let* $o, p_1, \ldots, p_m$ *be the centers of the Voronoï cells in its corona. The* m *vectors* $\vec{p}_1, \vec{p}_2, \ldots, \vec{p}_m$ *span an* $(n-k)$*-dimensional subspace* E^{n-k} *orthogonal to* $\mathbf{f}_k$, *so* $\mathbf{f}_k \cap E^{n-k}$ *is a single point* x_o, *and* $o, p_1, \ldots, p_m$ *lie on a sphere in* E^{n-k} *centered at* x_o.

Proof. Since $\mathbf{f}_k$ is the intersection of at least $n-k$ facets of $D(o)$, and since $D(o)$ is convex and compact, the corresponding facet vectors span an $(n-k)$-dimensional subspace of E^n. Thus $m \geq n-k$. By construction, these vectors and thus this subspace are orthogonal to $\mathbf{f}_k$. The intersection $E^{n-k} \cap \mathbf{f}_k$ is a single point x_o (otherwise the points of $\mathbf{f}_k$ could not all be equidistant from $o, p_1, \ldots, p_m$). Thus $o, p_1, \ldots, p_m$ lie on a sphere about x_o. □

Next we describe the Delone tiling Δ_Λ, obtained by connecting points of Λ. This tiling was in fact first introduced by Voronoï; later it was thoroughly studied by Delone. Today it is known as the Delone tessellation induced by Λ, except in Russian literature, where Delone tessellations are called L-tessellations, the name that Voronoï had given them.

Definition: Delone polytope The Delone polytope of a hole of Λ is the convex hull of the points of Λ that lie on its boundary.

From this definition it follows immediately that the set $\Delta_\Lambda = \Delta(x_i, r_i)$, $i \in \mathbb{Z}$ of Delone polytopes is a facet-to-facet tiling of E^n.

To make the connection with the Voronoï tiling induced by Λ, we remember that the center of any empty hole must be a vertex of the Voronoï tiling. For, the vertices y_j, $j = 1, \ldots, k$ of $\Delta(x_i, r_i)$, all points of Λ lie on the sphere about x_i and are the closest points of Λ to x_i. Thus x_i belongs to the all Voronoï cells $D(y_j)$. Since there are $n+1$ independent points among the y_j, $\cap_{j=1}^{k} D(y_j) = \{x_i\}$.

We will denote the Delone polytope associated with the vertex v of a Voronoï tiling by $\Delta(v)$, the set of vertices of $D(o)$ by $V(o)$, and the set of vertices of the Voronoï tiling $\mathcal{T}_\Lambda$ by $\mathcal{V}_\Lambda$.

Proposition 7 *For each $v \in V(o)$, the polytope $\Delta(v)$ is circumscribable, and so are its k-faces, $k = 0, \ldots, n-1$.*

Proof. The first statement follows immediately from the fact that the vertices of $D(o)$ lie on the boundary of an empty hole; the second is immediate since the intersection of a ball with a plane of lower dimension is again a ball. □

5.2.1 Primitive Delone sets

Definition: primitive (Delone set and Voronoï tessellation) A Delone set and the Voronoï tessellation it induces are said to be primitive if all of its Delone polytopes are simplices.

By the definition of the Delone polytope, we have

Proposition 8 *A Delone set is primitive if and only if every vertex of the Voronoï tessellation belongs to exactly $n+1$ Voronoï cells.*

More generally we have

Proposition 9 *In the Voronoï tessellation of a primitive Delone set, every k-face, $k = 0, \ldots, n-1$ belongs to exactly $n+1-k$ adjacent Voronoï cells.*

Proof. Voronoï proved this proposition for the case when Λ is a lattice but it is true more generally. If a k-face $\mathbf{f}_k$ is shared by exactly $n+1-k$ cells, then it lies in the intersection of exactly $n+1-k$ hyperplanes. Now let $\mathbf{f}_{(k+1)}$ be a $(k+1)$-face containing $\mathbf{f}_k$. It lies in the intersection of $m \leq n-k$ hyperplanes, and since it is $(k+1)$-dimensional, we must have $m = n-k$. □

Proposition 10 *Primitivity is generic.*

Proof. Since $n+1$ independent points determine a sphere in E^n, any additional points are redundant. □

Indeed in discrete geometry literature Delone tessellations are known as Delone triangulations. In addition to "most" lattices, many other important Delone sets are primitive.

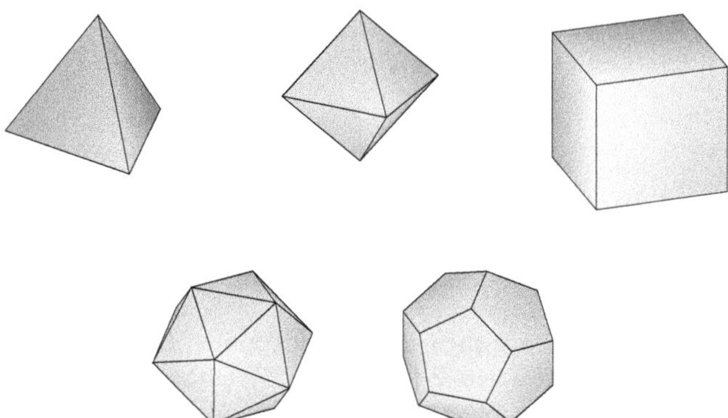

FIG. 5.10 – Examples of combinatorial duality of regular polyhedra. The tetrahedron is auto-dual. The cube and octahedron are combinatorially dual. The icosahedron and dodecahedron are also combinatorially dual.

5.3 Duality

We discussed dual lattices in Chapter 3. Here we introduce dual polytopes and dual tilings, for which duality has a different meaning.

Definition: combinatorially dual convex polytopes Two convex polytopes are said to be combinatorially dual if there is an inclusion-reversing bijection between the k-faces of one and the $(n-k)$-faces of the other.

For example, the cube and the regular octahedron are combinatorially dual, while the combinatorial dual of a tetrahedron is again a tetrahedron (see figure 5.10).

Definition: orthogonally dual polytopes Two combinatorially dual polytopes P and P' are said to be orthogonally dual if the corresponding k and $(n-k)$-faces are orthogonal.

Notice that we restrict these definitions to convex polytopes.

Duality for tilings is defined in an analogous way.

Definition: combinatorial and orthogonally dual tilings Two tilings by convex prototiles are combinatorially dual if there is an inclusion-reversing bijection between the k-faces of one and the $(n-k)$-faces of the other. When the corresponding k and $(n-k)$-faces are mutually orthogonal, the duality is said to be orthogonal.

Now we can formulate the duality relation between Voronoï and Delone tilings.

Proposition 11 *The tilings Δ_Λ and $\mathcal{T}_\Lambda$ are orthogonally dual.*

Proof. This is an immediate consequence of Proposition 6. We select a nested sequence of k-faces

$$D(o) \supset \mathbf{f} \supset \mathbf{f}_{n-2} \supset \cdots \supset \mathbf{f}_0 = v. \tag{5.3}$$

To construct an inclusion-reversing bijection ψ between $\mathcal{T}_\Lambda$ and Δ_Λ, we first set $\psi(D(o)) = o$. Let $D(p_1)$ be the Voronoï cell that shares $\mathbf{f}$ with $D(o)$. Then

$$op_1 = \bigcap_{v \in \mathbf{f}} \Delta(v) \tag{5.4}$$

and hence op_1 is an edge of Δ_Λ, so we set $\psi(\mathbf{f}) = op_1$. Next we set

$$\psi(\mathbf{f}_{n-2}) = \text{convex hull } \{o, p_1, \ldots, p_m\}, \tag{5.5}$$

where $D(p_2), \ldots, D(p_m)$ are the cells, in addition to $D(o)$ and $D(p_1)$, to which $\mathbf{f}_{n-2}$ belongs; this polygon is a 2-face of Δ_Λ. We continue in this way, taking for $\psi(\mathbf{f}_k)$ the $(n-k)$-face of Δ_Λ that is the convex hull of the points of Λ whose Voronoï cells share $\mathbf{f}_k$. Finally, the vertex v is associated to $\Delta(v)$. □

5.4 Voronoï and Delone cells of point lattices

5.4.1 Voronoï cells

When a Delone set Λ is a regular system of points (point lattice), its Voronoï tilings $\mathcal{V}_\Lambda$ is monohedral and we can speak of "the" Voronoï cell of the set. Thus by the Voronoï cell of a point lattice we will mean the Voronoï cell of the origin, $D(o)$. In this section we will discuss some of the fundamental properties of Voronoï cells of point lattices.

Since point lattices are orbits of translation groups, their Voronoï cells are parallelotopes. Since the Voronoï cell is the closure of a fundamental region for the translation subgroup of the symmetry group of the lattice, the volume of the Voronoï cell is equal to the volume of a lattice unit cell.

The Voronoï cell of a lattice is invariant under the lattice's point symmetry group.

Proposition 12 *The point symmetry group of a lattice L with fixed point o is also the symmetry group of the Voronoï cell $D(o)$; the full symmetry group of L is the symmetry group of the Voronoï tiling.*

Proof. This follows immediately from the definition of $D(o)$. □

Proposition 13 *$D(o)$ and its facets are centrosymmetric.*

Proof. Every lattice point is a center of symmetry for the lattice; thus $D(o)$ is centrosymmetric by construction. The midpoint between any pair of lattice points is also a center of symmetry for L; in particular if $\vec{f}$ is a facet vector, then $\frac{1}{2}\vec{f}$ is a center of symmetry for L. Thus it is the center of symmetry of $D(o) \cup D(f)$ and of $D(o) \cap D(f)$, and hence $\frac{1}{2}\vec{f}$ is the center of symmetry for the facet $\mathbf{f}$. □

Note: The k-faces of $D(o)$, $2 \le k \le n-2$, need not be centrosymmetric; for example, there are lattices in E^4 whose Voronoï cells have triangular or pentagonal 2-faces.

Since every point of $D(o)$ is a representative of a coset of L in $\mathbb{R}^n$, we can reformulate the definition of $D(o)$ in the following way.

Proposition 14 *The Voronoï cell of a lattice in E^n is the set of vectors $\vec{x} \in E^n$ of minimal norm in their L-coset $\vec{x} + L$:*

$$D_L = \{\vec{x} \in E^n | N(\vec{x}) \leq N(\vec{x} - \vec{\ell}),\ \forall \vec{\ell} \in L\}. \tag{5.6}$$

The interior points are unique in their coset but two or more boundary points may belong to the same coset: for example, if x is a point on the boundary $\partial D(o)$ of the Voronoï cell, then so is $-x$ and these points are congruent modulo L. This point x belongs to at least one intersection $D(o) \cap D(\vec{f})$, and translation by $-\vec{f}$ carries that intersection, and with it x, to $D(-\vec{f}) \cap D(o)$,

5.4.2 Delone polytopes

As in the case of general Delone sets, the tiles of the Delone tessellation induced by a lattice are convex polytopes whose vertices are the lattice points lying on the boundaries of empty spheres and the Delone and Voronoï tessellations are dual.

In general the Delone tiling has several prototiles. However, when $n = 2$, not only is the tiling monohedral, it is *isohedral*, i.e. the tiles form an orbit of the symmetry group of the tiling.

Proposition 15 *The Delone tiling associated to a lattice L in E^2 is isohedral.*

Proof. Since the midpoints of the edges of the Voronoï cell of L in E^2 are centers of symmetry for L, any pair of adjacent vertices can be interchanged by inversion in the center of the edge joining them. Thus all the vertices of the Voronoï cell are equivalent under the symmetry group of the lattice, from which it follows that the Delone cells corresponding to the vertices of $D(o)$ are equivalent too. □

5.4.3 Primitive lattices

Primitive Voronoï cells have received the most attention in the context of both lattices and quadratic forms. This is mainly due to the fact that the primitivity is generic. The relation to quadratic forms will be discussed in the next chapter. Here we describe several simple properties of primitive lattices.

Applying the definition of primitivity of Delone sets (see 5.2.1) to the lattice we get the following obvious statement:

Proposition 16 *A lattice L is primitive if and only if all its Delone cells are simplices.*

When $D(o)$ is primitive, exactly $n + k - 1$ Voronoï cells of the Voronoï tessellation share a given k-face, $k = 0, \ldots, n-1$.

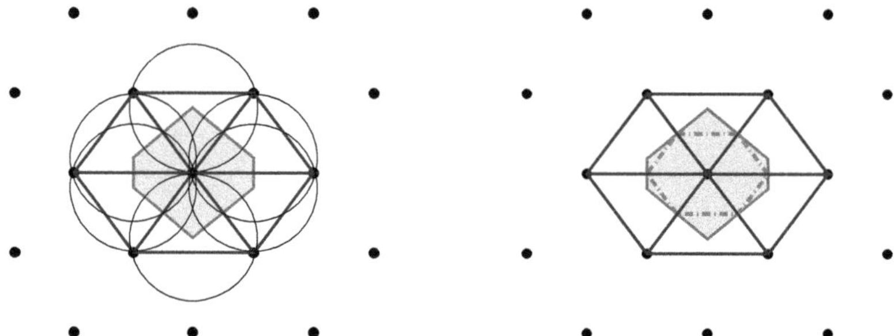

FIG. 5.11 – Illustration to Proposition 17. Left: Primitive two-dimensional lattice with its Voronoï cell whose vertices are situated in the centers of holes. Right: The same lattice with the Voronoï cell (shaded region with boundary), the dual to the Voronoï cell (dash line), and the Delone corona.

Definition: Delone corona The set of Delone cells that share the vertex o (the origin) is called the Delone corona of the lattice L.

Proposition 17 *If L is primitive, then the Delone corona of L is a scaled copy of the polytope dual to the Voronoï cell $D(o)$.*

This proposition is illustrated in figure 5.11.

We denote $\mathcal{V}_L$ the set of vertices of the Voronoï tiling of L. It is easy to check that $\mathcal{V}_L$ is a Delone set. The minimum distance between vertices of $D(o)$ can be taken as r_0, whereas R_0 can be chosen to be the length of the longest vertex vector of $D(o)$. Recall that when $D(o)$ is primitive, exactly $(n-k+1)$ Voronoï cells of the Voronoï tessellation share a given k-face, $k = 0, \dots, n-1$.

For each k, $0 \leq k \leq n$, the set of k-faces of a lattice Voronoï tessellation belongs to a finite number of orbits of the translation group of the lattice; in general, each Voronoï cell contains several elements of each orbit. Let $\mathbf{f}_k$ be a k-face of $D(o)$ and let $\{\vec{c}_m\}$, $1 \leq m \leq n-k$, be the set of vectors corresponding to the centers of the other $n-k$ Voronoï cells which share this k-face with $D(o)$. Each translation $-\vec{c}_m$ transforms $D(c_m)$ into $D(o)$ and therefore $\mathbf{f}_k$ into $\mathbf{f}_k - \vec{c}_m$, another k-face of the Voronoï cell $D(o)$. Conversely, if $\mathbf{f}'_k$ is a k-face of $D(o)$, where $\mathbf{f}'_k + \vec{t} = \mathbf{f}_k$ for some $\vec{t} \in L$, then $\mathbf{f}_k$ is a k-face of $D(t)$. Thus we have

Proposition 18 *Each k-face of a primitive Voronoï cell $D(o)$ is equivalent, under translations of L, to exactly $n-k$ other k-faces of $D(o)$.*

This means that the number of k-faces of a primitive Voronoï cell should be a multiple of $n-k+1$. In fact for $0 \leq k < n-1$, it should be proportional to $2(n-k+1)$ (see proposition 29).

The set $\mathcal{V}_L$ can be decomposed into L-orbits. Selecting one Delone cell from each orbit, we have the closure of a fundamental region of L, and so the volume of the union of these Delone cells must be equal to the volume of the lattice introduced in (3.2) as $\mathrm{vol}(L) = |\det(\ell_i)|$.

Moreover the set $C(o) = \cup_{v \in V(o)} \Delta(v)$ is the union of $n+1$ fundamental domains; hence the value of the invariant vol(L) for all primitive lattices is equal to $(n+1)$:

Proposition 19 *When L is primitive,*

$$\text{vol}(L) = \text{vol}C(o)/\text{vol}(D) = n + 1. \tag{5.7}$$

Any vertex of the Voronoï cell of a primitive lattice L belongs to exactly n facets of that cell; since the corresponding facet vectors are linearly independent, these vectors form a basis of E^n though they may generate only a sub-lattice L'. But there are many primitive lattices for which this set of vectors is a basis. For example, this is the case for the primitive lattices in E^2, E^3, and E^4.

Definition: principal primitive A primitive lattice, and its Voronoï cell, is said to be principal primitive if for each vertex of the cell, the facet vectors of the n facets meeting at this vertex form a basis of the lattice.

The Delone cells of principal primitive lattices are simplices whose edges issuing from 0 are the edges of a unit cell for L. Thus all these simplices have the same volume, vol(simplex$(x_0, \ldots, x_n)) = \det(L)/n!$.

Proposition 20 *A principal primitive Voronoï cell has* $(n+1)!$ *vertices.*

Proof. When all Delone cells have the same volume vol$\Delta(v)$, denoting the number of vertices of the Voronoï cell V by $N_0(V)$, we have

$$\frac{N_0(V)}{n+1} = \frac{\det(L)}{\text{vol}\Delta(v)} = n!. \tag{5.8}$$

Corollary 4 *A principal primitive Voronoï cell has* $(n+1)!\ n/2$ *edges.*

Proof. Exactly n edges of the cell meet at each vertex, and each edge has two vertices. □

Taking into account that the Euler characteristic for a n-dimensional polytope is $1-(-1)^n$ and it is expressed as an alternative sum of the numbers of k-faces of an n-polytope, $N_k(n)$, namely $\sum_{0 \leq k \leq n-1} (-1)^k N_k(n) = 1 - (-1)^n$, we can find immediately the number of faces for 3-dimensional principal primitive polytopes. The table 5.1 gives values of $N_k(n)$ for $n = 2, 3, 4$ for principal primitive polytopes. Note, that for $n = 2, 3, 4$ all primitive polytopes are principal. Additional topological restrictions on the numbers of k-faces for primitive higher dimensional polytopes will be discussed in the next chapter (see section 6.4).

5.5 Classification of corona vectors

In the geometry (and the algebra) of lattices, one is interested in the set of vectors that are (relatively) short. Historically, the vectors of minimum length have received the most attention. Here we consider three sets of "short"

TAB. 5.1 – Values of the number of k-faces, $N_k(n)$ for n-dimensional primitive polytopes for $n = 2, 3, 4$.

n	$N_0(n)$	$N_1(n)$	$N_2(n)$	$N_3(n)$
2	6	6		
3	24	36	14	
4	120	240	150	30

vectors, all defined in terms of the Voronoï cell of the lattice. We begin with the largest of these sets, the corona vectors of the lattice.

5.5.1 Corona vectors for lattices

The corona of a tile T in a tiling is defined in section 5.1. When T is the Voronoï cell $D(o)$ of a lattice then every tile in the corona is associated to a lattice vector.

Definition: corona vector The corona vectors of a lattice L are the vectors $\vec{c}$ from o to the centers c of the cells comprising the corona of the Voronoï cell $D(o)$.

Proposition 21 *A lattice vector $\vec{c} \in L$ is a corona vector if and only if $\frac{1}{2}\vec{c} \in \partial D(o)$.*

Proof. Let $\vec{c}$ be a corona vector. Let $I(o,c) = D(o) \cap D(c)$. Then $I(o,c) \neq \emptyset$, and it is convex because $D(o)$ and $D(c)$ are convex. The midpoint $\frac{1}{2}\vec{c}$ is a center of symmetry of the lattice that interchanges $D(o)$ and $D(c)$ and hence stabilizes $I(o,c)$, and again by convexity, $\frac{1}{2}\vec{c} \in I(o,c)$. The converse is immediate by the definition of the corona vector. □

Corollary 5 *$\vec{c} \in L$ is a corona vector if and only if $\frac{1}{2}\vec{c}$ is the center of symmetry of the nonempty intersection $D(o) \cap D(c)$.*

Corollary 6 *If a k-face of $D(o)$ does not contain a center of symmetry, then any tile that shares that k-face also shares one of higher dimension.*

We denote the set of corona vectors of L by $\mathcal{C}$.

Proposition 22 *The number of corona vectors is even.*

Proof. Since $D(o)$ is centrosymmetric, $\frac{1}{2}\vec{c} \in \partial D(o) \leftrightarrow -\frac{1}{2}\vec{c} \in \partial D(o)$. □

Theorem 7 *$\vec{c}$ is a corona vector of L if and only if it is a vector of minimal norm in its $L/2L$ coset.*

Proof. By definition, $\frac{1}{2}\vec{c}$ belongs to the Voronoï cell of o and so, by Proposition 21,

$$\vec{c} \in \mathcal{C} \leftrightarrow N\left(\frac{1}{2}\vec{c}\right) \leq N(\vec{c}/2 - \vec{\ell}), \quad \forall \vec{\ell} \in L. \tag{5.9}$$

Thus

$$\vec{c} \in \mathcal{C} \leftrightarrow N(\vec{c}) \leq N(\vec{c} - 2\vec{\ell}), \quad \forall \vec{\ell} \in L, \tag{5.10}$$

i.e. $\vec{c}$ is a vector of minimal norm in its $L/2L$ coset. □

Note that $\vec{c}$ and $\vec{c}\,'$ are two vectors of the same length and $\vec{c} - \vec{c}\,' \in 2L$ if and only if $\frac{1}{2}(\vec{c} - \vec{c}\,') \in L$. In this case, $\vec{c}\,'$ is the image of $\vec{c}$ through the center of symmetry $\frac{1}{2}(\vec{c} - \vec{c}\,')$. With this observation it is easy to prove (Theorem 8 below) that if $\pm\vec{c}$ are the only vectors of minimal length in their $L/2L$ coset, then they are facet vectors.

The corona vectors of a lattice are of special interest because they encode many of its properties.

Proposition 23 *A corona vector is the shortest lattice vector in its mL coset for all integers $m \geq 3$.*

Proof. If $\vec{c} \in \mathcal{C}$ and $\vec{x} \neq 0$, then $N(\vec{c} + 2\vec{x}) - N(\vec{c}) \geq 0$, so $(\vec{c}, \vec{x}) + N(\vec{x}) \geq 0$. Then

$$\begin{aligned} N(\vec{c} + m\vec{x}) - N(\vec{c}) &= 2m(\vec{c}, \vec{x}) + m^2 N(\vec{x}) \\ = m(2(\vec{c}, \vec{x}) + mN(\vec{x})) &= m(2((\vec{c}, \vec{x}) + N(\vec{x})) + (m-2)N(\vec{x})) \end{aligned} \tag{5.11}$$

which is positive for $m > 2$. Thus $N(\vec{c}) < N(\vec{c} + m\vec{x})$. □

Note that when $m > 2$, $\vec{c}$ and $-\vec{c}$ do not belong to the same mL coset.

Proposition 24 *The set $\mathcal{C}$ is the set of vertices (except o) of the Delone corona of o.*

Proof. The Delone corona of o is $\cup_{v \in V(o)} \Delta(v)$. For each such v, the vertices of $\Delta(v)$ are the centers of the Voronoï cells that meet at v, and thus by definition the vertices of $\Delta(v)$ are corona vectors. Conversely, every corona vector is a vertex of some $\Delta(v)$, $v \in V(o)$. □

5.5.2 The subsets $\mathcal{S}$ and $\mathcal{F}$ of the set $\mathcal{C}$ of corona vectors

We distinguish now two important subsets of the set $\mathcal{C}$ of corona vectors of a lattice L.

- The set $\mathcal{S}$ of vectors of minimal norm s in L, i.e. the set of shortest vectors.
- The set $\mathcal{F}$ of facet vectors of the Voronoï cell.

We have already noted the simple criterion for determining whether a lattice vector is an element of $\mathcal{F}$.

Theorem 8 *(Voronoï). The following conditions on $\vec{c} \in \mathcal{C}$ are equivalent:*

i) $\pm\vec{c}$ are the facet vectors;

ii) $\pm\vec{c}$ are the shortest vectors in their $L/2L$ coset;

iii) $(\vec{c}, \vec{v}) < N(\vec{v})\ \forall \vec{v} \in L,\ \vec{v} \neq 0,\ \vec{v} \neq \vec{c}$;

iv) the closed ball $B_{c/2}(|\frac{1}{2}\vec{c}|)$ *contains no points of* L *other than* o *and* c.

Proof.

ii) ⇒ i). Assume $\pm\vec{c}$ are the shortest vectors in their $L/2L$ coset. Let **f** be a facet containing $\frac{1}{2}\vec{c}$. Then the image of $\frac{1}{2}\vec{c}$ through the center of symmetry $\frac{1}{2}\vec{f}$ is $\frac{1}{2}\vec{c}\,' = \vec{f} - \frac{1}{2}\vec{c}$, and hence $N(\vec{c}\,') = N(\vec{c})$ and $\vec{c}\,' = 2\vec{f} - \vec{c}$. Thus our hypothesis implies $\vec{c}\,' = \vec{c}$ or $\vec{c}\,' = -\vec{c}$. If $\vec{c}\,' = \vec{c}$, $\frac{1}{2}\vec{c}$ was fixed by this symmetry and hence $\vec{c} = \vec{f}$. The case $\vec{c}\,' = -\vec{c}$ is impossible, since in that case $\frac{1}{2}\vec{c}$ and $-\frac{1}{2}\vec{c}$ would lie in the same facet of $D(o)$. Thus $\vec{c}$ is a facet vector, and the same argument works for $-\vec{c}$.

i) ⇒ ii). Conversely, assume that $\vec{c}$ is a facet vector. Then $\frac{1}{2}\vec{c}$ is the center of a facet and so is closer to $\vec{c}$ and to o than to any other points of L. That is,

$$\forall \vec{\ell} \in L, \vec{\ell} \neq 0, \quad N\left(\frac{1}{2}\vec{c}\right) < N\left(\frac{1}{2}\vec{c} - \vec{\ell}\right). \tag{5.12}$$

Again the same argument works for $-\vec{c}$, so $\pm\vec{c}$ are the shortest vectors in their $2L$ coset.

ii) ⇒ iii). This is equivalent to condition $N(\vec{c}) < N(\vec{c} - 2\vec{v})$, $\forall \vec{v} \in L$, $\vec{v} \neq 0, \neq \vec{c}$.

i) ⇒ iv). If $\vec{c}/2$ is the center of a facet, then it is equidistant from o and c and all other points of L are farther away. But any lattice point w in $B_{c/2}(|\frac{1}{2}\vec{c}|)$ would be at least as close, a contradiction. The converse is obvious. □

Corollary 7 $\mathcal{S} \subseteq \mathcal{F} \subseteq \mathcal{C}$.

Proof. By the definition of a facet vector, $\mathcal{F} \subseteq \mathcal{C}$. It is also obvious that $\mathcal{S} \subseteq \mathcal{C}$, since $\mathcal{S}$ vectors have minimal norm in L and hence also in the $L/2L$ cosets to which they belong. To show that $\mathcal{S} \subseteq \mathcal{F}$, we prove that no two $\mathcal{S}$ vectors $\vec{s}_1, \vec{s}_2$, $\vec{s}_1 \neq \pm\vec{s}_2$, can belong to the same $L/2L$ coset. Let θ be the angle between $\vec{s}_1$ and $\vec{s}_2$; we may assume $0 < \theta < \pi$. If $\vec{s}_1 = \vec{s}_2 + 2\vec{y}$ for some $\vec{y} \in L$, we have $(\vec{s}_1 - \vec{s}_2) = \vec{y} \in L$, and

$$N\left(\frac{\vec{s}_1 - \vec{s}_2}{2}\right) = \frac{s}{2}(1 - \cos\theta) < s, \tag{5.13}$$

where s is the norm of the vectors in $\mathcal{S}$. This is a contradiction. Thus $\mathcal{S} \subseteq \mathcal{F}$. □

There is exactly one planar lattice for which $\mathcal{S} = \mathcal{F} = \mathcal{C}$: the hexagonal lattice, whose Voronoï cell is a regular hexagon. Surprisingly, there are no examples in any higher dimension.

We next study some key properties of $\mathcal{F}$. The next two propositions are due to Minkowski [78].

Proposition 25 *(Minkowski).* $2n \leq |\mathcal{F}| \leq 2(2^n - 1)$.

Proof. The lower bound is implied by the centrosymmetry of $D(o)$. The 2^n cosets of $2L$ in L have as coset representatives the vectors $\{(\epsilon_1, \ldots, \epsilon_n)\}$ where $\epsilon_i \in \{0, 1\}$. Since if $\vec{f} \in \mathcal{F}$ the only other facet vector in its $L/2L$ coset is $-\vec{f}$, the maximum number of face vectors is twice the number of cosets, excluding of cause the 0-coset. □

Proposition 26 *(Minkowski).* $2(2^n - 1) \leq |\mathcal{C}| \leq 3^n - 1$.

Proof. The lower bound follows from the fact that every $L/2L$ coset contains at least two corona vectors. The upper bound is a corollary of Proposition 23 since there are 3^n cosets of $3L$ in L, one of which is represented by 0. □

The upper bound is attained in every dimension by the cubic lattice, whose Voronoï cell is the unit n-cube. To calculate the number of corona vectors for cubic lattices the notion of k-vector is useful.

Definition: k-vector. A vector $\vec{c} \in \mathcal{C}$ is a k-vector if $\frac{1}{2}\vec{c}$ lies in the interior of a k-face of $D(o)$, that is if it lies in the intersection of exactly $n-k$ independent facets.

For cubic lattices, the vector $\vec{c}$ is a k-vector if and only if $\epsilon_i = \pm 1$ for exactly $n - k$ values of i and is equal to 0 for all the others. Thus, since we do not include the 0-coset, the number of corona vectors for a n-dimensional cubic lattice is

$$|\mathcal{C}| = \sum_{k=0}^{n} 2^k \binom{n}{n-k} - 1 = 3^n - 1. \tag{5.14}$$

If L is primitive then $\mathcal{F} = \mathcal{C}$ and L has exactly $2(2^n - 1) < 3^n - 1$ corona vectors ($n \geq 2$).

Taking into account that $|\mathcal{F}|$ is maximal if and only if $\mathcal{F}$ contains a representative of every $L/2L$ coset (except 0), we get

Proposition 27 *$|\mathcal{F}|$ is maximal if and only if $\mathcal{F} = \mathcal{C}$.*

Lattices with maximal $|\mathcal{F}|$ are not necessarily primitive: if $D(o)$ has "few" vertices then some of them will be an intersection of more than n facets (this occurs first when $n = 4$, see example in subsection 6.4.1). However, if $|\mathcal{F}|$ is maximal and the number of facets at each vertex of $D(o)$ is minimal, then L is primitive. More precisely,

Proposition 28 *L is primitive if and only if $|\mathcal{F}|$ is maximal and exactly n facets of $D(o)$ meet at each vertex.*

Proof. Suppose that L is primitive. Then every Delone cell is a simplex. Since at least n facets of $D(o)$ must meet at every vertex, all of the vertices of the Delone cell, except o, correspond to facet vectors, so $|\mathcal{F}|$ is maximal. Conversely, if $|\mathcal{F}|$ is maximal and exactly n facets meet at each vertex, then every corona vector is a facet vector and hence the Delone cells are simplices. □

Denoting the number of k-faces of the Voronoï cell of an n-dimensional primitive lattice by $N_k(n)$, we have:

Proposition 29 *For a primitive lattice, $N_k(n)$ is a multiple of $2(n+1-k)$ for $0 \leq k < n-1$.*

Proof. Proposition 18 shows that this number is a multiple of $n+1-k$. Consider a k-face $\mathbf{f}_k$ and its image $\mathbf{f}'_k = -\mathbf{f}_k$ through the origin. If $\mathbf{f}_k$ and $\mathbf{f}'_k$ belong to the same translation orbit, there would be a translation $-\vec{c}$ carrying $\mathbf{f}_k$ into $\mathbf{f}'_k$. Then $\frac{1}{2}\vec{c} \in \mathbf{f}_k$. So $\vec{c} \in \mathcal{C}$, but $\vec{c} \notin \mathcal{F}$ since $k < n-1$. This is impossible, since $\mathcal{C} = \mathcal{F}$. So $\mathbf{f}_k$ and $\mathbf{f}'_k$ belong to two distinct translation orbits. Thus when a Voronoï cell is primitive, the k-faces belong to an even number of translation orbits, each containing $(n+1-k)$ k-faces of the cell. □

Proposition 30 *In any lattice, the vectors of norm less than $2s$ are facet vectors, where s is the minimal norm of the lattice.*

Proof. Let $N(\vec{v}_1) < 2s$; we will show that there is no $\vec{v}_2$ in the same $L/2L$ coset with norm $N(\vec{v}_2) < 2s$. Assume $N(\vec{v}_2) \leq 2s$ and $\vec{v}_1 - \vec{v}_2 = \vec{y}$, $\vec{y} \in L$. Then

$$N(\vec{v}_1 - \vec{v}_2) = N(2\vec{y}) = 4N(\vec{y}) \geq 4s, \tag{5.15}$$

so

$$4s \leq N(\vec{v}_1 - \vec{v}_2) = N(\vec{v}_1) + N(\vec{v}_2) - 1(\vec{v}_1, \vec{v}_2) \leq 4s - 1(\vec{v}_1.\vec{v}_2). \tag{5.16}$$

Choosing $\vec{v}_2$ so that $(\vec{v}_1, \vec{v}_2) > 0$ - that is replacing $\vec{v}_2$ by $-\vec{v}_2$ if necessary - we have a contradiction. Thus $\vec{v}_1$ is a facet vector. □

Corollary 8 *A vector of norm $2s$ is a corona vector.*

Proof. It follows from the proof of the preceding proposition that no vectors of norm $2s$ can be in the same $L/2L$ coset as a shorter vector. □

Corollary 9 *The vectors of norm $2s$ in the same $L/2L$ coset are pairwise orthogonal.*

Proof. Let $N(\vec{v}_1) = 2s$; we will show that if there is a $\vec{v}_2$ in the same $2L$ coset with norm $N(\vec{v}_2) = 2s$ and $\vec{v}_2 \neq \vec{v}_1$, then $(\vec{v}_1, \vec{v}_2) = 0$. Let $\vec{v}_1 - \vec{v}_2 = 2\vec{y}$, $\vec{y} \in L$. Then again (5.15) and (5.16) takes place. Since if $(\vec{v}_1, \vec{v}_2) \neq 0$ we can replace $\vec{v}_2$ by $-\vec{v}_2$ if necessary and to assure that $(\vec{v}_1, \vec{v}_2) > 0$, we must have $(\vec{v}_1, \vec{v}_2) = 0$. □

The following criterium, due to Venkov, allows us to distinguish the facet vectors among the vectors of norm $2s$.

Proposition 31 *A vector of norm $2s$ is a facet vector if and only if it is not a sum of two orthogonal vectors of $\mathcal{S}$.*

Proof. Let $\vec{s}_1, \vec{s}_2 \in \mathcal{S}$ where $(\vec{s}_1, \vec{s}_2) = 0$. Then the four vectors $\pm(\vec{s}_2 \pm \vec{s}_2)$ all have norm $2s$ and belong to the same $L/2L$ coset, so they cannot be facet vectors. Conversely, if $N(\pm\vec{c}_i) = 2s$ but $\pm\vec{c}_i \notin \mathcal{F}$ then there are vectors $\pm\vec{c}_j$ orthogonal to $\pm\vec{c}_i$ and in the same $2L$ coset. Then the four lattice vectors $\pm\frac{1}{2}(\vec{c}_1 \pm \vec{c}_2)$ are elements of $\mathcal{S}$ and form two orthogonal opposite pairs, and $\vec{c}_i = \frac{1}{2}(\vec{c}_i + \vec{c}_j) + \frac{1}{2}(\vec{c}_i - \vec{c}_j)$. □

The following obvious remark is also very useful:

Proposition 32 *$\mathcal{F}$ generates L.*

Proof. Since the Voronoï tesselation is facet-to-facet, we can pass from any cell, say $D(o)$ to any other, say $D(x)$, by a path that does not intersect the boundary of any face. This path defines a sequence of facet vectors from o to x. □

The set $\mathcal{S}$ of shortest vectors may not generate L, even if it spans the whole E^n. Also note that a generating set need not include a basis. For example, the integers 2 and 3 generate $\mathbb{Z}$ but neither 2 or 3 does. When $n > 9$, there exist lattices in E^n generated by $\mathcal{S}$ which have no basis in that set. The first example, in 11 dimensions, was found by Conway and Sloane [36].

5.5.3 A lattice without a basis of minimal vectors

Conway and Sloane have proved in [36] that the 11-dimensional lattice with Gram matrix

$$\begin{bmatrix} 60 & 5 & 5 & 5 & 5 & 5 & -12 & -12 & -12 & -12 & -7 \\ 5 & 60 & 5 & 5 & 5 & 5 & -12 & -12 & -12 & -12 & -7 \\ 5 & 5 & 60 & 5 & 5 & 5 & -12 & -12 & -12 & -12 & -7 \\ 5 & 5 & 5 & 60 & 5 & 5 & -12 & -12 & -12 & -12 & -7 \\ 5 & 5 & 5 & 5 & 60 & 5 & -12 & -12 & -12 & -12 & -7 \\ 5 & 5 & 5 & 5 & 5 & 60 & -12 & -12 & -12 & -12 & -7 \\ -12 & -12 & -12 & -12 & -12 & -12 & 60 & -1 & -1 & -1 & -13 \\ -12 & -12 & -12 & -12 & -12 & -12 & -1 & 60 & -1 & -1 & -13 \\ -12 & -12 & -12 & -12 & -12 & -12 & -1 & -1 & 60 & -1 & -13 \\ -12 & -12 & -12 & -12 & -12 & -12 & -1 & -1 & -1 & 60 & -13 \\ -7 & -7 & -7 & -7 & -7 & -7 & -13 & -13 & -13 & -13 & 96 \end{bmatrix} \tag{5.17}$$

has minimal norm 60, is generated by its 24 minimal vectors, but no set of 11 minimal vectors forms a basis.

We want just to use this example to illustrate relations between facet vectors and shortest vectors of the lattice. We note that for the lattice (5.17) all lattice vectors with norm less than 120 are facet vectors. In particular, the basis in which the Gram matrix is written is formed by facet vectors. Numerical calculations made by Engel (private communication) show that there are 2974 facet vectors. The maximal norm for facet vectors is 168. The minimal norm of lattice vectors which are not facet vectors is 122. There are 20 lattice vectors with norm 122 which are not facet vectors.

We do not touch here the question of existence of a basis of facet vectors conjectured by Voronoï and discussed later on several occasions [66, 52, 53].

Chapter 6

Lattices and positive quadratic forms

6.1 Introduction

Previous chapters were devoted to the study of lattices from the point of view of their symmetry and their Voronoï and Delone cells. This analysis was done essentially without explicit introduction of the basis in the ambient Euclidean space, E^n. Now we return to the study of lattices through associated positive quadratic forms. This approach requires us to introduce initially a lattice basis and to represent the translation lattice Λ^n in this basis

$$\Lambda^n := \{\mathbf{t}|\mathbf{t} = t_1\vec{b}_1 + \cdots + t_n\vec{b}_n,\ t_i \in \mathbb{Z}\}. \tag{6.1}$$

Here $\{\vec{b}_i\}$ is a basis of E^n. From the associated scalar products one can form the Gram matrix Q:

$$q_{ij} = (b_i, b_j); \quad Q = BB^\top = Q^\top. \tag{6.2}$$

Using the dual basis, defined in section 3.4 we obtain

$$Q(L^*) = Q(L)^{-1}. \tag{6.3}$$

We emphasize that the bases in the same orbit of the orthogonal group have the same Gram matrix; indeed $\forall S \in O_n$, $BS^\top(BS^\top)^\top = BS^\top SB^\top = BB^\top$, so Q describes the intrinsic lattice.

This symmetric matrix Q defines also a positive quadratic form $q(\vec{\ell})$ on E^n and, in particular on L, the lattice generated by the basis $\{\vec{b}_i\}$;

$$N(\vec{\ell}) = q(\vec{\ell}) := \sum_{i,j} \lambda_i q_{ij} \lambda_j, \tag{6.4}$$

where $\vec{\ell} = \sum_{i=1}^n \lambda_i \vec{b}_i$. Conversely, given the Gram matrix Q, one can reconstruct the intrinsic lattice.

Real quadratic forms in n variables and, equivalently, $n \times n$ real symmetric matrices Q, form a group under addition and they can be considered as elements of the vector space $\mathcal{Q}_n \sim \mathbb{R}^N$, where $N = n(n+1)/2$. This vector space $\mathcal{Q}_n$ carries a natural orthogonal scalar product $(Q, Q') = \mathrm{tr}QQ' = \mathrm{tr}Q'Q$. Since the sum of two positive quadratic forms is again a positive quadratic form, the set of n-variable positive quadratic forms is the interior $\mathcal{C}_+(\mathcal{Q}_n)$ of a convex closed cone $\overline{\mathcal{C}_+}(\mathcal{Q}_n)$. Notice that $\mathcal{C}_+(\mathcal{Q}_n)$ can be identified as the orbit space of the manifold $\mathcal{B}_n$ of bases under the action of the orthogonal group:

$$\mathcal{B}_n|O_n = GL_n(\mathbb{R}) : O_n = \mathcal{C}_+(\mathcal{Q}_n). \tag{6.5}$$

By a change of lattice basis, $\vec{b}'_i = \sum_j m_{ij}\vec{b}_j$, $M \in GL_n(\mathbb{Z})$, the Gram matrix Q is changed into the matrix:

$$Q \mapsto M.Q = MQM^\top. \tag{6.6}$$

So an intrinsic lattice corresponds to an orbit of $GL_n(\mathbb{Z})$ acting by (6.6) on $\mathcal{C}_+(\mathcal{Q}_n)$. The problem of choosing a fundamental domain for the $GL_n(\mathbb{Z})$ action on positive quadratic forms is equivalent to construction of the so called reduced forms. Also the overall scaling is unimportant for the study of intrinsic lattices. Therefore, it is possible to restrict analysis to appropriate sections of the cone, whose dimension is $n(n+1)/2 - 1$.

For two-dimensional lattices the corresponding cone of positive quadratic forms is three-dimensional, it can be easily visualized (see figure 6.1). Moreover, what we really need to look for in the case of quadratic forms in two variables is the two-dimensional section of the cone of positive quadratic forms represented, for example, in figure 6.2 where stratification of the cone is shown. Although the case of quadratic forms in two variables and associated two-dimensional lattices do not possess many complications arising for higher dimensional quadratic forms and lattices, it is quite instructive to study this particular case especially due to the possibility of visualization of corresponding structures.

6.2 Two dimensional quadratic forms and lattices

6.2.1 The $GL_2(\mathbb{Z})$ orbits on $\overline{\mathcal{C}}_+(\mathcal{Q}_2)$

The strata of the action of $GL_2(\mathbb{Z})$ on $\mathcal{C}_+(\mathcal{Q}_2)$ are the Bravais classes (see section 4.3 for initial definitions and chapter 8 for further details).

The three dimensional generic stratum represents the Bravais class $p2 = \mathbb{Z}_2$. After restriction to a section of the cone (see Figure 6.2) we see only a two-dimensional generic stratum.

Strata with stabilizers *p2mm* and *c2mm* are represented by one-dimensional lines on the section. On the whole cone of positive quadratic forms these strata are two-dimensional.

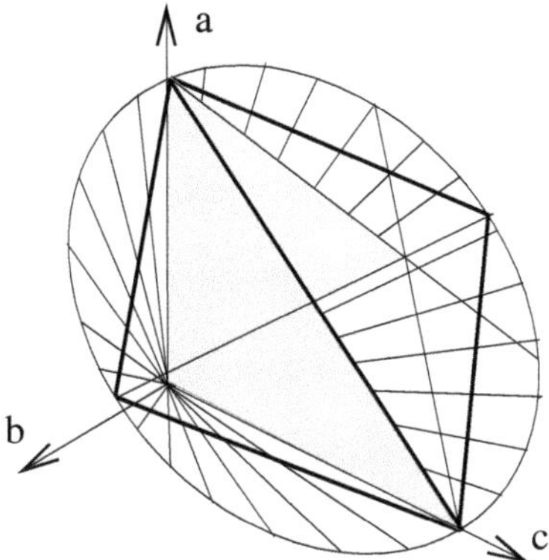

FIG. 6.1 – Representation of the cone of positive quadratic forms depending on two variables. Only interior points correspond to positive quadratic forms. The cone is divided into sub-cones with a given combinatorial type of Voronoï cell by planes passing through a vertex of the cone. One of such planes is shown by a dark shading. The cone is cut by the plane orthogonal to the axis. The traces of walls on this plane are shown by thick black lines. The number of walls is infinite and only a small number of walls is shown. Points on walls correspond to rectangular Voronoi cells. Generic points represent 2-dimensional lattices with the Voronoi cell being a parallelogon with six edges. Each generic region is further stratified by the action of the $GL(2, Z)$ group. The fundamental domain of $GL(2, Z)$ action consists of a sixth part of a generic domain together with its boundary. It is shown in figure as a lightly shaded region with its boundary.

From the partial ordering of Bravais classes (see section 4.4, Figures 4.6, 4.7) we know that $p4mm$ is generated by $p2mm$ and $c2mm$. Consequently, in Figure 6.2 the point at the intersection of $p2mm$ and $c2mm$ lines should correspond to a $p4mm$ Bravais class. For the 3d-cone, the $p4mm$ stratum is one-dimensional. It corresponds to intersections of the $p2mm$ and $c2mm$ planes. Similarly, the $p6mm$-invariant lattices appear at intersections of three $c2mm$ invariant strata. On Figure 6.2 the $p6mm$ stratum is shown as a system of isolated points whereas for the 3d-cone it is represented as a system of one-dimensional rays going through the cone vertex.

In order to construct the fundamental domain of the $GL_2(\mathbb{Z})$ action it is sufficient to choose one triangular domain (for example that shown in Figure 6.1 by light hatching) with its three boundaries but without a point belonging to the boundary of the cone.

Along with symmetry induced stratification of the cone of positive quadratic forms it is useful to look for a combinatorial classification of the Voronoï cells of corresponding lattices. We know that for two-dimensional lattices there are only two combinatorial types of Voronoï cells: hexagons for

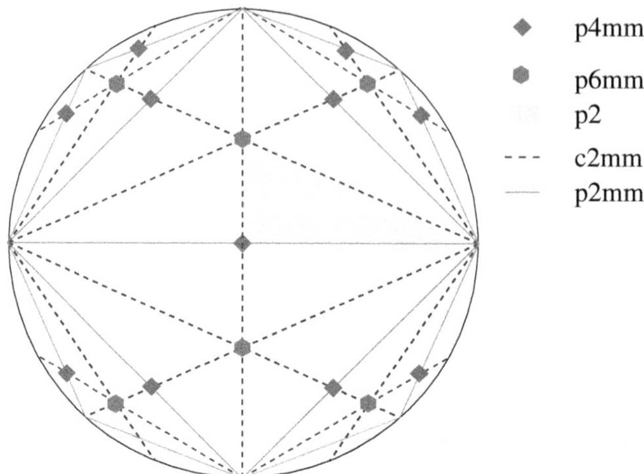

FIG. 6.2 – Representation of the section of the cone of positive quadratic forms depending on two variables. Stratification by the action of $GL(2, Z)$ into Bravais classes is shown. The fundamental domain includes a two-dimensional stratum ($p2$ lattices); two one-dimensional strata ($c2mm$ and $p2mm$); and two zero-dimensional strata ($p4mm$ and $p6mm$),

the generic primitive case and rectangles for non-primitive case. Rectangular Voronoï cells are compatible only with $p2mm$ and $p4mm$ symmetry. This means that from the point of view of combinatorial classification big triangular domains in Figure 6.2 formed by $p2mm$ boundary lines have in their interior points associated with primitive lattices (hexagon cells), whereas their boundaries (except vertices lying on the boundary of the cone) correspond to non-primitive lattices with rectangular Voronoï cells. Each such triangular domain consists of six fundamental regions of $GL_2(\mathbb{Z})$ action, intersecting at their boundaries.

6.2.2 Graphical representation of $GL_2(\mathbb{Z})$ transformation on the cone of positive quadratic forms

Remember that the action of a $GL_2(\mathbb{Z})$ element represented by matrix $B = \begin{pmatrix} b_{11} & b_{12} \\ b_{21} & b_{22} \end{pmatrix}$, satisfying condition $b_{11}b_{22} - b_{12}b_{21} = \pm 1$, on matrix $Q = \begin{pmatrix} q_{11} & q_{12} \\ q_{21} & q_{22} \end{pmatrix}$ is written as

$$Q \to Q' = BQB^{\top}, \tag{6.7}$$

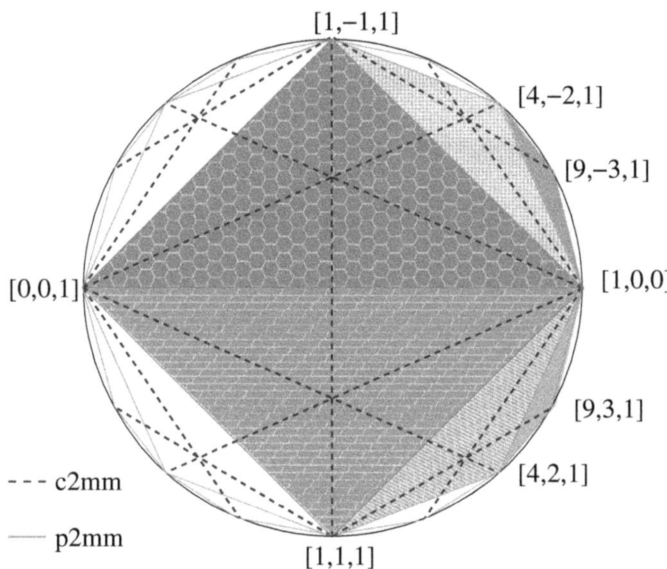

FIG. 6.3 – The action of B_1 transformation on the section of the cone of quadratic forms. Triangular domains shown by different shadings transform consecutively one into another in a clockwise direction around the point $[1,0,0]$ under B_1 action. Transformation of all other domains follows by applying the continuity arguments and invariance of combinatorial type under transformation. B_1^{-1} action corresponds to counterclockwise transformation of consecutive triangular domains around the same point $[1,0,0]$.

where $B^\top$ is the transposed matrix. The determinant of Q is invariant under $GL_2(\mathbb{Z})$ transformation. But on the representative section of the cone each point is denoted by the $[q_{11}, q_{12}, q_{22}]$ symbol which refers to the whole ray of quadratic forms with all possible determinants. The $[q_{11}, q_{12}, q_{22}]$ parameterization of points and lines used in Figures 6.3-6.5 is concretized in subsection 6.2.3 and Table 6.1.

$GL_2(\mathbb{Z})$ transformation is a continuous transformation of the disk representing the section of the cone of positive quadratic forms. Necessarily, it transforms each connected domain of one combinatorial (or symmetry) type into a domain of the same type and its boundaries into the respective boundaries. So to see the automorphism of the disk under the action of a concrete element of the $GL_2(\mathbb{Z})$ group, it is sufficient to study the transformation properties of special points being the vertices of domains of a given combinatorial type.

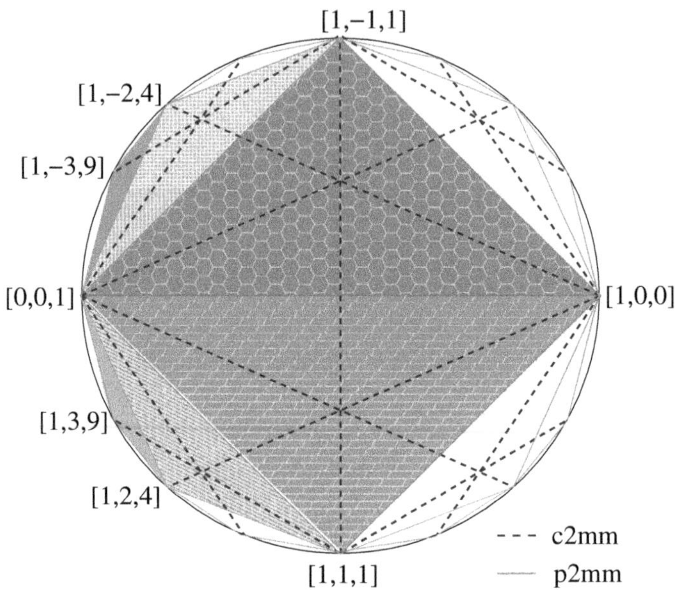

FIG. 6.4 – The action of B_2 transformation on the section of the cone of quadratic forms. Triangular domains shown by different shadings transform consecutively one into another in a clockwise direction around the point $[0, 0, 1]$ under B_2 action. Transformation of all other domains follows by applying the continuity arguments and invariance of combinatorial type under transformation.

Let us study the automorphism of the disk under the action of $B_1 = \begin{pmatrix} 1 & -1 \\ 0 & 1 \end{pmatrix}$ and its inverse $B_1^{-1} = \begin{pmatrix} 1 & 1 \\ 0 & 1 \end{pmatrix}$.

The point $[1, 0, 0]$ is invariant under B_1 action. The orbit of the point $[0, 0, 1]$ under the action of B_1 includes an infinite number of points which are obviously situated on the boundary of the disk

$$\begin{pmatrix} 1 & -1 \\ 0 & 1 \end{pmatrix}^K \begin{pmatrix} 0 & 0 \\ 0 & 1 \end{pmatrix} \begin{pmatrix} 1 & 0 \\ -1 & 1 \end{pmatrix}^K = \begin{pmatrix} K^2 & -K \\ -K & 1 \end{pmatrix}. \tag{6.8}$$

Expression (6.8) is valid for any integer K value, positive or negative. From this transformation formula we see immediately that, for example, the triangle $([1, 0, 0], [0, 0, 1], [1, 1, 1])$ transforms under the action of B_1 into triangle $([1, 0, 0], [1, -1, 1], [0, 0, 1])$, then under the repeated action to triangle $([1, 0, 0], [4, -2, 1], [1, -1, 1])$, next to triangle $([1, 0, 0], [9, -3, 1], [4, -2, 1])$, etc. Figure 6.3 shows schematically these transformations.

In a similar way we can study the automorphism of the disk under the action of $B_2 = \begin{pmatrix} 1 & 0 \\ 1 & 1 \end{pmatrix}$ and its inverse $B_2^{-1} = \begin{pmatrix} 1 & 0 \\ -1 & 1 \end{pmatrix}$.

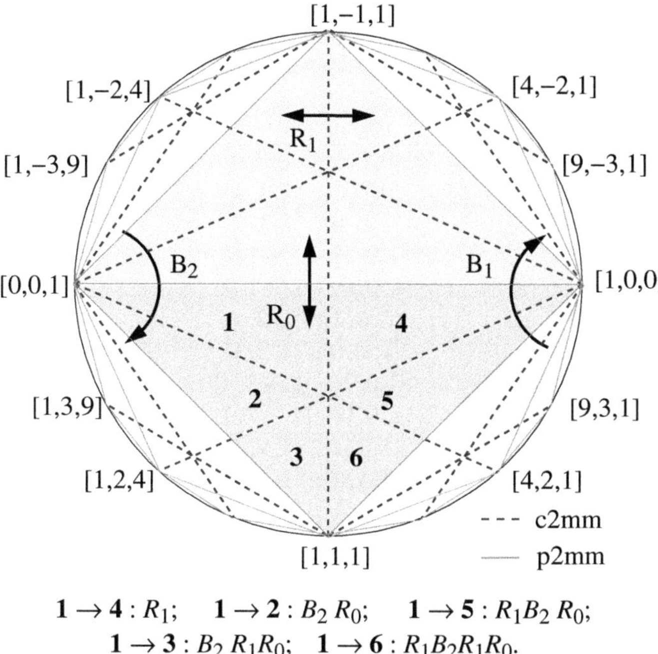

FIG. 6.5 – Examples of $GL_2(\mathbb{Z})$ elements realizing transformation between six equivalent sub-domains of the same connected combinatorial domain. Six sub-domains are labeled by big bold numbers **1**, **2**, **3**, **4**, **5**, and **6**.

Now the point $[0,0,1]$ is invariant under the B_2 action. The orbit of the point $[1,0,0]$ under the action of B_2 consists again in an infinite number of points situated on the boundary of the disk,

$$\begin{pmatrix} 1 & 0 \\ 1 & 1 \end{pmatrix}^K \begin{pmatrix} 1 & 0 \\ 0 & 0 \end{pmatrix} \begin{pmatrix} 1 & 1 \\ 0 & 1 \end{pmatrix}^K = \begin{pmatrix} 1 & K \\ K & K^2 \end{pmatrix}. \tag{6.9}$$

Expression (6.9) allows us to construct a graphical visualization of the B_2 transformation shown in Figure 6.4 and to see, in particular, that the triangle ($[0,0,1]$, $[1,0,0]$, $[1,-1,1]$) transforms under B_2 action into triangle ($[0,0,1], [1,1,1], [1,0,0]$), then under the repeated action to triangle ($[0,0,1]$, $[1,2,4]$, $[1,1,1]$), next to triangle ($[0,0,1]$, $[1,3,9]$, $[1,2,4]$), etc.

Along with transformation of points we can directly analyze transformation of lines. For example, we can find the image of the line $q_{12} = 0$ (corresponding to the $p2mm$ invariant boundary between generic combinatorial domains) under the action of B_2^K,

$$\begin{pmatrix} 1 & 0 \\ K & 1 \end{pmatrix} \begin{pmatrix} q_{11} & 0 \\ 0 & q_{22} \end{pmatrix} \begin{pmatrix} 1 & K \\ 0 & 1 \end{pmatrix} = \begin{pmatrix} q_{11} & Kq_{11} \\ Kq_{11} & K^2 q_{11} + q_{22} \end{pmatrix}. \tag{6.10}$$

TAB. 6.1 – $[q_{11}, q_{12}, q_{22}]$ parameterization of several lines on the section of cone of positive quadratic forms together with points lying on them and the combinatorial type of corresponding Voronoï cell.

Line	Points on line	Combinatorial type
$q_{12} = 0$	$[0, 0, 1]$; $[1, 0, 0]$	4-cell
$q_{12} - q_{11} = 0$	$[0, 0, 1]$; $[1, 1, 1]$	4-cell
$q_{12} - q_{22} = 0$	$[1, 0, 0]$; $[1, 1, 1]$	4-cell
$q_{12} + q_{11} = 0$	$[0, 0, 1]$; $[1, -1, 1]$	4-cell
$q_{12} + q_{22} = 0$	$[1, 0, 0]$; $[1, -1, 1]$	4-cell
$q_{12} - 2q_{11} = 0$	$[0, 0, 1]$; $[1, 2, 4]$	4-cell
$q_{12} - 2q_{22} = 0$	$[1, 0, 0]$; $[4, 2, 1]$	4-cell
$q_{12} + 2q_{11} = 0$	$[0, 0, 1]$; $[1, -2, 4]$	4-cell
$q_{12} + 2q_{22} = 0$	$[1, 0, 0]$; $[4, -2, 1]$	4-cell
$q_{11} - 3q_{12} + 2q_{22} = 0$	$[1, 1, 1]$; $[4, 2, 1]$	4-cell
$2q_{11} - 3q_{12} + q_{22} = 0$	$[1, 1, 1]$; $[1, 2, 4]$	4-cell
$q_{11} + 3q_{12} + 2q_{22} = 0$	$[1, -1, 1]$; $[4, -2, 1]$	4-cell
$2q_{11} + 3q_{12} + q_{22} = 0$	$[1, -1, 1]$; $[1, -2, 4]$	4-cell
$2q_{12} - q_{11} = 0$	$[0, 0, 1]$; $[4, 2, 1]$	6-cell
$2q_{12} - q_{22} = 0$	$[1, 0, 0]$; $[1, 2, 4]$	6-cell
$2q_{12} + q_{11} = 0$	$[0, 0, 1]$; $[4, -2, 1]$	6-cell
$2q_{12} + q_{22} = 0$	$[1, 0, 0]$; $[1, -2, 4]$	6-cell

This means that the line $q_{12} = 0$ transforms under the action of B_2^K into the line $q_{12} = Kq_{11}$. This allows us to easily label all boundaries between different combinatorial domains going through the $[0, 0, 1]$ fixed point of B_2 action.

Obviously, one can apply the same transformation to lines which are boundaries between different fundamental domains of $GL_2(\mathbb{Z})$ action but which correspond to the primitive combinatorial type ($c2mm$ invariant lines). For example, for the $q_{11} - 2q_{12} = 0$ line we get

$$\begin{pmatrix} 1 & 0 \\ K & 1 \end{pmatrix} \begin{pmatrix} 2q_{12} & q_{12} \\ q_{12} & q_{22} \end{pmatrix} \begin{pmatrix} 1 & K \\ 0 & 1 \end{pmatrix} = \begin{pmatrix} 2q_{12} & (2K+1)q_{12} \\ (2K+1)q_{12} & 2K(K+1)q_{12} + q_{22} \end{pmatrix}. \tag{6.11}$$

To see other important $GL_2(\mathbb{Z})$ transformations we need to add two reflections. The reflection $R_0 = \begin{pmatrix} 1 & 0 \\ 0 & -1 \end{pmatrix}$ corresponds to a reflection in the $q_{12} = 0$ line. It reverses the sign of q_{12}. Another reflection, $R_1 = \begin{pmatrix} 0 & 1 \\ 1 & 0 \end{pmatrix}$ interchanges q_{11} and q_{22}. It may be geometrically seen as reflection in the $q_{12} = 0$ line.

The action of four elements B_0, B_1, R_0, R_1 on the section of cone of quadratic forms is shown schematically in Figure 6.5. Using their geometrical visualization it is easy to find some simple sequences of transformations

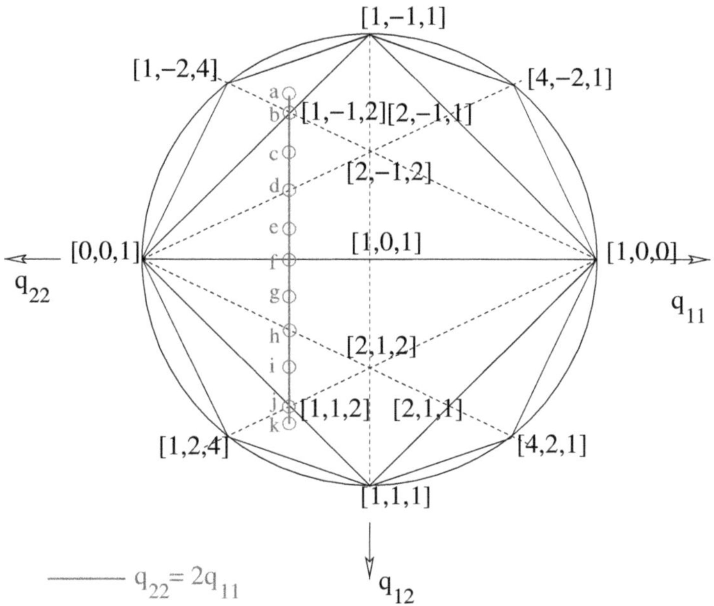

FIG. 6.6 – Section of the cone of quadratic forms with a path (a–k) along which evolution of lattices together with their Voronoï cell is shown in the next Figure 6.7. Notation for different lines is given in a separate Table 6.1.

which allow passage from one possible choice of fundamental domain to another one within the same domain of the combinatorial type. Examples of such transformations between six subdomains are given also in Figure 6.5.

6.2.3 Correspondence between quadratic forms and Voronoï cells

In order to see better the correspondence between points of the cone of positive quadratic forms and the corresponding Voronoï cell we take in figure 6.6 a series of points and represent in Figure 6.7 the evolution of the corresponding lattice and its Voronoï cell.

As we are interested not really in points of the cone but in rays, only two parameters are needed to define a ray. All matrices $Q = \begin{pmatrix} q_{11} & q_{12} \\ q_{12} & q_{22} \end{pmatrix}$ with different nonzero determinants but with the same ratio $q_{11} : q_{12} : q_{22}$ correspond to the same ray of the cone. Thus we can represent a ray by its projective coordinates $[q_{11}, q_{12}, q_{22}]$. Figure 6.6 shows stratification of the cone of positive quadratic forms in projective coordinates $[q_{11}, q_{12}, q_{22}]$. Equations for several lines corresponding to $c2mm$ and $p2mm$ strata are given in Table 6.1.

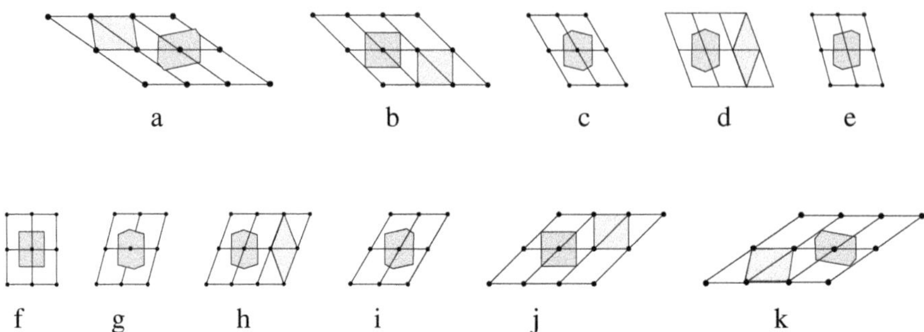

FIG. 6.7 – Lattices and their Voronoï cells associated with points on the section of the cone of positive quadratic forms shown in Figure 6.6.

As a path in the section of the cone of positive quadratic forms we take the line given by equation $q_{22} = 2q_{11}$. Along this line the 11 representative points $a, b, \ldots, k$ are chosen to cover different domains and to cross $c2mm$ and $p2mm$ strata. Lattices with their Voronoï cell for all these representative points are collected in Figure 6.7.

6.2.4 Reduction of two variable quadratic forms

To build a basis for a lattice L, we can start with any visible vector. We will choose a shortest vector $\vec{s}_1 \in S \subset L$; $\vec{s}_1$ defines a 1-sublattice $\{\mu\vec{s}_1;\ \mu \in \mathbb{Z}\}$. Then the 2-dimensional point lattice L becomes a union[1] $L = \cup_{\lambda\in\mathbb{Z}}\Sigma_\lambda$ of one-dimensional identical point lattices ("rangées") with $\Sigma_0 := \{\mu\vec{s}_1\}$ and $\Sigma_{\pm1}$ its nearest "rangées". The second basis vector $\vec{s}_2$ should belong to $\Sigma_{\pm1}$. These two rangées contain at least one vector whose orthogonal projection on the axis defined by $\vec{s}_1$ has the coordinate x which satisfies[2] $-\frac{1}{2} \leq x \leq 0$. When x satisfies the inequalities, we choose the corresponding vector as $\vec{s}_2$. The quadratic form defined by this basis is represented by the matrix with elements $q_{ij} = (\vec{s}_i, \vec{s}_j)$; these matrix elements satisfy exactly the conditions;

$$0 \leq -2q_{12} \leq q_{11} \leq q_{22}, \qquad 0 < q_{11}. \tag{6.12}$$

The set of quadratic forms defined by (6.12) is a fundamental domain of $\mathcal{C}_+(\mathcal{Q}_2)$: i.e. this domain contains one, and only one, quadratic form of each orbit of the $GL_2(\mathbb{Z})$ action on $\mathcal{C}_+(\mathcal{Q}_2)$.

[1] The arguments used here are those of [29]. Bravais wrote in French and used the words: "rangée, réseau, assemblage" for 1-, 2-, and 3-dimensional lattices, respectively. That makes his paper more colorful!

[2] The choice of the sign of x is arbitrary. We choose here the negative sign because this has a natural generalization to arbitrary n.

Determining a fundamental domain on $\mathcal{C}_+(\mathcal{Q}_n)$ is known as the problem of arithmetic reduction of quadratic forms. For $n = 2$ it was first solved by Lagrange [65].

Another approach to classification of lattices and associated quadratic forms was introduced by Voronoï ([94], p.157) and developed later by Delone [42]. Using Lagrange reduction, we can always choose a basis of a lattice such that the coefficients of the associated quadratic form $q_{11}x^2 + 2q_{12}xy + q_{22}y^2$ satisfy (6.12) $0 \leq -2q_{12} \leq q_{11} \leq q_{22}, 0 < q_{11}$. With the variables $\lambda = q_{11}+q_{12}$, $\mu = q_{22} + q_{12}$, $\nu = -q_{12}$, the quadratic form becomes a sum of squares:

$$\lambda x^2 + \mu y^2 + \nu(x-y)^2, \quad \lambda \geq 0,\ \mu \geq 0,\ \nu \geq 0. \quad \det(q_{ij}) = \lambda\mu + \mu\nu + \nu\lambda > 0. \tag{6.13}$$

As the value of the determinant shows, the quadratic form is positive if no more than one of the three parameters vanishes. We have the norms:

$$N\begin{pmatrix}1\\0\end{pmatrix} = \lambda + \nu; \quad N\begin{pmatrix}0\\1\end{pmatrix} = \mu + \nu; \quad N\begin{pmatrix}1\\1\end{pmatrix} = \lambda + \mu; \tag{6.14}$$

there is a complete syntactic symmetry among the parameters λ, μ, ν. The domain in $\mathcal{C}_+(\mathcal{Q}_2)$ associated with generic lattices possessing a primitive (hexagon) combinatorial type of Voronoï cell is invariant by the group of permutations S_3 of the three parameters λ, μ, ν. Indeed, it corresponds to the triangle $[0,0,1]$, $[1,0,0]$, $[1,1,1]$ of Figure 6.6 and S_3 permutes the six fundamental domains contained in the domain of (6.13).

It is straightforward to describe the five Bravais strata by studying them in the parameter space $\mathcal{C}_+(\mathcal{Q}_2)$ with λ, μ, ν parameterization. They can be labeled by an elegant symbol invented by Delone: the three parameters are represented by the three sides of a triangle.

First case: $\lambda\mu\nu \neq 0$:

- **i)** Generic Bravais class $p2$: represented by the Delone symbol △
- **ii)** When two parameters are equal, an order 2 symmetry appears: the invariance by R_1 in figure 6.6 (for instance $\lambda = \mu$); it exchanges the two equal sides of the triangle; it corresponds to the Bravais class $c2mm$:
- **iii)** When the three parameters are equal: we have the full symmetry S_3 of the triangle; with the inversion through the origin (=rotation by π), one describes the hexagonal Bravais class $p6mm$:

Second case: one of the three parameters vanishes[3].

[3] If we choose $\nu = 0$ the quadratic form is diagonal (invariant by R_0 in Figure 6.6). The cases $\lambda = 0$ and $\mu = 0$ are obtained from the preceding one by transforming the quadratic form by the $SL_2(\mathbb{Z})$ matrices $\begin{pmatrix}1 & 0\\-1 & 1\end{pmatrix}$ and $\begin{pmatrix}1 & -1\\0 & 1\end{pmatrix}$ respectively.

i) The two other parameters are different: Bravais class $p2mm$ 

ii) The two other parameters are equal: Bravais class $p4mm$

An extension of Delone classification to higher dimensional lattices and quadratic forms results in more fine classification of lattices than simply combinatorial or symmetry (Bravais) classification. (See Delone classification of three-dimensional lattices in Chapter 8, section 8.5 and the representation of combinatorial types of lattices by graphs in section 6.7.)

6.3 Three dimensional quadratic forms and 3D-lattices

The set of 3-dimensional quadratic forms $\{q\}$ (corresponding to symmetric real 3×3 matrices Q) forms a 6-dimensional real vector space $\mathbb{R}^6$, with the scalar product $(Q.Q') = \mathrm{tr}QQ'$. The 6-dimensional submanifold of positive forms, $\mathcal{C}_+(\mathcal{Q}_3)$, is the interior of a convex, homogeneous, self-dual[4] cone. Since each positive quadratic form represents a 3-dimensional Euclidean lattice, modulo position, it is interesting to partition $\mathcal{C}_+(\mathcal{Q}_3)$ both, into the 14 domains of Bravais classes, and the 5 domains of combinatorial types of Voronoï cells.

This would be very redundant, however, because the representation of an Euclidean lattice by a quadratic form depends on the choice of basis vectors, as we have seen during the analysis of a more simpler case of 2-dimensional quadratic forms in the preceding section.

To study the set of 3-dimensional lattices one has to consider only a fundamental domain of $\mathcal{C}_+(\mathcal{Q}_3)$ for the $GL_3(\mathbb{Z})$ action. To choose such a domain was a classical problem: the first solution was given by Seeber in 1831 [84]. The interior of such a domain can be chosen, using the main conditions for obtuse forms, to be:

$$0 < q_{11} \leq q_{22} \leq q_{33}, \quad i \neq j : \; q_{ij} \leq 0; \quad 2|q_{ij}| \leq q_{ii}; \quad 2|q_{12}+q_{13}+q_{23}| \leq q_{11}+q_{22}. \tag{6.15}$$

On the boundary of that domain there occur only non-generic Bravais classes with still some redundancy, which are solved by the auxiliary conditions[5]. That domain is unbounded. Since we are interested in lattices up to a dilation, we can consider only a five dimensional (bounded) domain of the group $GL_3(\mathbb{Z}) \times \mathbb{R}^\times_+$. The most natural way to do it is to choose the intersection of the domain (6.15) by the hyperplane $\mathrm{tr}Q = c$, with c a positive constant. We shall choose $\mathrm{tr}Q = 3$ and call $T\mathcal{C}_+(\mathcal{Q}_3)$ this 5 dimensional bounded domain. However it is still difficult to draw its picture! For studying a

---

[4] Both Q and Q^{-1} are in the cone.

[5] $|q_{23}| \leq |q_{13}|$ if $q_{11} = q_{22}$; $|q_{13}| \leq |q_{12}|$ if $q_{22} = q_{33}$; $q_{12} = 0$ if $2|q_{23}| = q_{22}$; $q_{12} = 0$ if $2|q_{13}| = q_{11}$; $q_{13} = 0$ if $2|q_{12}| = q_{11}$; $q_{11} \leq |q_{12} + 2q_{13}|$ if $2|q_{12} + q_{13} + q_{23}| = q_{11} + q_{22}$.

3-dimensional picture, we have to restrict ourselves to a section of $T\mathcal{C}_+(\mathcal{Q}_3)$ by a well chosen 4-dimensional subspace of $\mathcal{C}_+(\mathcal{Q}_3)$. To check the dimension arguments we note that the $\mathcal{C}_+(\mathcal{Q}_3)$ space is 6-dimensional. If we intersect 6-dimensional space by a 5-dimensional $(T\mathcal{C}_+(\mathcal{Q}_3))$ and by a 4-dimensional subspaces, generically the intersection of 5-dimensional and 4-dimensional subspaces is 3-dimensional.

How to cut the maximal number of different Bravais class domains? There are four maximal Bravais classes:

$$Pm\bar{3}m, \quad Fm\bar{3}m, \quad Im\bar{3}m, \quad P6/mmm.$$

For the partial ordering of the set of Bravais classes there is a unique largest element (i.e. with largest symmetry), smaller than these four maximal classes; that is the Bravais class Mono C $= C2/m$, whose domain has dimension 4. We choose a group G belonging to the conjugacy class of the $C2/m$ subgroups of $GL_3(\mathbb{Z})$. We denote by $\mathcal{H} = \mathcal{Q}_3^G$ the 4-dimensional subspace of the G-invariant quadratic forms. Its intersection with the hyperplane of the trace 3 quadratic forms will define the Euclidean 3-plane of our model (Figures 6.8, 6.9). Figure 6.8 shows a fundamental domain of the Mono C $= C2/m$ Bravais class. Its boundary shows, with some redundancy the fundamental domains of the 10 Bravais classes which have a larger symmetry. Moreover, the model shows simultaneously parts of the 5 domains of combinatorial types of Voronoï cell represented in Figure 6.9.

6.3.1 Michel's model of the 3D-case

We start by describing the stratification of the suggested above 3-dimensional model into different strata corresponding to different Bravais classes and into different domains associated with different combinatorial types of Voronoï cell. Note that this 3D-model was designed by Louis Michel during his visits and lecturing in Smith College, Northampton (USA) and Technion, Haifa (Israel).

We give now the description of the model and reserve some hints for its construction till the end of this section.

The model is the tetrahedron $ABCD$ (see Figure 6.8). Four vertices, five edges (except for the edge $\overline{AD}$) and the facet ABC correspond to points representing quadratic forms with $\det Q = 0$. All internal points, internal points of the facet ABC and of the edge $\overline{AD}$ represent positive quadratic forms.

Stratification of the tetrahedron $ABCD$ into Bravais classes for three-dimensional lattices is shown in Figure 6.8. There are 0-, 1-, 2-, and 3-dimensional strata for eleven Bravais classes (among 14 existing for the 3D-case). They are summarized in the following table

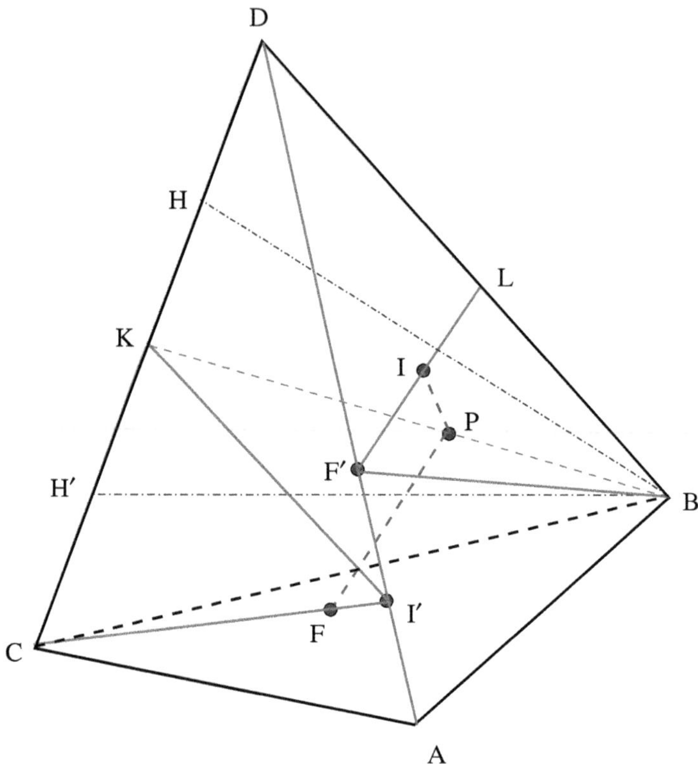

FIG. 6.8 – Partial model of the stratification of the cone of positive quadratic forms into Bravais classes for three-dimensional lattices. Strata of Bravais classes. Notice that $\det Q = 0$ on the facet ABC and on five edges of the tetrahedron, except for the edge $\overline{AD}$.

Mono C:	$C2/m$	interior of tetrahedron except intervals $\overline{PI}, \overline{PF}$;
Ort C:	$Cmmm$	facet BCD except $\overline{BH}, \overline{BH'}, \overline{BK}$;
Ort F:	$Fmmm$	facet BDA except $\overline{LF'}, \overline{BF'}$;
Ort I:	$Immm$	facet ACD except $\overline{KI'}, \overline{I'C}$;
Tet P:	$P4/mmm$	$\overline{BP}, \overline{PK}$;
Tet I:	$I4/mmm$	$\overline{BF'}, \overline{F'I}, \overline{IL}, \overline{DF'}, \overline{F'I'}, \overline{I'A}, \overline{KI'}, \overline{I'F}, \overline{FC}$;
Trig R:	$R\bar{3}m$	$\overline{PF}, \overline{PI}$;
Hex P:	$P6/mmm$	$\overline{BH}, \overline{BH'}$;
Cub P:	$Pm\bar{3}m$	P;
Cub F:	$Fm\bar{3}m$	F, F';
Cub I:	$Im\bar{3}m$	I, I'.

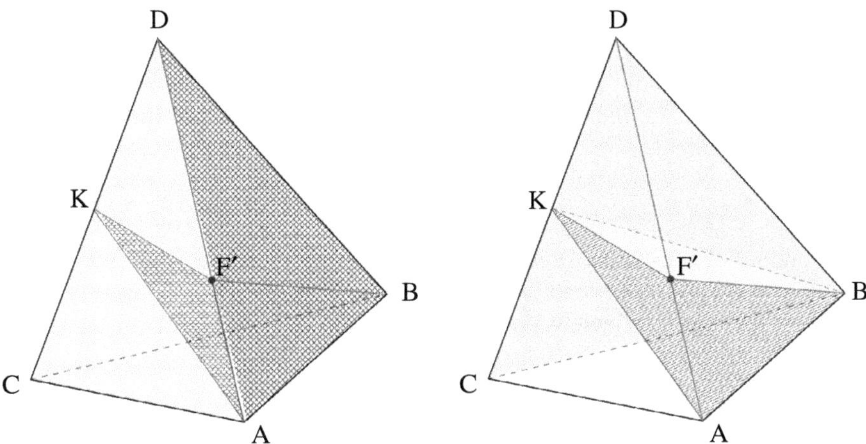

FIG. 6.9 – Partial model of the partition of the cone of positive quadratic forms into sub-cones of different combinatorial Voronoï cells. Notice that $\det Q = 0$ on the facet ABC and on five edges of the tetrahedron, except for the edge $\overline{AD}$. Left: Stratification of the facets ADB and ACD of the tetrahedron. Right: Strata non-visible on the left figure.

In order to visualize stratification of the tetrahedron $ABCD$ into domains of different combinatorial types we use in Figure 6.9 two images of the same tetrahedron and keep in this figure only points and lines important for stratification into combinatorial types. All points shown in Figure 6.9 are equally present in Figure 6.8, but some lines and planes present in Figure 6.8 are absent in Figure 6.9 because they have no specific combinatorial meaning. Remember that the lines and points absent in Figure 6.9 but present in Figure 6.8 are important to see the topology of the space of orbits (redundancy).

The stratification of the tetrahedron by different combinatorial types of Voronoï cell is given in the following table

14.24: interior of $DBF'K$ and $ABF'K$;
interior of DBF', $BF'A$, and $F'KA$;
interval $\overline{AF'}$

12.18: interior of $BACK$,
interior of $BF'K$, DKF', and CKA;
intervals $\overline{BF'}$ and $\overline{F'D}$

12.14: interior of ABK;
intervals $\overline{KF'}$ and $\overline{KA}$;
point F'

8.12: facet BCD except $\overline{BK}$

6.8: interval $\overline{BK}$

In order to see better the relation of the 3D-model to the six-dimensional cone of positive quadratic forms of three variables we recall below the relevant data for different combinatorial types of Voronoï cell and also on the dimension of these domains in the five-dimensional domain of positive quadratic forms with a given trace:

short notation	**14.24**	**12.18**	**12.14**	**8.12**	**6.8**
number of facet vectors	14	12	12	8	6
number of non-facet corona vectors	0	4	6	12	20
dimension of the domain in $TC_+(\mathcal{Q}_3)$	5	4	3	3	2
dimension of the domain in model	3	3	2	2	1

6.3.2 Construction of the model

Now we return briefly to some points important for the construction of the described above model.

The 4 element group $G = \mathbb{Z}_2(r) \times \mathbb{Z}_2(-I)$, generated by the two matrices:

$$R = \begin{pmatrix} 0 & 1 & 0 \\ 1 & 0 & 0 \\ 0 & 0 & 1 \end{pmatrix}, \quad -I = \begin{pmatrix} -1 & 0 & 0 \\ 0 & -1 & 0 \\ 0 & 0 & -1 \end{pmatrix} \tag{6.16}$$

is a realization in $GL_3(\mathbb{Z})$ of the point symmetry of the monoclinic $C2/m$ lattices. Its invariant quadratic forms form the 4-dimensional space:

$$\mathcal{H} :\equiv \mathcal{Q}_3^G = \left\{ Q = \begin{pmatrix} u & x & y \\ x & u & y \\ y & y & v \end{pmatrix}, \quad u, v, x, y \in \mathbb{R} \right\}. \tag{6.17}$$

In $\mathcal{H}$, the hyperplane of the trace 3 quadratic forms is:

$$\mathcal{H}' := \left\{ Q(x,y,z) = \begin{pmatrix} 1-z & x & y \\ x & 1-z & y \\ y & y & 1+2z \end{pmatrix} \right\}, \tag{6.18}$$

i.e. $2u + v = 3$, $v - u = 3z$. Given two quadratic forms $q, q' \in \mathcal{H}'$, their Euclidean distance is the square root of:

$$\operatorname{tr}(Q - Q')^2 = 2\big((x - x')^2 + 2(y - y')^2 + 3(z - z')^2\big). \tag{6.19}$$

The positive quadratic forms of $\mathcal{H}'$ form a bounded domain whose boundary is given by the condition for the quadratic forms of (6.18) to be positive:

$$-\frac{1}{2} < z < 1, \quad -(1 - z) < x < 1 - z, \quad y^2 < (1 + x - z)(1 + 2z)/2. \tag{6.20}$$

In $\mathcal{H}'$ this is a convex domain $\mathcal{K}$ bounded by three planes and one sheet of a (two sheet) hyperbolic quadric.

By construction of $\mathcal{H}$, all G-invariant lattices are represented in it: those are all lattices of the Bravais classes $\geq$ Mono C $= C2/m$. We denote by N the stabilizer of $\mathcal{H}$ in the linear action of $GL_3(\mathbb{Z})$ on the space $\mathcal{Q}_3$. It is easy to prove that N is the normalizer $N_{GL_3(\mathbb{Z})}(G)$, i.e. the largest subgroup of $GL_3(\mathbb{Z})$ containing G as the invariant subgroup. The lattices of the Bravais classes $\geq$ Mono C are represented by the orbits of N inside $\mathcal{H} \cap \mathcal{C}_+(\mathcal{Q}_3)$. The represented Bravais classes correspond to the strata of this action; the stratum representing the smallest class, Mono C, is open dense and we want to choose a fundamental domain in it. For this we have first to determine N.

We notice that G is in the center of N. Since $G \triangleleft N$, every $n \in N$ has to conjugate the 4 matrices of G into each other; since the matrices of G have different traces, n commutes with them. So N is the centralizer of G in $GL_3(\mathbb{Z})$:

$$N = C_{GL_3(\mathbb{Z})}(G). \tag{6.21}$$

To compute this centralizer, it is sufficient to find the integral matrices n which satisfy $nr = rn$, $r \in G$, and require their determinant to be ± 1:

$$n = \begin{pmatrix} \alpha & \beta & \delta \\ \beta & \alpha & \delta \\ \delta' & \delta' & \gamma \end{pmatrix}, \quad \det n = (\alpha - \beta)\big(\gamma(\alpha + \beta) - 2\delta\delta'\big). \tag{6.22}$$

Each factor of the determinant should be ± 1:

$$\varepsilon^2 = 1, \quad \eta^2 = 1, \quad \alpha - \beta = \varepsilon, \quad \gamma(\alpha + \beta) - 2\delta\delta' = \eta. \tag{6.23}$$

One can prove that N is generated by the matrices

$$-I,\ R,\ S = \begin{pmatrix} 1 & 0 & 0 \\ 0 & 1 & 0 \\ 0 & 0 & -1 \end{pmatrix}, \ D = \begin{pmatrix} 1 & 0 & 1 \\ 0 & 1 & 1 \\ 0 & 0 & 1 \end{pmatrix}, \ D' = \begin{pmatrix} 1 & 0 & 0 \\ 0 & 1 & 0 \\ 1 & 1 & 1 \end{pmatrix}. \tag{6.24}$$

The matrices $-I, R, S$, generate a group of the Bravais class Ort C $= Cmmm$. Each of the matrices D, D' generates an infinite cyclic group ($\sim \mathbb{Z}$). Since the stabilizer of any lattice is finite, the orbits of N in $\mathcal{H} \cap \mathcal{C}_+(\mathcal{Q}_3)$ are infinite. In general the action of $g \in N$ on $\mathcal{H}$ does not preserve the trace of quadratic forms; so we deduce the action of N on $\mathcal{H}'$ from the action on $\mathcal{H}$ by adding the stereographic projection normalizing the trace.

By construction, the matrices $-I, R$ act trivially on $\mathcal{H}$; the matrix S changes y into $-y$ (both in $\mathcal{H}$ and $\mathcal{H}'$); so from now on we make the convention:

$$\text{convention}: \quad y \leq 0. \tag{6.25}$$

In $\mathcal{H}'$, the intersection of the positivity domain (6.20) with the 2-plane $y = 0$ is chosen to be part of the boundary of our fundamental domain; its points represent lattices of the Bravais class $Cmmm$ or greater ones.

Finite subgroups of N are crystallographic point groups; therefore each one containing G as a strict subgroup will have a linear manifold of fixed points containing a domain of a larger Bravais class. To find the finite subgroups of N, we must first determine its elements of finite order. As for $GL_3(\mathbb{Z})$ their order can be only 1, 2, 3, 4, or 6. Elements of order 3 must have as eigenvalues the three cubic roots of 1, so their trace, $\tau := \mathrm{tr}\, n = 2\alpha + \gamma$, must be 0. That is impossible since we know that γ is odd [see (6.18)]. Hence N *has no elements of order 3 or 6* (the square of an element of order 6 would be of order 3). The equation $n^2 = 1$ yields the following conditions in addition to those of (6.23), and combined with them:

$$\gamma^2 + 2\delta\delta' = 1, \quad 2\alpha(\alpha - \varepsilon) + \delta\delta' = 0, \quad \delta(\tau - \varepsilon) = 0 = \delta'(\tau - \varepsilon). \tag{6.26}$$

Since the eigenvalues of these matrices are ± 1, their trace can be either -3 or ± 1. In the former case we find easily that $n = -I$. When the trace $\tau = \pm 1$ we must have $\tau + \det n = 0$ so

$$\tau = 2\alpha + \gamma = -\varepsilon\eta. \tag{6.27}$$

That, with the first two conditions of (6.26), yields $\eta = -1$. Notice that for elements of N which are four fold, $\varepsilon = \eta = 1$, so there are no elements of order 4 in N. That proves that in N, all non-trivial elements of finite order are of order 2. Hence, all finite subgroups of N have the structure $\mathbf{Z}_2^k$, and we know from the study of the finite subgroups of $GL_3(\mathbb{Z})$ that $k \leq 3$.

It is easy to verify that the largest finite subgroups of N represent three of the four conjugacy classes of $\mathbb{Z}_2^3$ subgroups in $GL_3(\mathbb{Z})$; explicitly, they can be generated by the matrices[6]:

$$Cmmm : \langle R, S, -I \rangle, \quad Fmmm : \langle R, W^\top, -I \rangle, \quad Immm : \langle R, W, -I \rangle, \tag{6.28}$$

with

$$W = \begin{pmatrix} 0 & 1 & 0 \\ 1 & 0 & 0 \\ -1 & -1 & -1 \end{pmatrix}. \tag{6.29}$$

The domains of these 3 Bravais classes are two-dimensional. We determine those invariants by the three matrix groups chosen in (6.28); they belong to the boundary of the fundamental domain that we have chosen to represent the Bravais class $C2/m$.

We recall now that given a subgroup G of $GL_3(\mathbb{Z})$ it is easy to verify that the linear map on the orthogonal space $\mathbb{R}^6$ (see [11], Chapter 7.3):

$$\mathcal{C}_+(\mathcal{Q}_3) \ni Q \;\mapsto\; |G|^{-1} \sum_{g \in G} g^\top Q g \tag{6.30}$$

[6] Among the different method for distinguishing the two point groups $Fmmm$ and $Immm$, the fastest one is the computation of their fixed points (i.e. their cohomology group $H^0(P, L)$) by their action on the lattice L: $Fmmm$ has four and $Immm$ two fixed points per unit cell.

is an orthogonal projector over the subspace of G-invariant quadratic forms. From (6.30) we obtain the equations of the 2-planes supporting boundaries of the fundamental domain in $\mathcal{H}'$ invariant by matrix groups in (6.28). The boundary of the positivity domain (6.20) also has a facet supported by a 2-plane. We define the 2-planes by

$$f_1 : y = 0; \quad f_2 : 1 - z + x + 2y = 0; \quad f_3 : 1 + 2z + 2y = 0; \quad f_4 : 1 - z - x = 0.$$

So the fundamental domain we have chosen in $\mathcal{H}'$ is a tetrahedron $ABCD$ whose facets are

$$Cmmm = BCD \subset f_1; \quad Fmmm = ABD \subset f_2; \quad Immm = ACD \subset f_3;$$
$$\text{positivity boundary } = ABC \subset f_4. \tag{6.31}$$

The coordinates x, y, z of its vertices are:

$$A = \left(\frac{3}{4}, -\frac{3}{4}, \frac{1}{4}\right), \quad B = (0,0,1), \quad C = \left(\frac{3}{2}, 0, -\frac{1}{2}\right), \quad D = \left(-\frac{3}{2}, 0, -\frac{1}{2}\right). \tag{6.32}$$

Notice that on the facet ABC and on five edges of the tetrahedron, $\det Q = 0$. This is not true for the edge $\overline{AD} = ABD \cap ACD$, so it represents the Bravais class Tet I $= I4/mmm$ or higher.

Now we pass to the analysis of the Bravais class domains of dimension 1 and 0 in $\mathcal{H}'$.

Besides the four orthorhombic Bravais classes[7] the Bravais class Trig R $=$ $R\bar{3}m$ is also a minimal supergroup[8] of $C2/m$. Its domain has dimension 1; indeed in $GL_3(\mathbb{Z})$ there are two groups of the conjugacy class $R\bar{3}m$ which contains $G \sim C2/m$ defined in (6.16); these groups are generated by the matrices:

$$R\bar{3}m = \langle R, -I, T\rangle, \ R\bar{3}m' = \langle R, -I, S^{-1}TS\rangle, \text{ with } T = \begin{pmatrix} 0 & 1 & 0 \\ 0 & 0 & 1 \\ 1 & 0 & 0 \end{pmatrix}, \tag{6.33}$$

and the corresponding invariant subspaces in $\mathcal{H}'$ are defined by $z = 0$ and $x = y$ or $x = -y$, respectively. Hence in our figure (we want $y < 0$) the trigonal Bravais class $R\bar{3}m$ is represented by two open segments inside the tetrahedron: in the subspace $z = 0$,

$$-\frac{1}{3} < x < 0 \quad \text{when } x = y; \quad 0 < x < \frac{1}{2} \quad \text{when } x = -y. \tag{6.34}$$

Their boundary is made of 3 points representing the 3 minimal supergroups of $R\bar{3}m$, i.e. the three cubic Bravais classes: we call these points:

$$P = (0,0,0), \quad I = \left(-\frac{1}{3}, -\frac{1}{3}, 0\right), \quad F = \left(\frac{1}{2}, -\frac{1}{2}, 0\right). \tag{6.35}$$

---

[7] Ort P $= Pmmm$ is not represented on the figure; this is also the case of the two other Bravais classes: Mono P $= P2/m$ and Tric $= \bar{1}$.

[8] i.e. there is no Bravais class X which satisfies $C2/m < X < R\bar{3}m$.

Notice that the point I is invariant by the group $R\bar{3}m$, the point F by $R\bar{3}m'$ and the point P by both groups.

There are two Bravais classes directly greater than the Bravais class Ort C = $Cmmm$; those are Tet P = $P4/mmm$ and Hex P = $P6/mmm$. The representative domain of each is 1-dimensional and has to belong to the facet BCD of the tetrahedron.

The stabilizer in $GL_3(\mathbb{Z})$ of the 2-plane $y = 0$ is the normalizer $N_{GL_3(\mathbb{Z})}(Cmmm) = P4/mmm$ which belongs to the Bravais class Tet P. Since $Cmmm$ acts trivially, its normalizer (which is a subgroup of $O_3(\mathbb{Z})$) acts only through the quotient

$$(P4/mmm)/Cmmm \sim \mathbb{Z}_2.$$

This action must be the orthogonal symmetry through an axis and this invariant axis represents the Bravais class Tet P. To realize the action of this quotient we can choose for instance the diagonal matrix $\begin{pmatrix} -1 & 0 & 0 \\ 0 & 1 & 0 \\ 0 & 0 & 1 \end{pmatrix}$, (in $P4/mmm$ but not in $Cmmm$); it changes x into $-x$ and leaves z invariant. So Tet P = $P4/mmm$ is represented by:

$$P4/mmm \mapsto x = y = 0, \quad -\frac{1}{2} < z < 0 < z < 1 \equiv]\overline{BK}[\setminus o. \qquad (6.36)$$

Note that the point $x = y = z = 0 \equiv o$ is represented in Figure 6.8 as point P.

In a similar way for Hex P = $P6/mmm$ we have

$$P6/mmm \mapsto y = 0, \quad x = \pm(1-z)/2, \quad -\frac{1}{2} < z < 1 \equiv]\overline{BH}[\cup]\overline{BH'}. \qquad (6.37)$$

The positions of the specified points H, H', K are given below

$$H = \left(-\frac{3}{4}, 0, -\frac{1}{2}\right), \quad H' = \left(\frac{3}{4}, 0, -\frac{1}{2}\right), \quad K = \left(0, 0, -\frac{1}{2}\right). \qquad (6.38)$$

We emphasize the redundancy in the facet BCD: when $x \neq 0$, the points $(\pm x, 0, z)$ represent the same lattice. In the boundary of the open segments defined in (6.36), only one point represents a Bravais class; that is $x = y = z = 0$ representing the Cub P class. This point (given in (6.35)) is common to the boundaries of the domains representing Tet P and Trig R, the two Bravais classes directly smaller than Cub P.

We noticed in 6.3.1 that no vertices and only one of the six edges of the tetrahedron represents a Bravais class: it is $]\overline{AD}[=]ABD \cap ACD[$, corresponding to $Fmmm \cap Immm$, which represents Tet P = $I4/mmm$.

This edge must also carry two points F' and I' representing the two Bravais classes Cub F and Cub I directly greater than Tet I. To find these points we can use again the same method as for the facet BCD representing Ort C = $Cmmm$.

The facet BDA represents Ort F $= Fmmm$; its stabilizer is

$$N_{GL_3(\mathbb{Z})}(Fmmm) = Fm\bar{3}m$$

belonging to the Cub F Bravais class. It acts on the plane as a linear representation of the quotient $Fm\bar{3}m/Fmmm \sim \mathcal{S}_3$. We can take as representative of this quotient in $Fm\bar{3}m$ a subgroup conjugate to $R\bar{3}m$ defined in (6.33); it is generated by the matrices:

$$R\bar{3}m'' = \langle -I, \quad R' = M^{-1}RM, \quad T' = M^{-1}TM \rangle, \tag{6.39}$$

with $M = \begin{pmatrix} 0 & 1 & -1 \\ -1 & 0 & 1 \\ 1 & 0 & 0 \end{pmatrix}$. Then, using (6.30) for this group we obtain the point representing Cub F:

$$F' = (0, -\frac{1}{2}, 0) \in \overline{AD}. \tag{6.40}$$

Using (6.33), we verify

$$Q_{F'} = M^\top S^\top Q_F SM. \tag{6.41}$$

The same group transforms the segment $\overline{AD}$ into two other ones $\overline{A'D'}$, $\overline{A''D''}$ defined by:

$$\overline{A'D'} \ : \ A' = (0, -\frac{3}{5}, -\frac{1}{5}), \quad D' = (0,0,1) = B; \tag{6.42}$$

$$\overline{A''D''} \ : \ A'' = (-1,0,0), \quad D'' = (1,-1,0). \tag{6.43}$$

The segment parts $\overline{A'F}$ and $\overline{F'D''}$ are in $\mathcal{H}'$ but outside the tetrahedron. The segment $\overline{A''F'}$ contains the point I defined in (6.34). The orbit of this point for the group $R\bar{3}m''$ contains the two other points:

$$I' = (\frac{3}{10}, -\frac{3}{5}, \frac{1}{10}) \in \overline{AD}; \quad I'' = (0, -\frac{6}{11}, -\frac{1}{11}). \tag{6.44}$$

The points I' and F' are on the edge $\overline{AD}$, (which represents Tet I). The point I'' does not belong to the tetrahedron.

Similarly, the stabilizer of the facet ACD, which represents the Bravais class Ort I $= Immm$, has normalizer $N_{GL_3(\mathbb{Z})}(Immm) = Im\bar{3}m$ which belongs to the Bravais class Cub I. This class is represented by the $Im\bar{3}m$ invariant point I' (defined in (6.44)). Moreover that normalizer transforms $\overline{AD}$ into two other segments whose intersections with the facet ACD are $\overline{KI'}$ and $\overline{CI'}$.

Finally, similar to the case of the facet BCD representing the class Ort C, we notice the same type of redundancy for the facets BAD and CAD

representing respectively the Bravais classes Ort F and Ort I. Indeed the intermediate groups

$$Fmmm < I4/mmm < N_{GL_3(\mathbb{Z})}(Fmmm) = Fm\bar{3}m, \tag{6.45}$$

$$Immm < I4/mmm < N_{GL_3(\mathbb{Z})}(Immm) = Im\bar{3}m, \tag{6.46}$$

belong to the Bravais class Tet I $= I4/mmm$; they respectively leave invariant the segments $\overline{BF'} \subset BAD$ and $\overline{KI'} \subset CAD$ which both represent Tet I. We notice that the interior of the triangles $DLF' \subset BDA$, $I'CA \subset CDA$ are not redundant.

All the obtained information is used for the construction of Figure 6.8.

To take into account all redundancies for points on the boundary of the tetrahedron $ABCD$ and to see the topology of the fundamental domain, the following identification of domains of the boundary of the tetrahedron should be done:

- Triangle BKD should be identified with BKC.
- Triangle $F'LB$ should be identified with $F'LD$.
- Triangle $I'KC$ should be identified with $I'KD$.

This implies that the following identification of 1-dimensional and 0-dimensional subsets on the boundary of $ABCD$ should be done:

- BH should be identified with BH'.
- $I'F'$ should be identified with $I'F$.
- CF should be identified with DF' and with BF'.
- F should be identified with F'.

6.4 Parallelohedra and cells for N-dimensional lattices.

In this section we give a brief description of some important new features related to the combinatorial classification of lattices and to the associated cone of positive quadratic forms which appear for lattices in higher dimensional $d \geq 4$ space as compared to the cases of planar $d = 2$ and space $d = 3$ lattices studied earlier in this chapter.

First of all it is necessary to make the definition of the combinatorial type of polytopes and their labeling for arbitrary dimension more precise.

The k-faces of a polytope P are partially ordered with respect to inclusion. Together with the empty set $\{\emptyset\}$ the k-faces form the *face lattice* $\mathcal{L}(P)$. (See the definition of a lattice as a partial ordered set in appendix A.) For any two faces F and F' of $\mathcal{L}(P)$, the least upper bound is given by the k-face $F_\vee \supset F \cup F'$ having the least k. The k-face $F_\vee$ is unique because otherwise there would exist a face $\tilde{F} := F_\vee \cap F'_\vee \supset F \cup F'$ and thus, k would not be minimal. The greatest lower bound is given by the l-face $F_\wedge = F \cap F'$.

Definition: combinatorial type Two polytopes P and P' are combinatorially equivalent, $P' \overset{\text{comb}}{\simeq} P$, and belong to the same combinatorial type, if there exist a combinatorial isomorphism $\tau : \mathcal{L}(P) \to \mathcal{L}(P')$.

The combinatorial type of P is denoted by the *short symbol* $\mathbf{N}_{(n-1)}.\mathbf{N}_0$. For different combinatorial types having the same short symbol, additional letter/number symbols A, a, B, b, ... are added to distinguish them. In particular, we denote by n_h the number of 2-faces of P which are hexagons. In many cases the short symbol with the addition of -n_h uniquely characterizes special sets of parallelotopes [11]. More generally, for any k, $1 < k < n$, let $d_i^{(k)}$ be the number of k-faces of P which have $f_i^{(k)}$ subordinated $(k-1)$-faces, $i = 1, \ldots, r$. The k-subordination symbol is defined by

$$f^{(k)}_{1_{d_1^{(k)}}} f^{(k)}_{2_{d_2^{(k)}}} \cdots f^{(k)}_{r_{d_r^{(k)}}},$$

with $f_1^{(k)} < f_2^{(k)} < \cdots < f_r^{(k)}$. We give a few easy examples. The 2-subordination symbol of the 3-dimensional cubooctahedron is $4_6 6_8$, which means that there are six quadrilateral facets (2-faces) and eight hexagonal facets (2-faces). The 4-dimensional cube has the 3-subordination symbol 6_8 (there are eight facets (3-faces) possessing each six 2-faces) and the 2-subordination symbol 4_{24} (there are 24 quadrilateral 2-faces).

In order to verify combinatorial equivalence, the k-subordination symbols are determined for $k = (n-1), \ldots, 2$. The concatenation of these k-subordination symbols is called a *subordination scheme.* The subordination scheme does not characterize a polytope uniquely in dimension $d \geq 3$, but it is sufficient for parallelotopes in $\mathbb{R}^n$ for at least $n \leq 7$. A unique characterization of a polytope obtained by the unified polytope scheme is described in [48].

As we have introduced in section 5.4, each vertex of a primitive parallelotope in E^n is determined by the intersection of n facets. Let $\{\vec{f}_{l_1}, \ldots, \vec{f}_{l_n}\}$, be the set of the corresponding facet vectors. These vectors are linearly independent and determine a sublattice of the lattice L of index $\omega(v)$. It was shown by Voronoï[94], §66 that the upper bound for the number of vertices is reached exactly if, for each vertex v of a primitive parallelotope, $\omega(v) = 1$. Ryšhkov and Baranovskii [83] gave upper bounds for the index $\omega(v)$.

Theorem 9 *For dimensions* $n = 2, 3, 4, 5,$ *and* 6 *the maximal values of the index* $\omega(v)$ *are* 1, 1, 1, 2, *and* 3, *respectively.*

The index $\omega(v)$ has direct correlation to the number of vertices N_0 of a primitive parallelotope P. The primitive parallelotope with $\omega(v) = 1$ for each of its vertices is called the principal primitive. Voronoï have shown [94] that the number of k-faces N_k, $0 \le k \le d$ of a parallelotope in E^d is

$$N_k \le (d+1-k)\sum_{l=0}^{d-k}(-1)^{d-k-l}\binom{d-k}{l}(1+l)^d. \qquad (6.47)$$

For the number of facets ($k = d-1$) equation (6.47) becomes an equality *for all primitive* parallelohedra

$$N_{d-1} = 2(2^d - 1) \quad \text{for primitive parallelohedra} \qquad (6.48)$$

and coincides with the upper bound in the inequality for the number of facets vectors given by Minkowski [78] for a d-dimensional parallelohedron:

$$2d \le |\mathcal{F}| \le 2(2^d - 1). \qquad (6.49)$$

The equality sign in (6.47) holds for principal primitive parallelohedra for any k.

In particular, from (6.47) we immediately have the following estimations for the number of vertices N_0, edges N_1 and $(d-2)$-faces $N_{(d-2)}$, related to the number of belts, for d-dimensional parallelohedra

$$N_0 \le (d+1)!, \quad N_1 \le \frac{d}{2}(d+1)!, \quad N_{(d-2)} \le 3\left(1 - 2^{(d+1)} + 3^d\right). \qquad (6.50)$$

The equalities in (6.50) hold only for principal primitive parallelohedra. We note here that primitive parallelohedra contain sixfold belts only. This allows the number of belts N_b for primitive parallelohedra to be expressed as $N_b = N_{(d-2)}/6$.

Non-principal primitive parallelohedra exist for $d \ge 5$. They have the same number of facets as principal primitive parallelohedra but the number of k-faces with $k \le d-2$ is less (for some k) than the maximal possible value for principal primitive parallelohedra.

The number of combinatorial types of primitive parallelohedra in E^d increases rapidly with increasing dimension d. In dimensions 2 and 3 there exists only one combinatorial type of primitive parallelohedra. In $d = 2$ this is a hexagon and in $d = 3$ this is a truncated octahedron. In $d = 4$ there are three combinatorially different parallelohedra which are all principal primitive. In $d = 4$ there is also one non-primitive parallelohedron which has the same maximal number of faces as primitive ones. In dimension 5 as found by Engel [47], there are 222 combinatorially different types of primitive parallelohedra among which there are 21 non-principal. In dimension 6 only the lower bounds for the number of primitive parallelohedra are known [25]. There are at least 567613632 combinatorial types among which there are 293517383 non-principal ones.

It is interesting to see the recently found results on the numbers N_k of k-faces of primitive parallelohedra [25]. They are reproduced in Table 6.2

TAB. 6.2 – The numbers N_k of k-faces of primitive parallelohedra in E^d, $2 \leq d \leq 6$. Different sets of numbers N_k of k-faces for six-dimensional non-primitive parallelohedra correspond to sixteen different values of $t = 1, 2, \ldots, 16$. The table is based on the numerical data given in [25].

d	N_0	N_1	N_2	N_3	N_4	N_5	Belts
2	6	6					6_1
3	24	36	14				6_6
4	120	240	150	30			6_{25}
5	720	1800	1560	540	62		6_{90}
	708	1770	1536	534	62		6_{89}
6	5040	15120	16800	8400	1806	126	6_{301}
	$5040 - 28t$	$15120 - 84t$	$16800 - 90t$	$8400 - 40t$	$1806 - 6t$	126	6_{301-t}

in a slightly different manner which explicitly shows that for non-principal primitive parallelohedra the $d+1$ dimensional vector of numbers N_k, $k = 0, 1, \ldots, d$ can be written as a linear function of only one auxiliary parameter chosen in Table 6.2 as t and taking for $d = 5$ only one value $t = 1$ and for $d = 6$ taking 16 consecutive values $t = 1, \ldots, 16$.

The origin of this linear dependence on only one auxiliary parameter remains unexplained for non-principal primitive parallelohedra. Several linear relations between numbers of k-faces are known for a larger class of convex polytopes, namely for simple polytopes.

Definition: simple polytope A d-dimensional polytope P is called *simple* if every vertex v of P belongs to exactly d facets of P.

The class of simple polytopes is larger than the class of primitive polytopes defined in terms of primitive tilings. For example the d-dimensional cube is simple but not the primitive polytope. For a simple d-dimensional polytope the system of linear relations between numbers of k-faces (known as Dehn-Sommerville relations) consists of $\lfloor (d+1)/2 \rfloor$ relations, where $\lfloor x \rfloor$ is the integer part of x. The simplest way to introduce this relationship is to use the so called h-vectors of the polytope [2].

Definition: h-vector Let P be a d-dimensional simple polytope and $N_k(P)$ be the number of k-dimensional faces of P (we agree that $f_d(P) = 1$). Let

$$h_k(P) = \sum_{i=k}^{d} (-1)^{i-k} \binom{i}{k} N_i(P) \quad \text{for} \quad k = 0, \ldots, d. \tag{6.51}$$

The $(d+1)$-tuple $(h_0(P), \ldots, h_d(P))$ is called the h-vector of P.

It can be proved that the numbers of k-faces, N_k, can be uniquely determined from $h_k(P)$:

$$N_i(P) = \sum_{k=i}^{d} \binom{k}{i} h_k(P) \quad \text{for} \quad i = 0, \ldots, d. \tag{6.52}$$

Now we formulate without proof the following important proposition.

Proposition 33 (Dehn-Sommerville relations). *Let P be a simple d-dimensional polytope. Then*

$$h_k(P) = h_{d-k}(P) \quad \text{for} \quad k = 0, \dots, d. \tag{6.53}$$

and

$$1 = h_0 \leq h_1 \leq \dots \leq h_{\lfloor d/2 \rfloor}. \tag{6.54}$$

For centrally symmetric simple d-polytopes Stanley [18, 90] improved inequality (6.54), namely:

$$h_i - h_{i-1} \geq \binom{d}{i} - \binom{d}{i-1}, \quad \text{for} \quad i \leq \lfloor d/2 \rfloor. \tag{6.55}$$

For primitive parallelohedra we can apply Dehn-Sommerville relations together with the explicit expression (6.48) for the number of facets of primitive parallelohedra and the upper bound for the number of k-faces of primitive parallelohedra given by Voronoï (6.47). Also we take into account that the number of k-faces of primitive parallelohedra should be a multiple of $2(d-k+1)$ for $k \leq n-1$ (see proposition 29).

For $d = 2$ the only Dehn-Sommerville relation coincides with Euler characteristic of the polytope. Together with $N_1 = 6$ (6.48) this determines the unique vector of the numbers of faces ($N_1 = 6, N_0 = 6$) for the primitive 2-dimensional polytopes.

For $d = 3$ the second Dehn-Sommerville relation appears which can be written in a form applicable for any $d \geq 3$,

$$dN_0(P) = 2N_1(P) \quad \text{for} \quad d \geq 3. \tag{6.56}$$

Applying two Dehn-Sommerville relations to three-dimensional simple polytopes we get for the numbers of faces expression

$$(N_0 = 2N_2 - 4, \quad N_1 = 3N_2 - 6, \quad N_2), \tag{6.57}$$

which includes one free parameter, N_2. For primitive $3d$-polytope the number of facets is $N_2 = 14$ (6.48) and we get the unique possible set of numbers of faces for $3d$-primitive parallelohedron: ($N_0 = 24, N_1 = 36, N_2 = 14$).

The same two general linear Dehn-Sommerville relations exist for $4d$-simple polytopes. This means that we can express the numbers of k-faces for four dimensional simple polytopes in terms of two free parameters, say N_3 and N_2:

$$(N_0 = N_2 - N_3, \quad N_1 = 2N_2 - 2N_3, \quad N_2, \quad N_3). \tag{6.58}$$

It follows that for primitive 4-polytopes after imposing $N_3 = 30$ and $N_2 = 150 - 6\alpha$, we get for the number of faces and for the components of h-vector the following expressions which depend on one free parameter α:

$$N_0 = 120 - 6\alpha, N_1 = 240 - 12\alpha, N_2 = 150 - 6\alpha, N_3 = 30, N_4 = 1; \tag{6.59}$$

$$h_0 = 1 = h_4, \quad h_1 = 26 = h_3, \quad h_2 = 66 - 6\alpha. \tag{6.60}$$

Applying relation (6.54) we get immediately that α can take only a small number of values, namely $\alpha = 0, 1, 2, 3, 4, 5, 6$. But among these values only $\alpha = 0$ and $\alpha = 5$ give the number of vertices divisible by 10 and among these two possible values only $\alpha = 0$ gives the number of edges divisible by 8. Consequently, we get that the only possible set of the numbers of faces for primitive four-dimensional parallelohedra is $(N_0 = 120,\ N_1 = 240,\ N_2 = 150,\ N_3 = 30,\ N_4 = 1)$.

For five dimensional simple polytopes there are three Dehn-Sommerville linear relations.

$$N_0 - N_1 + N_2 - N_3 + N_4 - 2 = 0; \tag{6.61}$$
$$N_1 - 2N_2 + 3N_3 - 5N_4 + 10 = 0; \tag{6.62}$$
$$N_2 - 4N_3 + 10N_4 - 20 = 0. \tag{6.63}$$

For primitive parallelohedra $N_4 = 62$ and we can express N_{d-2} as $N_3 = 540 - 6\alpha$ taking into account that primitive parallelohedra have only six-fold belts (i.e. N_{d-2} should be divisible by 6). This allows us to express all numbers of faces in terms of one free parameter α and to explain the linear relation between numbers of faces for 5d-primitive parallelohedra with 90 and 89 belts given in Table 6.2. Namely we get

$$\begin{gathered}(N_0 = 720 - 12\alpha, \quad N_1 = 1800 - 30\alpha, \quad N_2 = 1560 - 24\alpha, \\ N_3 = 540 - 6\alpha, \quad N_4 = 62) \quad \text{with} \quad \alpha = 0, 1, \ldots .\end{gathered} \tag{6.64}$$

This expression fits numerical results listed in Table 6.2, but the restriction of α to only two possible values $\alpha = 0, 1$ remains unexplained. The inequality (6.55) allows only to state that $0 \leq \alpha \leq 40$.

For six-dimensional simple polytopes there are again three linear Dehn-Sommerville relations. Together with $N_5 = 126$ this gives for six-dimensional primitive polytopes expressions for the number of faces depending on two free parameters.

$$\begin{gathered}N_5 = 126; \quad N_4 = 1806 - 6\alpha; \quad N_3 = 8400 - 8\beta; \\ N_2 = 16800 + 30\alpha - 24\beta; \quad N_1 = 15120 + 36\alpha - 24\beta; \\ N_0 = 5040 + 12\alpha - 8\beta.\end{gathered} \tag{6.65}$$

We see that for any integer α, β the N_4 is divisible by 6, the N_3 is divisible by 8, the N_1 is divisible by 12. At the same time N_2 becomes a multiple of 10 only for $\beta = 5\gamma$, with $\gamma = 0, 1, 2, \ldots$. Replacing β by 5γ we get

$$\begin{gathered}N_5 = 126; \quad N_4 = 1806 - 6\alpha; \quad N_3 = 8400 - 40\gamma; \\ N_2 = 16800 + 30\alpha - 120\gamma; \quad N_1 = 15120 + 36\alpha - 120\gamma; \\ N_0 = 5040 + 12\alpha - 40\gamma.\end{gathered} \tag{6.66}$$

But we still need to check that N_0 is divisible by 14. This is equivalent to the requirement for $(3\alpha - 10\gamma)$ to be a multiple of 7. This is possible only for $\alpha = 0, \gamma = 0, 7, 14, \ldots$; $\alpha = 1, \gamma = 1, 8, 15, \ldots$; $\alpha = 2, \gamma = 2, 9, 16, \ldots$, etc. More generally we should have $\gamma - \alpha = 7k$.

Taking into account that for any set of two free parameters, α, γ, the numbers of faces cannot exceed their values for principal primitive parallelohedra we get general restrictions on possible values of free parameters $0 \leq 3\alpha \leq 10\gamma$. Together with the divisibility constraint $\gamma = \alpha + 7k$, with k being any integer, it follows that for $\gamma = 0$ the only possible value of the second parameter is $\alpha = 0$. Similarly, for $\gamma = 1$ we should have $\alpha = 1$ and for $\gamma = 2$, $\alpha = 2$. Only starting from $\gamma = 3$, several values of the second parameter are possible, in particular formal solutions are $(\gamma = 3, \alpha = 3)$ and $(\gamma = 3, \alpha = 10)$. Numerical results given by Baburin and Engel [25] correspond to face vectors with $\alpha = \gamma = 0, 1, \ldots, 16$. The fact that for six-dimensional primitive parallelohedra the whole observed set of face vectors can be described as only one-parameter family should be related to additional properties of primitive parallelohedra which are not taken into account in the present analysis.

It is clear that with increasing dimension the number of free parameters for the face vectors obtained within the adopted above scheme increases. For 7-dimensional parallelohedra we still have two free parameters but for 8-dimensional there are three such parameters, etc. The question whether the exact solution for face vectors of primitive parallelohedra in arbitrary dimension can be described by a one parameter family or a multi-parameter family is an interesting open problem.

6.4.1 Four dimensional lattices

This section illustrates correspondence between description of the four-dimensional lattices in terms of combinatorial types of parallelohedra and in terms of the subdivision of the cone of positive quadratic forms.

In four-dimensional space E^4 there exist three types of primitive parallelohedra which are principal (i.e. have the maximal numbers of k-faces for all k, namely $N_3 = 30, N_2 = 150, N_1 = 240, N_0 = 120$). Corresponding quadratic forms fill on the 10-dimensional cone of positive quadratic forms in four variables the 10-dimensional generic domains. Along with three primitive parallelohedra there exist one combinatorial type which is not primitive but has the maximal number of facets. The face vector for this non-primitive but maximal type is $(N_3 = 30, N_2 = 144, N_1 = 216, N_0 = 102)$. The quadratic forms associated with this non-primitive parallelohedron form a 9-dimensional domain.

Starting from these four maximal parallelohedra all other combinatorial types can be obtained by a consecutive application of the zone contraction. There are two zone-contraction/extension families consisting in 35 and 17 combinatorial types respectively. These two families are shown in Figures 6.10 and 6.11. The whole list of different combinatorial types of 4-dimensional parallelohedra was given initially by Delone [41] who found 51 types and

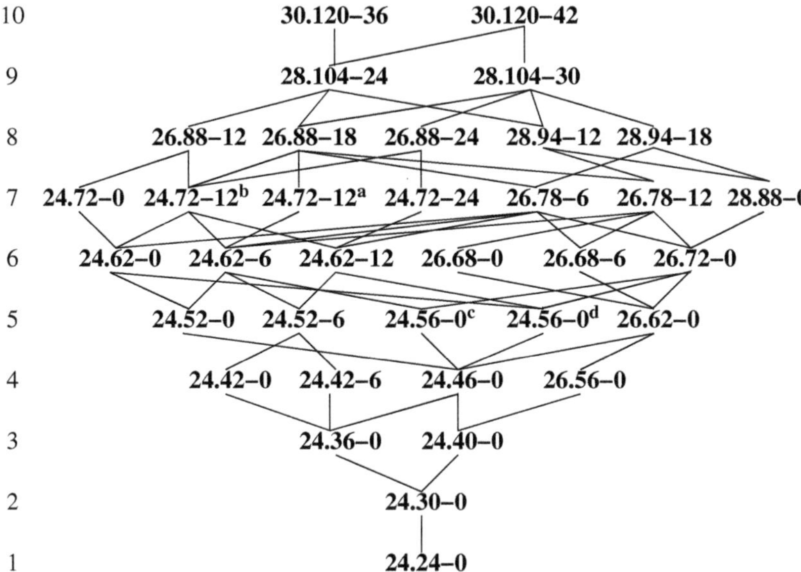

FIG. 6.10 – Zone contraction/extension family of Voronoï cells in E^4 consisting of 35 combinatorial types including two primitive cells, **30.120**-42 and **30.120**-36 and **24.24**-0 cell (F_4). Each cell is denoted by a $\mathbf{N_3}.\mathbf{N_0}$-$n_6$ symbol where n_6 is the number of hexagonal 2-faces. When this symbol is insufficient for a unique definition of the cell we give as a footnote the 3-subordination symbol: a - $8_{12}10_{12}$; b - $8_{14}10_8 12_2$; c - $8_{16}10_8$; d - $8_{18}10_4 12_2$. The dimension of the corresponding regions within the ten-dimensional cone of positive quadratic forms is indicated on the left. Note that some minor modifications have been introduced into the original figure taken from [11]. The modifications are justified by an explicit graphical correlation discussed in the next section.

was corrected by Shtogrin [87], adding one missed type. The organization of combinatorial types into two families was studied by Engel [11, 49]. (For a more detailed recent analysis see [32, 91, 44]. We will discuss briefly this organization using graphical representation in the next section 6.7.)

Each of the three primitive parallelohedra are associated with a 10-dimensional domain on the cone of the positive quadratic cone bounded each by 10 hyperplanes (walls). Schematic representation of these generic domains is given in Figure 6.12. (We return to the more profound discussion of this figure in section 6.8 after introducing graphical representation.) We use in these figures an abbreviated notation for primitive parallelohedra used by Engel [11], namely **30.120**-60 is denoted by "2"; **30.120**-42 is denoted by "3"; and **30.120**-36 is denoted by "4". All walls between **30.120**-60 and **30.120**-42 (i.e. between "2" and "3") are of **28.96**-40 type. It is important that the

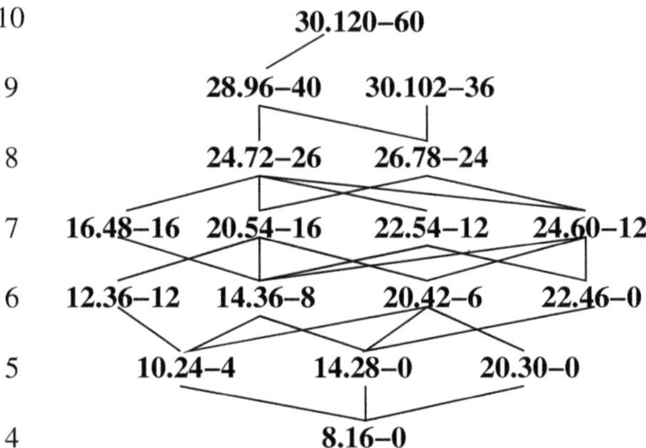

FIG. 6.11 – Zonohedral contraction/extension family of Voronoï cells in E^4 consisting of 17 cells. Notation is explained in caption to figure 6.10. The dimension of the corresponding regions within the ten-dimensional cone of positive quadratic forms is indicated on the left.

passage from "2" to the **28.96**-40 wall corresponds to the contraction of the **30.120**-60 parallelohedron whereas there is no contraction/extension transformation between "3", i.e. **30.120**-42 and the same wall **28.96**-40. All walls between disconnected domains of "3" type (i.e. **30.120**-42) are of **28.104**-30 type. They correspond to contraction of the cell "3".

Each isolated domain of **30.120**-36 type (i.e. of type "4") has nine walls of **28.104**-24 type separating "4" and "3" and associated with contraction from both sides and one wall between two disconnected domains of the same type "4". This wall is of **30.102**-36 type. It corresponds to a non-primitive paralelohedron with maximal number of facets and there is no contraction leading from the region "4" to that wall.

Finally the domain "3" (i.e. **30.120**-42) has six walls with similar disconnected domains of the same type "3", three walls with domains of type "4" (i.e. **30.120**-36) and one wall with domain "2" (i.e. **30.120**-60).

6.5 Partition of the cone of positive-definite quadratic forms

We describe now in slightly more detail the algebraic structure of the cone of positive-definite quadratic forms in n variables. Special attention will be paid now to the evolution of combinatorial type along a path in the space of positive quadratic forms going from one generic domain to another different (or equivalent by $GL_n(\mathbb{Z})$ transformation) domain by crossing the wall.

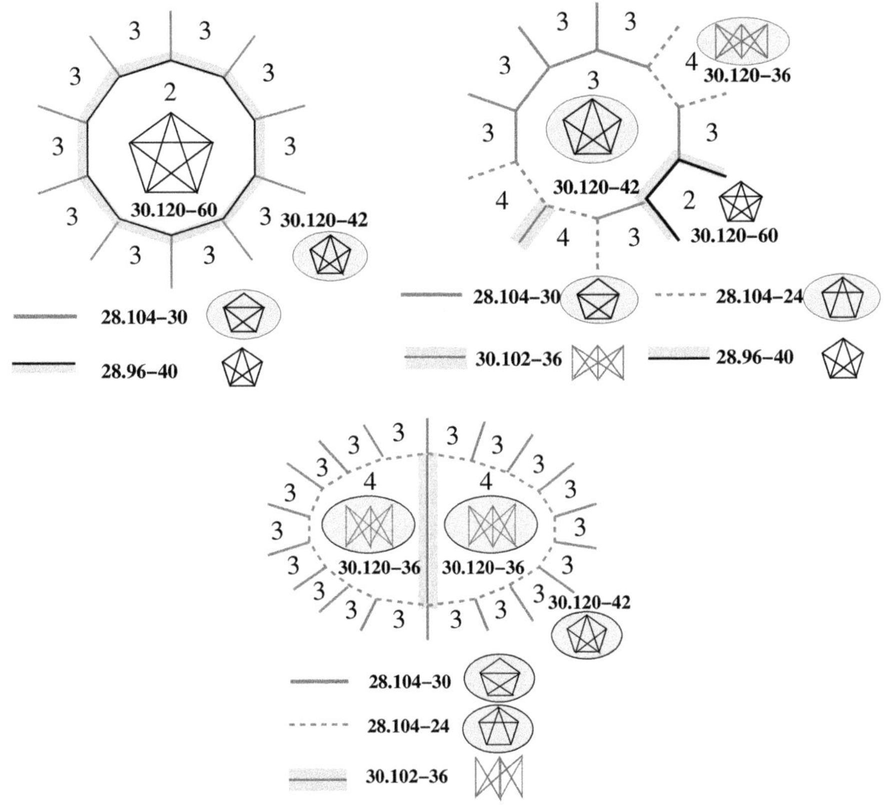

FIG. 6.12 – Schematic representation of local arrangements of generic subcones of the cone of quadratic forms for $d = 4$. Ten-dimensional domains with nine-dimensional boundaries are represented by two-dimensional regions with one-dimensional boundaries. **30.120**-36 type is abbreviated as "4", **30.120**-42 type as "3", and **30.120**-60 type as "2" in accordance with notation used by Engel [11]. For comments on graphical visualization see section 6.8.

A quadratic form is defined by $\varphi(\vec{x}) := \vec{x}^{\mathrm{t}} Q \vec{x}$. We denote by

$$\mathcal{C}^+ := \left\{ Q \in \mathbb{R}^{\binom{n+1}{2}} | \varphi(\vec{x}) > 0, \forall \vec{x} \in E^n \setminus \{0\} \right\} \tag{6.67}$$

the cone of positive-definite quadratic forms. Its dimension is $\binom{n+1}{2} = \frac{n(n+1)}{2}$. The closure of the cone is denoted by $\mathcal{C} := clos(\mathcal{C})^+$, and its boundary by $\mathcal{C}^0 := \mathcal{C} \setminus \mathcal{C}^+$.

Given an orthonormal basis $\vec{e}_1, \ldots, \vec{e}_n$ of $\mathbb{R}^n$, a basis of $\mathbb{R}^{n \times n}$ is obtained by the tensor products $\vec{e}_{ij} := \vec{e}_i \otimes \vec{e}_j$, $i, j = 1, \ldots, n$, with $\vec{e}_{ij}\vec{e}_{kl} = \delta_{ik}\delta_{jl}$. Since Q is symmetric, $Q = Q^{\mathrm{t}}$, it follows that the cone $\mathcal{C}^+$ can be restricted to a subspace $\mathbb{R}^{\binom{n+1}{2}}$ of dimension $\binom{n+1}{2}$, defined by $\vec{e}_{ij} = \vec{e}_{ji}$, $i \leq j = 1, \ldots, n$.

In $\mathbb{R}^{n\times n}$, the Gram matrix Q is represented by a vector $\vec{q}$ with components q_{ij}, $1 \leq i,j \leq n$. Each zone vector $\vec{z}^*$ has a representation in $\mathbb{R}^{n\times n}$ by $\vec{z}^* \otimes \vec{z}^*$.

One can study the symmetry of a lattice L by investigating the symmetry of its Gram matrix in the cone $\mathcal{C}^+$. For any $A \in GL_n(\mathbb{Z})$, $Q' = AQA^{\mathrm{t}}$ is arithmetically equivalent to Q. Thus, $\vec{q}\,' = A^{\mathrm{t}} \otimes A^{\mathrm{t}}\vec{q}$ is arithmetically equivalent to q. If $S \in GL_n(\mathbb{Z})$ fixes Q, $Q = SQS^{\mathrm{t}}$, then $S^{\mathrm{t}} \otimes S^{\mathrm{t}}$ fixes $\vec{q}$.

For any vector $\vec{v}^* = v_1\vec{a}_1^*, \ldots, v_n\vec{a}_n^*$ in dual space, the tensor product $\vec{l} = \vec{v}^* \otimes \vec{v}^*$ is denoted to be a ray vector. Since $\det(\vec{v}^* \otimes \vec{v}^*) = 0$, it follows that the ray vector $\vec{l}$ lies on the boundary $\mathcal{C}^0$. Let $\vec{c}$ be the representation of the identity matrix in $\mathbb{R}^{n\times n}$. Then $\lambda\vec{c}$ is the axis of the cone $\mathcal{C}$, because for any ray vector $\vec{l}$, the cone angle ψ satisfies

$$\cos\psi = \frac{\vec{c}\cdot\vec{l}}{|\vec{c}||\vec{l}|} == \frac{v_1^2 + v_2^2 + \ldots + v_n^2}{\sqrt{n}\sqrt{v_1^4 + 2v_1^2v_2^2 + \ldots + v_n^4}} = \frac{1}{\sqrt{n}}. \tag{6.68}$$

Thus $\mathcal{C}$ is a cone of rotation with rotation axis $\lambda\vec{c}$. For $n = 2$ the cone angle ψ is $\pi/4$ (see Figure 6.1). For large dimensions n, the cone angle ψ is close to $\pi/2$. The cone $\mathcal{C}$ is intersected by subspaces of dimensions $\binom{k+1}{2}$, $k < n$.

Let us now study partition of the cone $\mathcal{C}$ into domains of non-equivalent combinatorial types.

Definition: domain of combinatorial type In the cone $\mathcal{C}$, the domain of combinatorial type of a parallelohedron P is the connected open subcone of Gram matrices

$$\Phi^+(P) = \{Q \in \mathcal{C}^+ | P(Q) \overset{\text{comb}}{\simeq} P\}. \tag{6.69}$$

By $\Phi = clos(\Phi^+)$ we denote its closure, and its boundary is given by $\Phi^0 = \Phi \setminus \Phi^+$.

Theorem 10 *The domain Φ of the combinatorial type of a parallelohedron P is a polyhedral subcone of $\mathcal{C}$.*

Proof. We have to show that the border between two neighboring domains of parallelohedra of different combinatorial type are flat walls. It is sufficient to do that for generic domains, the walls are then hyperplanes in $\mathcal{C}$. We give the condition for the existence of a wall $W \subset \Phi$. Let Φ be a generic domain. The length of at least one edge of P diminishes for some $Q \in \Phi^+$ approaching the boundary Φ^0, and when Q hits Φ^0, both vertices subordinated to that edge coincide. By this coincidence at least $n+1$ facets meet in the common vertex v. If a facet F_i contains the vertex v then the corresponding facet vector $\vec{f}_i$ fulfills the equation

$$\vec{v}\,^{\mathrm{t}}Q\vec{f}_i = \frac{1}{2}\vec{f}_i^{\,\mathrm{t}}Q\vec{f}_i, \quad i = 1, \ldots, n+1. \tag{6.70}$$

As a sufficient condition that $n+1$ facets meet in the vertex v, we have that the determinant

$$\begin{vmatrix} \sum q_{1j} f_{1j} & \dots & \sum q_{nj} f_{1j} & \vec{f_1}^{\,t} Q \vec{f_1} \\ \cdot & & \cdot & \cdot \\ \cdot & & \cdot & \cdot \\ \cdot & & \cdot & \cdot \\ \sum q_{1j} f_{nj} & \dots & \sum q_{nj} f_{nj} & \vec{f_n}^{\,t} Q \vec{f_n} \\ \sum q_{1j} f_{n+1,j} & \dots & \sum q_{nj} f_{n+1,j} & \vec{f}_{n+1}^{\,t} Q \vec{f}_{n+1} \end{vmatrix} = 0. \tag{6.71}$$

Since $\vec{f_1}, \dots, \vec{f_n}$ form a basis of a sublattice of L of index ω, it follows that

$$\vec{f}_{n+1} = \alpha_1 \vec{f_1} + \dots + \alpha_n \vec{f_n}, \quad \alpha_i, \in \mathbb{Z}/\omega\mathbb{Z}. \tag{6.72}$$

Hence, the determinant can be transformed to

$$\begin{vmatrix} \sum q_{1j} f_{1j} & \dots & \sum q_{nj} f_{1j} & \vec{f_1}^{\,t} Q \vec{f_1} \\ \cdot & & \cdot & \cdot \\ \cdot & & \cdot & \cdot \\ \cdot & & \cdot & \cdot \\ \sum q_{1j} f_{nj} & \dots & \sum q_{nj} f_{nj} & \vec{f_n}^{\,t} Q \vec{f_n} \\ 0 & \dots & 0 & A \end{vmatrix} = 0, \tag{6.73}$$

where

$$A := \sum_{i=1}^{n} \alpha_i(\alpha_i - 1) \vec{f_i}^{\,t} Q \vec{f_i} + 2 \sum_{i=1}^{n-1} \sum_{j=i+1}^{n} \alpha_i \alpha_j \vec{f_i}^{\,t} Q \vec{f_j}. \tag{6.74}$$

We set

$$\Delta_n = \begin{vmatrix} \sum q_{1j} f_{1j} & \dots & \sum q_{nj} f_{1j} \\ \cdot & & \cdot \\ \cdot & & \cdot \\ \cdot & & \cdot \\ \sum q_{1j} f_{nj} & \dots & \sum q_{nj} f_{nj} \end{vmatrix}. \tag{6.75}$$

The determinant thus becomes

$$A\Delta_n = A \ \det(Q) \ \det(\vec{f_1}, \dots, \vec{f_n}) = 0. \tag{6.76}$$

This product gives, in terms of the Gram matrix Q, the condition that the $n+1$ facets meet in the vertex v. Either factor can be zero.

- First consider the case $A = 0$. The term A is linear in the q_{ij} and hence, it determines a flat wall $W \subset \Phi$.
- The case $\det(Q) = 0$, or $\det(\vec{f_1}, \dots, \vec{f_n}) = 0$ means that $Q \in \mathcal{C}^0$ and the lattice L^n degenerates to L^k, $k < n$.

The parallelohedron P has only a finite number of edges, and therefore Φ is bounded by a finite number of hyperplanes. Thus Φ is a rational polyhedral subcone of $\mathcal{C}$. □

Since ω is finite, the term A can be represented by integral numbers h_{ij}, and thus the coincidence condition becomes

$$h_{11}q_{11} + h_{12}q_{12} + \ldots + h_{nn}q_{nn} = 0. \tag{6.77}$$

The wall normal

$$\vec{n} = h_{11}\vec{e}_{11} + h_{12}\vec{e}_{12} + \ldots + h_{nn}\vec{e}_{nn} \tag{6.78}$$

is orthogonal to the wall W.

In general, the wall W separates two domains of different combinatorial type. The wall itself is an open domain $\overline{\Phi}^+$ for some limiting type.

The edges of Φ are the extreme forms of Φ, and are referred to as edge forms. An edge form is either

- a ray vector lying on the boundary $\mathcal{C}^0$ which has a representation as a tensor product $\vec{z}^* \otimes \vec{z}^*$ with zero determinant, where $\vec{z}^*$ is a vector of a closed zone of P.
- a generic inner edge form of $\mathcal{C}^+$ having positive determinant.
- a non-generic inner edge form of $\mathcal{C}^+$ having zero determinant, i.e. it is a generic inner edge form of a cone $\overline{\mathcal{C}}^+$ of a lower dimension $\binom{k+1}{2}$, $k < n$. Inner edge forms occur only in dimensions $n \geq 4$.

An effective numerical algorithm to determine the walls and the edge forms is discussed in [25].

6.6 Zonotopes and zonohedral families of parallelohedra

After looking at the system of different combinatorial types of parallelohedra and their organization in families for four-dimensional lattices we return to some systematic classification of combinatorial types of parallelohedra for arbitrary dimension. We start with the definition of the Minkowski sum of polytopes.

Definition: Minkowski sum The vector sum or Minkowski sum of two convex polytopes P and P' is the polytope

$$P + P' = \{x + x' | x \in P, x' \in P'\}. \tag{6.79}$$

Equivalently, we can describe $P + P'$ as the convex sum of all combinations of their vertices. Let $V(P)$ and $V(P')$ be the set of vertices of P and P', then

$$P + P' = \text{conv}\{v + v' | v \in V(P), v' \in V(P')\}. \tag{6.80}$$

This can be generalized to any finite number of summands in an obvious way.

Now we define one special but very important class of polytopes.

Definition: Zonotope A zonotope is a finite vector sum of straight line segments.

We recall that a zone Z of a parallelohedron P is the set of all 1-faces (edges) E that are parallel to a zone vector $\vec{z}^*$,

$$Z := \{E \subset P | E \| \vec{z}^*\}. \tag{6.81}$$

In each edge at least $d - 1$ facets meet. The zone vector $\vec{z}^*$ is the outer product of the corresponding facet vectors. In the dual basis, $\vec{z}^*$ has integer components

$$\vec{z}^* = z_1 \vec{a}_1^* + \ldots + z_n \vec{a}_n^*, \quad z_i \in \mathbb{Z}. \tag{6.82}$$

With respect to any zone vector $\vec{z}^*$ we can classify the lattice vectors in layers

$$L_i(\vec{z}^*) := \{\vec{t} \in L^n | \vec{t}\vec{z}^* = i, \quad |i| = 0, 1, \ldots\}. \tag{6.83}$$

A zone Z is referred to as being closed if every 2-face of P contains either two edges of Z, or else none. Otherwise Z is denoted as being open.

The zone contraction is the process of contracting every edge of a closed zone by the amount of its shortest edges. As a result, the zone becomes open, or vanishes completely, but the properties of a parallelohedron are maintained and the result of the zone contraction is a parallelohedron of a new combinatorial type. If a d-dimensional parallelohedron P collapses under a zone contraction, then the resulting P' parallelohedron has dimension $d - 1$.

A parallelohedron P_c is referred to as being totally contracted, if all its zones are open. It is relatively contracted, if each further contraction leads to a collapse into a parallelohedron of a lower dimension. A parallelohedron P_m is maximal, if it cannot be obtained by a zone contraction of any other parallelohedron in the same dimension.

Note that a polytope P is a zonotope if and only if all its k-faces are centro-symmetric. In its turn, a zonotope is a parallelohedron if and only if all its belts have 4 or 6 facets. This is a consequence of Theorem 5.

The parallelohedra which are at the same time zonotopes have a particular simple combinatorial structure. They are named zonohedral parallelohedra.

For zonohedral parallelohedra P the following two conditions are equivalent:

i) each zone of P has edges of the same length;

ii) each zone of P is closed.

In each dimension there exists a unique family of parallelohedra which contains all zonohedral parallelohedra, and which is named a zonohedral family. In dimensions $d \leq 3$ all parallelohedra are zonohedral and belong to the unique family. The zonohedral family for $d = 4$ consists of 17 members shown in Figure 6.11.

In dimensions $d \geq 4$ the zonohedral family includes several maximal zonotopes. For $d = 4$ (see figure 6.11), for example, the zonotopes **30.120** $- 60$ and **30.102** $- 36$ are maximal. Each zonohedral family has one main zone-contraction lattice corresponding to the maximal zonohedral parallelohedron $P_m(\mathbf{A}_n^*)$ of the root lattice $\mathbf{A}_n^*$ which is a primitive principal and generic, i.e. fills $d(d+1)/2$-dimensional domain of the cone of positive quadratic forms. This main zone-contraction sub-family of the zonohedral family includes all parallelohedra which can be obtained from the $P_m(\mathbf{A}_n^*)$ zonotope by zone contraction. For dimension 4 (see again Figure 6.11) the main zone contraction sub-family consists of all zonotopes except one, namely **30.102** $- 36$. One contraction is necessary to transform **30.102** $- 36$ to a parallelohedron belonging to the main zone-contraction sub-family. Each maximal zonotope can be characterized by the number of zone-contraction steps needed to attain the main zone-contraction family. In dimension $d = 4$, the **30.102** $- 36$ parallelohedron is distanced from the main zone-contraction family by one step (contraction till **26.78** $- 24$). The zonohedral family for $d = 5$, for example, includes 81 zonotopes (see section 6.7), among which there are four maximal, with the maximal distance from main zone-contraction sub-family consisting of three contraction steps.

The minimal member of the zonohedral family has combinatorial type of a parallelepiped (hypercube) and occupies a d-dimensional domain on the cone of positive quadratic forms.

Apart from the zonohedral family in each dimension $d \geq 4$ there exist a number of parallelohedra which can be represented as a finite Minkowski sum of a totally zone contracted parallelohedron and a zonotope [51]. In dimension $d = 4$ the family consisting of 35 parallelohedra (see figure 6.10) can be constructed by applying a zone extension operation to the totally contracted 24-cell parallelohedron **24.24** $- 0$ associated with F_4 lattice.

Not every totally contracted parallelohedron can be extended by applying a Minkowski sum with a segment (without extending the dimension of the parallelohedron). The maximal and simultaneously totally contracted parallelohedron, for example, exists in $d = 6$. It is related to the E_6^* lattice [50, 57].

6.7 Graphical visualization of members of the zonohedral family

The fact that all members of the zonohedral family can be represented as a vector sum of a certain number of segments (vectors) allows us to construct relatively simple visualization of different combinatorial types of zonohedral lattices using graphs in such a way that each segment generating the

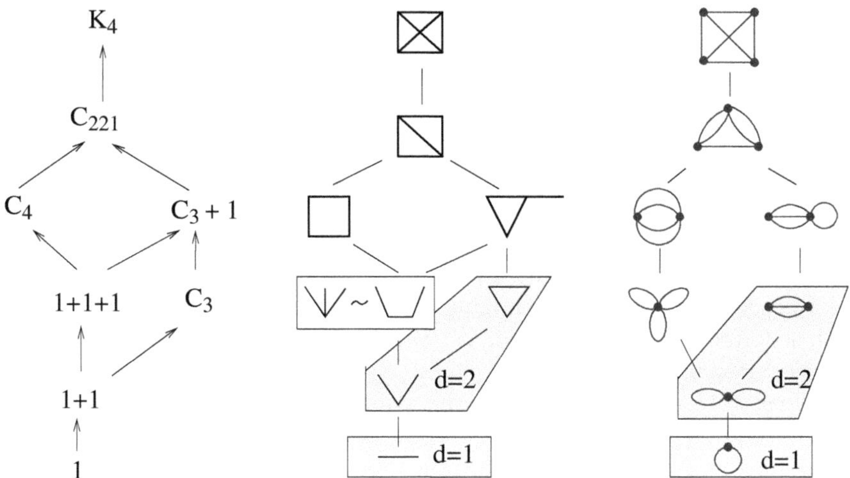

FIG. 6.13 – Graphical (center) and cographical (right) representations of zonohedral parallelotopes in dimensions $d = 1, 2, 3$ together with zone extension relations between them. The left diagram gives the notation of graphs used for graphical representation, as introduced in [32].

Minkowski sum is represented by a segment whereas linear dependencies between vectors corresponds to cycles of the graph. We cannot enter here into detailed mathematical theory of such a correspondence which is based on the matroid theory (for introduction see [23]). We hope that the more or less self-explaining correspondence shown in Figure 6.13 for dimensions $d = 1, 2, 3$ and in further figures for dimension $d = 4$ and $d = 5$ will stimulate the interest of the reader to study the corresponding mathematical theory.

The so called graphical representation for d-dimensional zonotopes consists in constructing connected graphs with $d+1$ nodes without loops and multiple edges. For $d = 1$ we obviously have one graph, for $d = 2$ there are two graphs (see Figure 6.13, center). In dimension $d = 3$ we need to introduce the equivalence relation between graphs, namely, for edges with one free end we should allow the other end of the same edge to move freely from one node to another. This means that all "tree-like" graphs or subgraphs should be treated as equivalent (see the equivalence between two three-edge graphs for $d = 3$ in Figure 6.13, center). This gives five inequivalent graphs for $d = 3$. For the notation of graphs (see left subfigure in 6.13) we follow the style used in the book [23].[9] The most important for further applications is the notation

[9] In [35], Conway and Sloane use these five graphs among different alternative versions of graphical visualizations and indicate as inconvenience the absence of symmetry transformations for this presentation. We note, however, that looking at these graphs up to topological equivalence, including 2-isomorphism [97] removes this inconvenience.

K_4 and its natural generalization to K_{d+1}, $d \geq 3$, which means the complete graph on four (or more generally on $d+1$) nodes. For dimension three, all five combinatorial types correspond exactly to all five connected graphs on four nodes (taking into account the introduced equivalence of graphs).

It is easy to see that the correlation between graphs (shown by connecting lines) corresponds to removing or adding one edge, when this correlation is within different graphs of the same number of nodes, i.e. between zonotopes of the same dimension. Removing one edge corresponds to zone contraction and all subgraphs of K_4 with four nodes can be obtained from K_4 by successively removing edges. Removing an edge with a free end leads to a graph with a lower number of nodes, i.e. we go to lower dimension with such a transformation.

Along with the graphical representation for 1-,2,-3-dimensional zonotopes we can equally use so called cographical representation which consists of replacing the graphical representation by a dual graph. To construct a dual graph, the original graph should be planar, i.e. when drawing a graph on paper (plane) no intersection or touching points between edges are allowed (except at the nodes). All graphs in Figure 6.13 are planar. (It is sufficient to deform graph K_4 to avoid the intersection of two edges.) To construct for a planar graph the dual graph, we need to associate with one connected domain of the plane a node and with each edge of the original graph an edge of the dual graph crossing this edge and relating nodes associated with left and right domains separated by an edge. (It may occur that the domain is the same and we get a loop.) The following simple examples give an intuitive understanding of the construction of the dual graph.

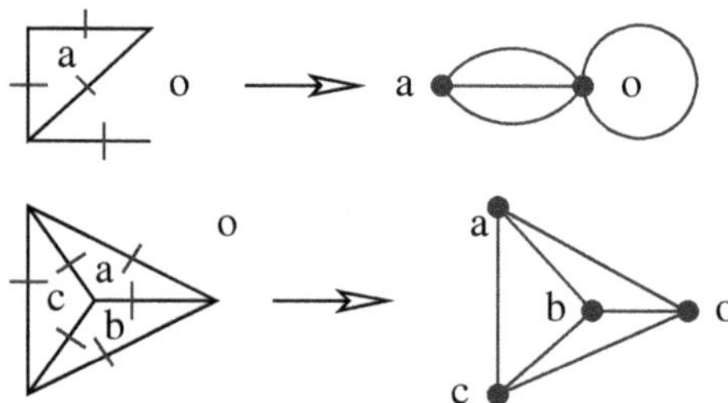

We see that a loop at one node and multiple edges between pairs of nodes appear naturally for a dual graph. Also we see that K_4 is self-dual. Elimination of one edge for graphical representation corresponds to shrinking of one edge by identifying two nodes for the corresponding dual graph. Increasing dimension for the graphical representation by adding one edge with a free end (adding an extra node) corresponds to adding a loop in the cographical representation. A cographical representation for three dimensional combinatorial types of lattices is given in Figure 6.13, right.

A very interesting and new situation (as compared with the three-dimensional case) appears for 4-dimensional zonotopes.

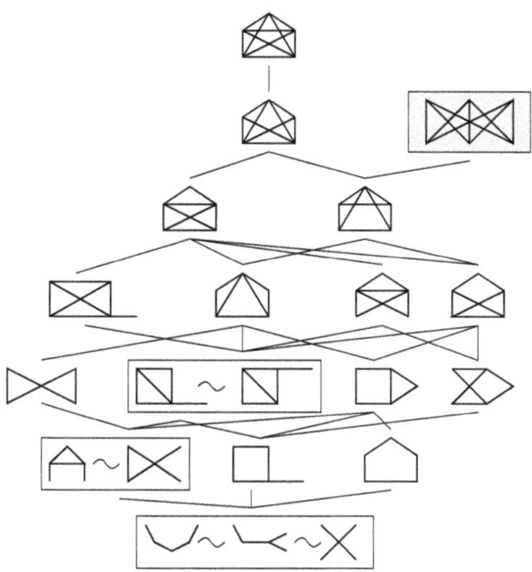

FIG. 6.14 – Zonohedral family in $d = 4$. A graphical representation is used for all zonohedral lattices except for the $K_{3,3}$ one. The arrangement of zonotopes reproduces the zonohedral family of lattices given in Figure 6.11.

Let us extend our visualization approach to the 4-dimensional case. Figure 6.14 gives a graphical representation for all zonohedral lattices for $d = 4$ (with exception of one case corresponding to maximal non-primitive **30.102**-36). In fact it is sufficient to construct all connected subgraphs of the complete graph K_5 with 5 vertices possessing 10 edges and to take into account certain equivalence relations. (The notations of graphs are summarized in Figure 6.15.) Certain equivalence relations in the graphical representation are shown in Figure 6.14. Namely, for C_3+1+1 and for $C_{2,2,1}+1$ graphs the edge with one free end can be attached to any node. To keep the figure more condensed we do not show for $C_3 + 1 + 1$ graph the isomorphism with the graph formed by a chain of length 2 attached to a 3-cycle. Starting from the complete graph K_5 we easily construct the zonohedral family consisting of 16 elements (except $K_{3,3}$ shown in Figure 6.14 in the special rectangle). To understand the logic of its appearance we need to study along with the graphical representation and the cographical one. First let us note that the K_5 graph is not planar and we cannot construct a dual for this graph. At the same time for all proper subgraphs of K_5 the dual graphs can be constructed. Figure 6.16 shows the result of cographical representations for all proper subgraphs of K_5. But this family naturally includes one extra graph, $K_{3,3}$ which can be obtained by an extension (point splitting) operation applied to the $K_5 - 1 - 1$ cographical representation. Point splitting is an inverse operation to edge contraction for the cographical representation. It allows us to find an

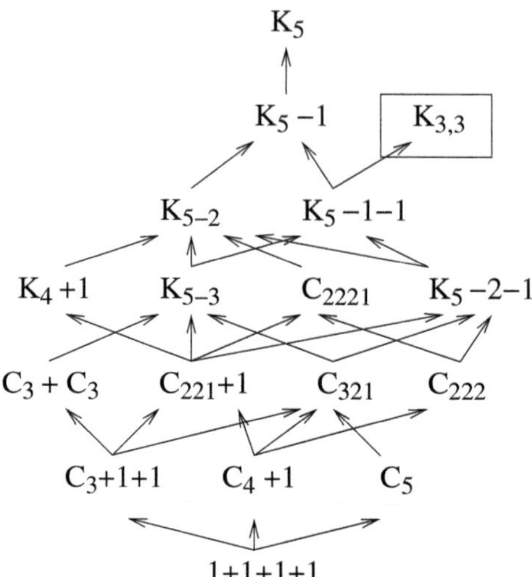

FIG. 6.15 – Conway notation [32] for zonohedral family in $d = 4$. (Note the misprints in [32]: K_4 used by Conway should be replaced by $C_{221} + 1$, whereas K_4 corresponds to the primitive combinatorial type of the three-dimensional lattice.)

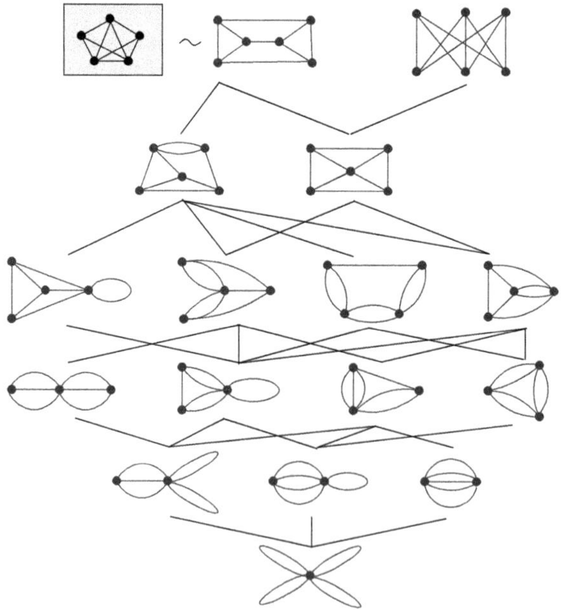

FIG. 6.16 – Cographical representation for 4-dimensional zonotopes.

additional zonotope belonging to a zonohedral family for four-dimensional lattices. Returning now to the semi-ordered set of zonotopes shown in Figure 6.14 we can explain the correlation between $K_5 - 1 - 1$ and $K_{3,3}$ as follows. From the graphical representation of $K_5 - 1 - 1$ we pass to the cographical representation and next realize point splitting of the only four-valence vertex. As a result we have two answers (depending on the type of rearrangement of edges during the point splitting), one is dual to $K_5 - 1$, another is $K_{3,3}$, for which we only have a cographical representation. These transformation are graphically summarized in the following symbolic equation.

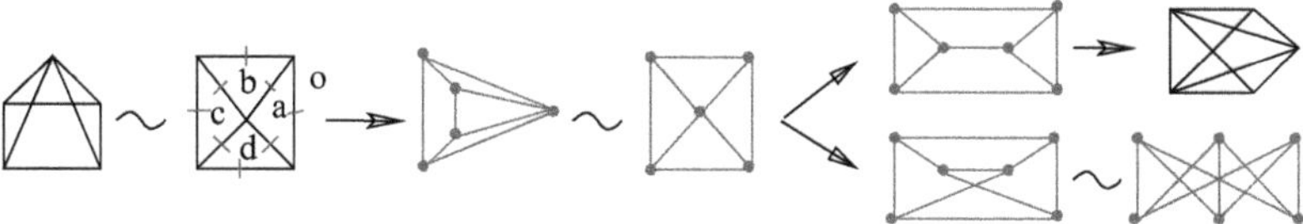

We use the four-dimensional case to introduce still one more representation of graphical zonotopal lattices. Namely, instead of plotting the graph which is a subgraph of K_5, we can simply plot the complement, i.e. the difference between K_5 and the subgraph. The only useful convention now is to keep all nodes explicitly shown. Such a representation is given in Figure 6.17. This representation becomes interesting when studying subgraphs with a number of edges close to the maximal possible value, i.e for subgraphs close to a complete graph K_5 and in higher dimensional cases close to K_{d+1}, or in other words for

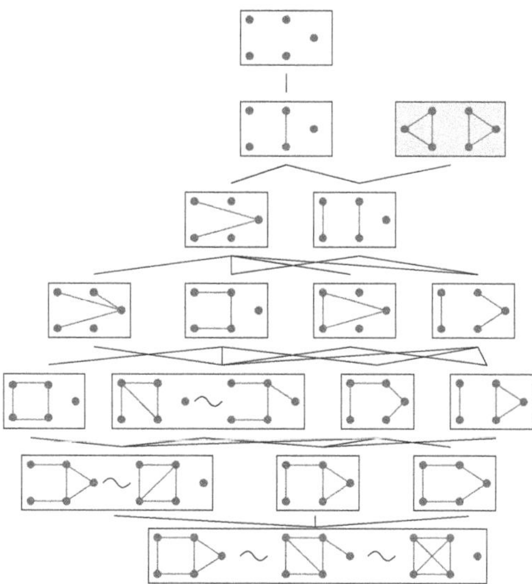

FIG. 6.17 – Representation of zonotopes through complement to graphical representation within the complete K_5 graph.

graphs with a small number of edges absent from the complete graph. This representation allows us to easily find equivalence between different graphical representations taking into account the topological equivalence of a complement to a graph. In contrast, for graphs with a small number of edges it is easier to see equivalence by looking directly at the graphical representation.

The five dimensional zonohedral family

To show the interest in the application of graphical visualization of zonohedral lattices we give now the application to five-dimensional lattices. The zonohedral family of five-dimensional lattices has been described by Engel [53], who has found 81 members of the family among which eight do not belong to the principal sub-family corresponding to the complete graph K_6 and its subgraphs. Engel characterizes members of the zonohedral family by symbols $N_{facets}.N_{vertices} - N_{hexagonal\ 2-faces}$ and gives the correlation between them corresponding to zone contraction. Figure 6.18 reproduces Engel's diagram with additional distinction between zonotopes belonging to

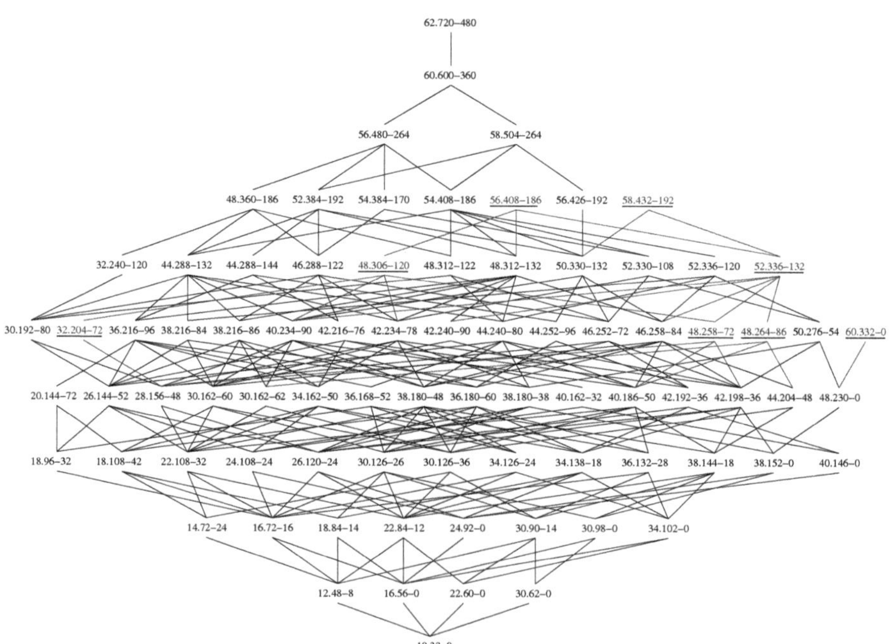

FIG. 6.18 – A representation of a zonohedral family in five-dimensional space made by Engel [53]. Lattices for which graphical representation is not available are underlined.

the main K_6-subfamily and correlations between members of this subfamily and zonotopes for which graphical representation is not available. Note that for the five-dimensional zonohedral lattices there exists one example of a lattice, namely **60.332**-0, which has neigher graphical nor cographical representation.

Figure 6.19 keeps the same organization of zonohedral lattices as that shown in Figure 6.18 but now each graphical lattice is given by its graph. Eight zonotopes which do not belong to the main family of subgraphs of K_6 are described by cographical representation or do not possess neigher graphical nor cographical representations. Their symbols are replaced in Figure 6.19 by an shaded rectangle. These lattices and their correlations with graphical lattices are discussed separately below.

To simplify the visualization for graphical representations we use graphs only when the number of edges is less than or equal to 10, whereas for graphs with the number of edges being more than or equal to 10 we use the representation of a complement to the graph with respect to the K_6 complete graph. For graphs with 10 edges both direct graphical and complement to graphical representations are given to clarify the correspondence.

Let us now give some comments about zonohedral lattices which do not appear as subgraphs of the complete graph K_6. First consider **60.332**-0, which is a special R_{10} graph introduced by Seymour [86], or E_5 used by Danilov and Grishukhin [38]. This graph cannot be described as belonging to the graphical or cographical representations. It can be considered as an extension of the $K_{3,3}$ cographical four-dimensional lattice by adding one loop. Seymour [86] uses for R_{10} the presentation of the type

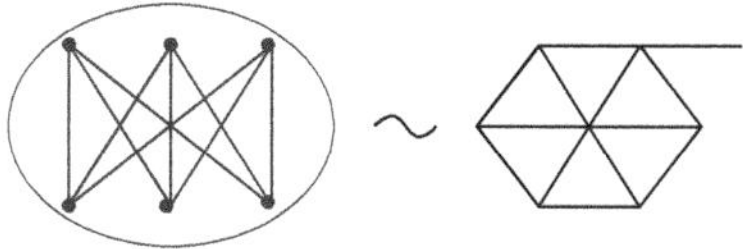

Note that the graphical presentation of **48.230**-0 as a subgraph of K_6 (which in fact equivalent as a graph to the $K_{3,3}$ representation) assumes that this graph corresponds to a five-dimensional lattice rather than the graph $K_{3,3}$ considered earlier and representing a four-dimensional zonohedral lattice. Because of that it is more natural to use $K^*_{3,3}$ notation for the four-dimensional lattice **30.120** $-$ 30.

Let us now turn to cographical representation of seven zonohedral lattices which are not subgraphs of K_6. These seven cographical lattices are shown in Figure 6.20.

Figure 6.21 demonstrates using the example of the **58.432**-192 polytope how to realize different contractions. It is possible to make two contractions for **58.432**-192. One consists in the contracting edge between nodes 1 and 2. (Numbering is given in figure 6.21.) It leads to cographical representation of the **52.336**-132 polytope. Another contraction (15) leads to cographical

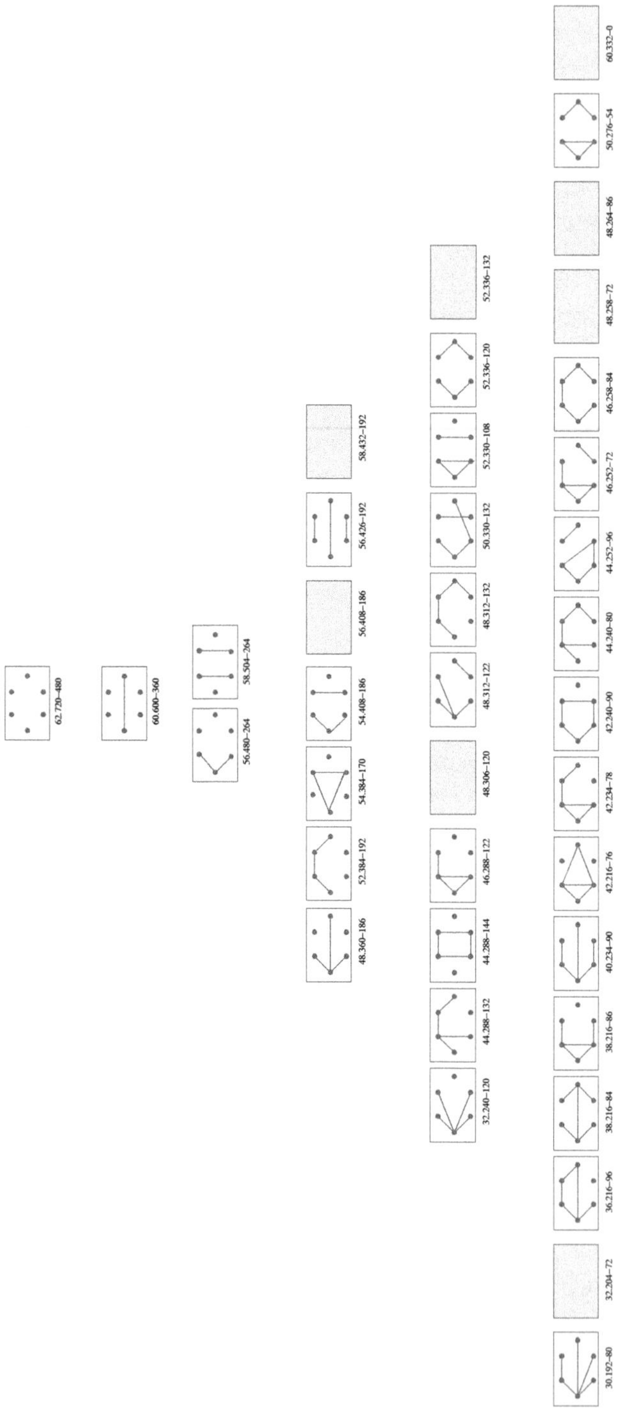

62.720–480

60.600–360

56.480–264
58.504–264

48.360–186
52.384–192
54.384–170
54.408–186
56.408–186
56.426–192
58.432–192

32.240–120
44.288–132
44.288–144
46.288–122
48.306–120
48.312–122
48.312–132
50.330–132
52.330–108
52.336–120
52.336–132

30.192–80
32.204–72
36.216–96
38.216–84
38.216–86
40.234–90
42.216–76
42.234–78
42.240–90
44.240–80
44.252–96
46.252–72
46.258–84
48.258–72
48.264–86
50.276–54
60.332–0

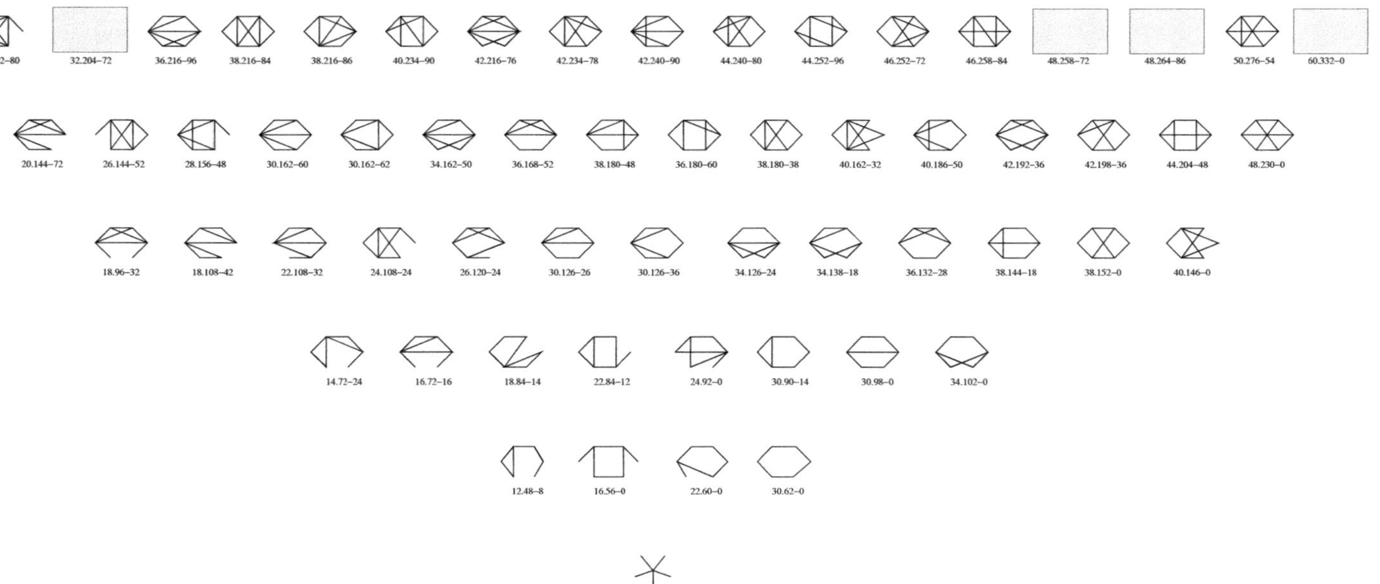

FIG. 6.19 – A representation of zonohedral family for five-dimensional lattices. For graphs with no more than 10 edges the direct representation is used, whereas complement graphs are drawn for graphs with 10 and more edges. Both direct and complement representations are shown for graphs with 10 edges in order to simplify the comparison between both representations. Graphs which are not subgraphs of K_6 are not shown in this figure. They are discussed separately. Their places in the original diagram by Engel are left free (shaded rectangle).

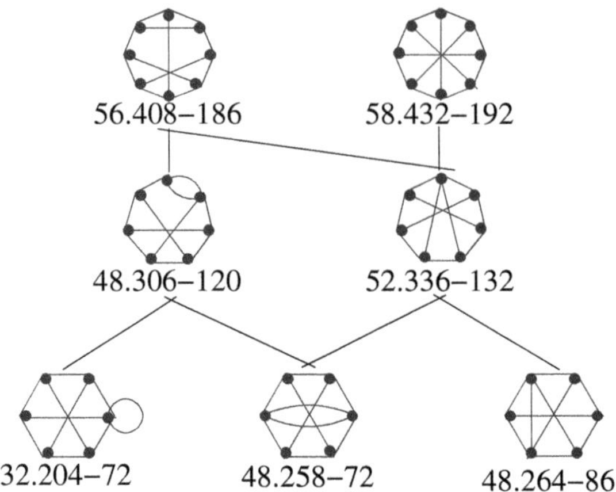

FIG. 6.20 – A cographical representation for seven zonohedral lattices associated with two maximal ones, **56.408**-186 and **58.432**-192. They are not shown in Figure 6.19.

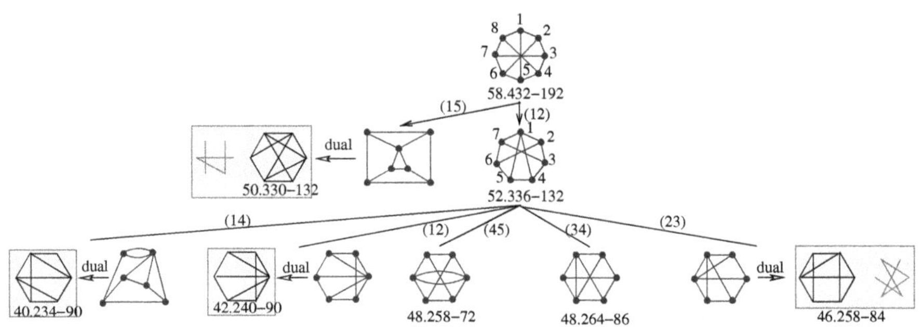

FIG. 6.21 – Different contractions of the **58.432**-192 zonohedral polytope shown in the cographical representation and transformed into the graphical representation for subgraphs of K_6.

representation which can be transformed to a dual graphical representation showing that the result is the polytope **50.330**-132. The complementary graph is given for **50.330**-132 along with the image of the graph itself in order to simplify the identification of the graph.

In its turn the cographical representation of the **52.336**-132 polytope shows that five different contractions are possible. Two among these contractions, namely (34) and (45) lead to two lattice zonotopes possessing only cographical presentation. Three other contractions (14), (12), and (23) lead to cographical presentation of zonotopes possessing graphical presentation and being subgraphs of K_6. Their transformation to graphical presentation through construction of a dual graph is explicitly shown in Figure 6.21.

Graphical representation of subgraphs of K_6 gives us an opportunity to see explicitly an application of important notion of 2-isomorphism of graphs introduced by Whitney [97]. Namely, for the **30.162**-60 zonotope two apparently different graphs can be assigned, but nevertheless these two graphs are 2-isomorphic as the following graphical equation demonstrates.

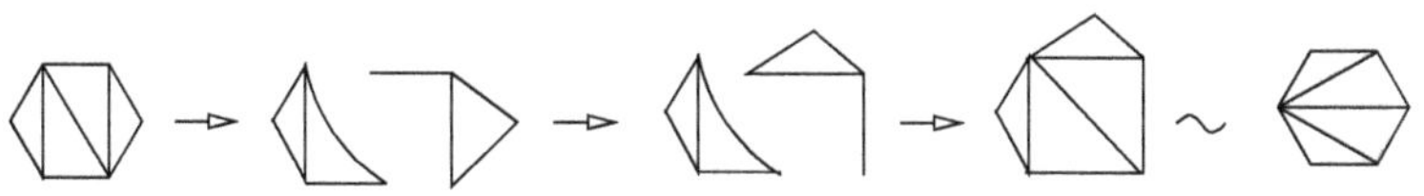

6.7.1 From Whitney numbers for graphs to face numbers for zonotopes

Simple visualization of zonohedral lattices by graphs would be much more interesting if it is possible to find zonotopes characteristics directly form graphs. And this is indeed possible. Face numbers of zonotopes can be expressed through rather elementary formulae in terms of topological invariants of graphs, the so called Whitney numbers [96, 82, 56]. A short guide to the calculation of Whitney numbers for simple graphs is given in appendix B. Here we simply give several explicit expressions for face numbers of 3- and 4-dimensional zonotopes in terms of doubly indexed Whitney numbers of the first and second kind.

For three dimensional zonotopes, i.e. for all combinatorial types of three-dimensional lattices we have

$$N_0 = w_{00}^+ + w_{01}^+ + w_{02}^+ + w_{03}^+; \tag{6.84}$$

$$N_1 = w_{11}^+ + w_{12}^+ + w_{13}^+; \tag{6.85}$$

$$N_2 = w_{22}^+ + w_{23}^+; \tag{6.86}$$

$$N_2^{(6)} = 4w_{02}^+ - 2w_{12}^+; \tag{6.87}$$

$$N_2^{(4)} = 4w_{12}^+ - 6w_{02}^+. \tag{6.88}$$

where w_{ij}^+ are absolute value of doubly indexed Whitney numbers of the first kind.

In fact, the total number of k-faces can be expressed more generally for arbitrary dimension d as [56]

$$N_k = \sum_{j=k}^{d} w_{kj}^+. \tag{6.89}$$

For four-dimensional zonotopes we add several expressions for particular types of k-faces.

$$
\begin{aligned}
N_2^{(6)} &= 2w_{13} + 4w_{23} + 2W_{13};\\
N_2^{(4)} &= -2w_{13} - 6w_{23} - 2W_{13};\\
N_3^{(6)} &= 2w_{01}w_{02} + 12w_{02} - 56w_{03} + 4w_{12} - 32w_{13} - 24w_{23} - 8W_{13};\\
N_3^{(8)} &= -4w_{01}w_{02} - 24w_{02} + 96w_{03} - 8w_{12} + 50w_{13} + 36w_{23} + 14W_{13};\\
N_3^{(12)} &= 4w_{01}w_{02} + 24w_{02} - 78w_{03} + 8w_{12} - 36w_{13} - 24w_{23} - 10W_{13};\\
N_3^{(14)} &= -2w_{01}w_{02} - 12w_{02} + 36w_{03} - 4w_{12} + 16w_{13} + 10w_{23} + 4W_{13}.
\end{aligned}
$$

Although these expressions are slightly complicated because they include one quadratic term, the existence of such expressions clearly supports the tight relation between zonohedral lattice and representative graph.

6.8 Graphical visualization of non-zonohedral lattices.

We have noted earlier in section 6.4.1, that in dimension four there exist two families of parallelohedra, the zonohedral family and the family obtained from the 24-cell polytope by making zone extension. This 24-cell family was represented in figure 6.10 taken (with minor modifications) from Engel's book [11]. In spite of the fact that these two families are often considered as completely independent and not related, there is a tight relation between them. The origin of this relation is the fact that all members of the 24-cell family can be constructed as a Minkowski sum of the 24-cell, P_{24} = **24.24**-0 and a zonotope which we denote $Z(U)$ and which in its turn can be constructed as a Minkowski sum of one, two, three, or four vectors. Thus, we can try to associate with each non-zonohedral polytope a zonotope (one-, two-, three-, or four-dimensional) which after making a Minkowski sum with the 24-cell leads to a required polytope. We need however to mention here a very important remark made by Deza and Grishukhin [44]. For a zonotope $Z(U)$ itself it is not important whether the summing vectors are orthogonal or not. A parallelepiped and a cube have the same combinatorial type. But the orthogonality of summing vectors in $Z(U)$ influences heavily the combinatorial type of the sum $P_{24} + Z(U)$. This means that the number of different types of $P_{24}+Z(U)$ can be larger than the number of different $Z(U)$ and we need to introduce an additional index characterizing orthogonality or non-orthogonality of vectors in the sum associated with a zonotope $Z(U)$. Nevertheless the contraction/extension relation between different non-zonohedral polytopes should respect the corresponding contraction/extension relation between zonotopal contributions. This allows us to represent all non-zonohedral polytopes (or lattices) in a way similar to zonohedral ones. Figure 6.22 is a graphical

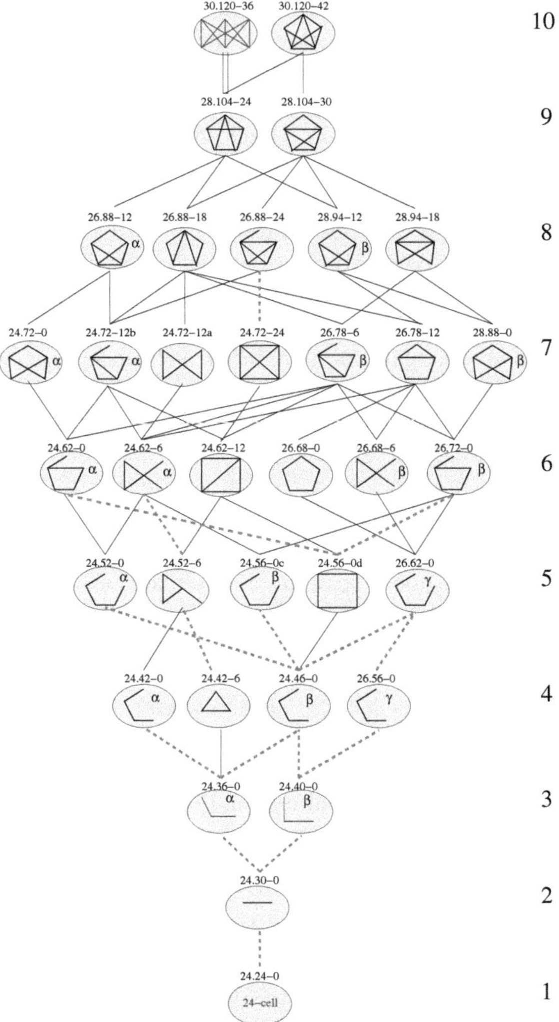

FIG. 6.22 – Graphical visualization of four-dimensional parallelohedra represented as a Minkowski sum $P_{24}+Z(U)$ of a zonotope $Z(U)$ and the 24-cell, $P_{24} = \mathbf{24.24}-0$. Shaded elliptic disks symbolize the P_{24} cell. Graphs for zonotopes coincide with those used to visualize zonohedral lattices. Symbols α, β, γ make further distinction between zonotope contributions $Z(U)$. Depending on the number of mutually orthogonal vectors in the Minkowski sum for a zonotope $Z(U)$, this additional index characterizes the cases with no orthogonal edges, with a pair of orthogonal edges and with three mutually orthogonal edges. A single thin line corresponds to elimination/addition of one edge without changing the number of points. A double thin line symbolizes transformation to the dual representation. A thick dash line corresponds to elimination of one edge together with one point.

visualization of the organization of the 24-cell family shown in 6.10. In Figure 6.22 the 24-cell contribution to the sum is symbolized by an elliptic shaded disk. The zonotope contribution is represented inside the shaded disk in a way similar to the representation of zonotopes discussed in the preceding subsection. The additional index is added when it is necessary to distinguish between the Minkowski sum with the same zonotope contribution but with special orthogonality between vectors forming the zonotope. This additional index is shown on the disk and takes values α, β, γ. It is useful equally to make the distinction between correlations (contraction/extension) associated with elimination of one edge without changing the number of points, i.e. within the zonotopes of the same dimension, and with elimination of the edge together with one point. The correlations associated with modification of the dimension of a zonotope are represented by a thick dash line. Graphical correlation allows us to localize small misprints in the figure representing a partially ordered set of non-zonohedral lattices in book [11], Figure 9-7. Namely, in the notation used in [11] it is necessary to change the line 26-7—24-14 by the line 26-7—26-3; the line 26-6—26-3 should be changed into 26-6—24-14; the line 28-2—26-7 should be changed into 28-2—26-6.

Using the discussed above graphical visualization of non-zonohedral lattices we can better understand the system of the organization of walls between generic domains for a cone of positive quadratic forms (see figure 6.12). The wall between the 30-2 and 30-3 domains is of **28.96**-40 type represented by $(K_5 - \mathbf{1})$ graph. Taking into account that the 30-2 domain corresponds to the K_5 zonotope graph and the 30-3 domain corresponds to the $P_{24}+(K_5-\mathbf{1})$ non zonotopal graph it is clear that going from domain 30-2 to the wall **28.96**-40 is a simple zone contraction graphically visualized as removing one edge. At the same time going from domain 30-3 to the same wall is not a zone contraction transformation. This transformation can be described as "elimination of the P_{24} contribution".

In a similar way going from the 30-4 domain to the wall **30** – **102**-36 has the same type. This transformation is again associated with "elimination of the P_{24} contribution" and is not of a standard contraction type. All other walls between generic domains are of simple contraction type, which are graphically represented by removing one edge from the graph.

The comparison of graphical representations of zonohedral lattices (Figure 6.23) and non-zonohedral ones (Figure 6.22) clearly indicates that there are similar transformations with "elimination of the P_{24}-contribution" during passage from lower dimensional subcones to their walls. For example the non-zonohedral lattice **28.104**-24 represented as $P_{24} + (K_5 - 2 \times \mathbf{1})$ and filling a 9-dimensional subcone can have as one of its 8-dimensional boundaries the zonohedral lattice **24.78**-24 which is graphically represented as $(K_5 - 2 \times \mathbf{1})$. Going from $P_{24} + (K_5 - 2 \times \mathbf{1})$ to $K_5 - 2 \times \mathbf{1}$ is not of a zone-contraction transformation but the "P_{24} elimination".

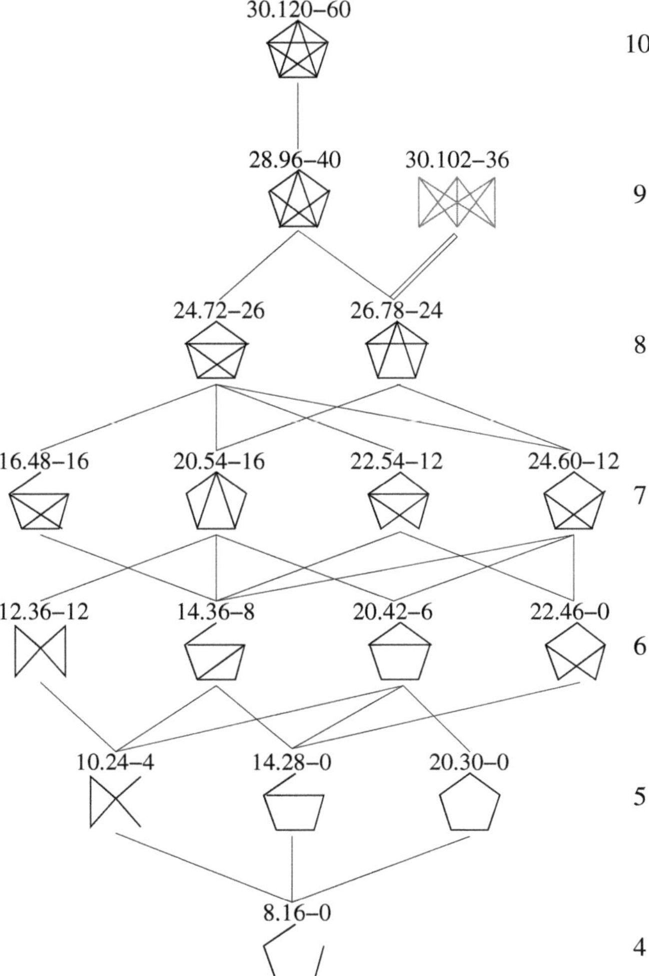

FIG. 6.23 – Graphical visualization of zonohedral four-dimensional parallelohedra.

To conclude the discussion of graphical representations of non zonohedral polytopes we note that this approach can be generalized to higher dimensional spaces. In order for the reader to follow this rather active direction of research we mention the recent paper [88] (and the most important of its predecessors [93, 38, 57, 58]). In [88] the description of six-dimensional polytopes represented in a form of $P_V(E_6) + Z(U)$ is studied. The $P_V(E_6)$ is the parallelotope associated with the root lattice E_6. (See chapter 7 of this book for an initial discussion of root lattices.)

6.9 On Voronoï conjecture

When discussing parallelohedra associated with facet-to-facet tiling of the space and corresponding lattice we have not stressed the difference between parallelohedra and Voronoï cells of lattices. It is clear that any Voronoï cell is a parallelohedron but the inverse is generally wrong.

In his famous paper [95] Voronoï formulated an important question: "Is an arbitrary parallelohedron affinely equivalent to the Dirichlet domain for some lattice?". Now the term "Dirichlet domain" is more often replaced by the "Voronoï cell" but the positive answer to this question is still absent and the affine equivalence between Voronoï cells of lattices and arbitrary parallelohedra is known as Voronoï's Conjecture.

Voronoï himself gave a positive answer to his question in the case when the parallelohedron P is primitive, i.e. when every vertex of corresponding tiling belongs to exactly $(d+1)$ copies of the d-dimensional parallelohedron P, or, in other words, each belts of P contains 6 facets. Since then, some progress has been made by extending Voronoï's Conjecture to a larger class of parallelohedra. The most serious steps are the following:

Delone [41] demonstrated that the conjecture is valid for all parallelohedra in dimensions $d \leq 4$.

Zhitomirskii [99] relaxed the condition of primitivity of parallelohedra for which Voronoï's Conjecture was proved to be valid. According to [99], a parallelohedron P is called k-primitive if each of its k-faces are primitive, i.e. every k-face of the corresponding tiling belongs to exactly $(d+1-k)$ copies of P. In particular, if each belt of P consists of 6 facets the parallelohedron is $(d-2)$-primitive. Zhitomirskii [99] extended the result of Voronoï on $(d-2)$-primitive parallelohedra.

Another class of parallelohedra for which the Voronoï's Conjecture was also proved [54] includes zonotopal parallelohedra.

Engel checked the Voronoï's Conjecture for five-dimensional parallelohedra by computer calculations [50, 51]. He enumerated all 179372 parallelohedra of dimension 5 and gave a Voronoï polytope affinely equivalent to each of the found parallelohedra.

Assuming the existence of an affine transformation that maps a parallelohedron onto a Voronoï polytope, Michel et al. [77] have shown that in the primitive case and in few other cases these mappings are uniquely determined up to an orthogonal transformation and scale factor.

Chapter 7

Root systems and root lattices

7.1 Root systems of lattices and root lattices

A hyperplane H_r of a n-dimensional vector space E_n is a $(n-1)$-dimensional subspace. It is completely characterized by a normal vector $\vec{r}$.

Definition: reflection through a hyperplane. The reflection σ_r through the hyperplane H_r is a linear involution of E_n which leaves the points of H_r fixed and transforms $\vec{r}$ into $-\vec{r}$.

Reflection through a hyperplane is an automorphism of E_n completely characterized by $\sigma_r^2 = I_n$, $\mathrm{Tr}\ \sigma_r = n-2$. Explicitly[1],

$$\forall \vec{x} \in E_n, \quad \sigma_r(\vec{x}) = \vec{x} - 2\frac{(\vec{x}, \vec{r})}{N(\vec{r})}\vec{r}. \tag{7.1}$$

Assume now that σ_r is a symmetry of the n-dimensional lattice L, i.e.

$$\forall \vec{\ell} \in L, \quad \sigma_r(\vec{\ell}) \in L. \tag{7.2}$$

Then

$$\vec{\ell} - \sigma_r(\vec{\ell}) \in L \Leftrightarrow 2\frac{(\vec{\ell}, \vec{r})}{N(\vec{r})}\vec{r} \in L, \tag{7.3}$$

which shows that the 1-dimensional vector subspace $\{\lambda\vec{r}\}$ contains a 1-sublattice[2] of L. From now on we choose $\vec{r}$ to be a generator of this 1-sublattice, so it is a visible vector[3]. This implies that the coefficient of the vector $\vec{r}$ in (7.3) is an integer. We call these vectors the **roots** of the lattice; their set is called the lattice **root system**,

$$\mathcal{R}(L) = \{\vec{r} \in L,\ \vec{r}\ \text{visible} :\ \forall \vec{\ell} \in L,\ 2\frac{(\vec{\ell}, \vec{r})}{N(\vec{r})} \in \mathbb{Z}\}. \tag{7.4}$$

[1] As we should expect from the definition of a reflection, the expression of σ_r is independent of the normalization of $\vec{r}$; in particular $\sigma_r \equiv \sigma_{-r}$.

[2] Concept defined in section 3.3.

[3] There are no shorter collinear vectors in the lattice.

We write $\mathcal{R}$ instead of $\mathcal{R}(L)$ when L is understood. Notice that $\vec{r} \in \mathcal{R} \Leftrightarrow -\vec{r} \in \mathcal{R}$ and that the dilation $L \mapsto \lambda L$ of the lattice changes simply $\mathcal{R}$ into $\lambda\mathcal{R}$. Moreover different pairs $\pm\vec{r}$ of roots correspond to distinct reflections.

We denote by $G_{\mathcal{R}}$ the group generated by the $|\mathcal{R}|/2$ reflections σ_r, $\vec{r} \in \mathcal{R}$; it is a subgroup of the Bravais group P_L^z of L. We know that $|P_L^z|$ is finite, so $|\mathcal{R}|$ is also finite.

We can write $\mathcal{R} = \cup_i \mathcal{R}_i$ where the $\mathcal{R}_i$ are the different orbits of P_L^z. The reflections σ_r, $\vec{r} \in \mathcal{R}_i$ form a conjugacy class of this group; we denote by $G_{\mathcal{R}_i}$ the subgroup they generate. Since in a finite group G, any subgroup generated by one (or several) conjugacy classes is an invariant subgroup of G, we have:

$$G_{\mathcal{R}_i} \triangleleft P_L^z, \quad G_{\mathcal{R}} \triangleleft P_L^z. \tag{7.5}$$

When $G_{\mathcal{R}}$ is $\mathbb{R}$-irreducible, any of its orbit spans the space E_n (if it were not true, $G_{\mathcal{R}}$ would leave invariant the subspace spanned by the orbit, and that contradicts its irreducibility). So each $\mathcal{R}_i$ spans E_n; that is also true of the short vectors $\mathcal{S} = \mathcal{S}(L)$.

Proposition 34 *When $G_{\mathcal{R}}(L)$ is $\mathbb{R}$-irreducible, the norm of any root satisfies $N(\vec{r}) < 4s(L)$.*

Proof: The proposition is true for roots in $\mathcal{S}$. Let $\vec{r}$ be a root not belonging to $\mathcal{S}$. Since $\mathcal{S}$ spans the space we can choose $\vec{s} \in \mathcal{S}$ such that $(\vec{r}, \vec{s}) > 0$. The transformed vector $\vec{s}_r = \sigma_r(\vec{s}) = \vec{s} - \mu\vec{r}$, with $0 < \mu = 2(\vec{s}, \vec{r})/N(\vec{r})$, is also in $\mathcal{S}$ since it has the same norm as $\vec{s}$. Since $\vec{r}$ is visible and $N(\vec{r}) > N(\vec{s})$, Schwarz's inequality

$$|(\vec{s}, \vec{r})| < \sqrt{N(\vec{s})N(\vec{r})} \tag{7.6}$$

implies $|(\vec{s}, \vec{r})| < N(\vec{r})$. Thus $\mu = 1$, i.e. $\vec{r} = \vec{s} - \vec{s}_r$. Thus $N(\vec{r}) < 4N(\vec{s})$. □

Definition: root lattice. A root lattice is a lattice generated by its roots.

As a trivial example, any one dimensional lattice $L = \{n\vec{r}, n \in \mathbb{Z}\}$ is a root lattice; indeed $\sigma_r(n\vec{r}) = -n\vec{r}$. We recall that any one dimensional lattice can be scaled to I_1.

Proposition 35 *The vectors of norm 1 and 2 of an integral lattice are roots of the lattice.*

Proof: In an integral lattice $\vec{\ell}, \vec{v} \in L \Rightarrow (\vec{\ell}, \vec{v}) \in \mathbb{Z}$. Assume $N(\vec{v}) = 1$ or 2; so $\vec{v}$ is visible. Then $2(\vec{\ell}, \vec{v})/N(\vec{v})$ is an integer, so $\vec{v}$ is a root. □

As we will see, this proposition gives important information on the symmetry of the lattice. From the definition of the root lattice we obtain:

Proposition 36 *An integral lattice L generated by its vectors of norm 1 and 2 is a root lattice which is the orthogonal sum $L = L_1 \overset{\perp}{\oplus} L_2$ where $L_1 = I_k$ is generated by the norm 1 vectors and L_2 is generated by its shortest vectors of norm 2.*

Proof. From Proposition 35, L is a root lattice. If $\vec{s}_i$, $\vec{s}_j$ are two linearly independent norm 1 vectors of L, Schwarz's inequality (7.6) implies $(\vec{s}_i, \vec{s}_j) = 0$. Let k be the number of mutually orthogonal pairs $\pm\vec{s}_i$; these short (norm 1) vectors generate a lattice I_k. Let $\vec{r}$ be a root of norm 2; the value of $\varepsilon = (\vec{r}, \vec{s}_i)$ is either ± 1 or 0. In the former case $N(\vec{r} - \varepsilon\vec{s}_i) = 1$ so $\vec{r} = \varepsilon\vec{s}_i + \vec{s}_j$, and $\vec{r}$ visible requires that it is the sum of two orthogonal short roots, i.e. $\vec{r} \in I_k = L_1$. Obviously, the norm 2 roots $\vec{r}$ orthogonal to all lattice vectors of norm 1, generate L_2. □

Note that an integral lattice which has no vectors of norm 1 and 2, may contain a root lattice; a trivial example is given by a non reduced integral lattice, i.e. the lattice $\sqrt{m}L$, $m \in \mathbb{Z}$ with $m \geq 3$ where L is an integral lattice with minimal norm $s(L) = 1$.

Let $\vec{r}$ and $\vec{r}\,'$ be two linearly independent roots of L and ϕ the angle between them. From (7.4) we obtain:

$$4\cos^2(\phi) = \frac{4(\vec{r}, \vec{r}\,')^2}{N(\vec{r})N(\vec{r}\,')} \in \mathbb{Z}. \tag{7.7}$$

Thus

$$4\cos^2(\phi) = 0, 1, 2, 3 \;\Rightarrow\; \phi = \frac{\pi}{2}; \frac{\pi}{3}, \frac{2\pi}{3}; \frac{\pi}{4}, \frac{3\pi}{4}; \frac{\pi}{6}, \frac{5\pi}{6}. \tag{7.8}$$

Since $\sigma_r\sigma_{r'}$ and its inverse $\sigma_{r'}\sigma_r$ are rotations by the angle 2ϕ in the 2-dimensional space spanned by $\vec{r}, \vec{r}\,'$, we have

$$(\sigma_r\sigma_{r'})^m = I, \quad \text{where } m = 2,\ 3,\ 4,\ 6. \tag{7.9}$$

The groups whose relations between generators are given by these equations are called *Weyl groups*. They are studied in the next subsection. To write explicitly the integer $4(\vec{r}, \vec{r}\,')^2/(N(\vec{r})N(\vec{r}\,'))$ as a function of m we use the Boolean function $m \mapsto (m = 6)$ whose values are 1 when $m = 6$ and 0 when $m \neq 6$. Then

$$\frac{4(\vec{r}, \vec{r}\,')^2}{N(\vec{r})N(\vec{r}\,')} = m - 2 - (m = 6). \tag{7.10}$$

Application to dimension 2

We have seen in section 4.3 that there are two maximal Bravais classes: *p4mm* (square lattices L_s) and *p6mm* (hexagonal lattices L_h). Their groups are irreducible (over $\mathbb{C}$). So we can consider the two integral lattices. Since their shortest vectors satisfy $s(L_s) = 1$, $s(L_h) = 2$, and generate the lattice, Proposition 35 shows that L_s and L_h are root lattices. For each one, the root system has two orbits of roots; one of them is the set of short vectors of the lattice. We use the value i of the root norm as an index for the root orbit $\mathcal{R}_i$. In the next equations we list the roots by giving their coordinates in the basis defined by the Gram matrix $Q(L)$.

For the Bravais class *p4mm* (square lattice),

$$Q(L_s) = I_2 \quad \mathcal{R}_1(L_s) = \mathcal{S}(L_s) = \left\{\pm\begin{pmatrix}1\\0\end{pmatrix}, \pm\begin{pmatrix}0\\1\end{pmatrix}\right\}, \quad |\mathcal{R}_1(L_s)| = 4;$$

$$\mathcal{R}_2(L_s) = \left\{\pm\begin{pmatrix}1\\1\end{pmatrix}, \pm\begin{pmatrix}1\\-1\end{pmatrix}\right\}, \quad |\mathcal{R}_2(L_s)| = 4. \tag{7.11}$$

For the Bravais class $p6mm$ (hexagonal lattice),

$$Q(L_h) = \begin{pmatrix}2 & -1\\-1 & 2\end{pmatrix}, \; \mathcal{S} = \mathcal{R}_2(L_h) = \left\{\pm\begin{pmatrix}1\\0\end{pmatrix}, \pm\begin{pmatrix}0\\1\end{pmatrix}, \pm\begin{pmatrix}1\\1\end{pmatrix}\right\}, \; |\mathcal{R}_2(L_h)| = 6,$$

$$\mathcal{R}_6(L_h) = \left\{\pm\begin{pmatrix}2\\1\end{pmatrix}, \pm\begin{pmatrix}1\\2\end{pmatrix}, \pm\begin{pmatrix}1\\-1\end{pmatrix}\right\}, \quad |\mathcal{R}_6(L_h)| = 6. \tag{7.12}$$

Finally, the lattices of the other two non-generic Bravais classes, $p2mm$ and $c2mm$, have the same point symmetry, $2mm \sim \mathbb{Z}_2^2$, which is reducible. The lattices of the Bravais class $p2mm$ are root lattices; those of $c2mm$ are not. For the latter Bravais class, depending on the lattice, there might be 4 or 2 shortest vectors; in the latter case, these two shortest vectors are roots. The generic lattices (Bravais class $p2$) have no roots.

7.1.1 Finite groups generated by reflections

We will give in this subsection the list of irreducible finite groups generated by reflections, for short *finite reflection groups*. Those which satisfy equation (7.9) were introduced by H. Weyl in 1925 in his study of the finite-dimensional representations of the semi-simple Lie groups and they were listed by E. Cartan [30] (p. 218-224). Here we shall give the results of Coxeter, who established the complete list of finite reflection groups[4] [37]. A finite reflection group G acting linearly on the orthogonal vector space E_n is defined by n' generators and the relations:

$$1 \leq i \leq n', \qquad (\sigma_{\vec{r}_i}\sigma_{\vec{r}_j})^{m_{ij}} = I_n, \quad m_{ii} = 1, \quad i \neq j,\; 2 \leq m_{ij} \in \mathbb{Z} \tag{7.13}$$

and this abstract group is realizable as a finite subgroup of O_n with the generators $\sigma_{\vec{r}_i}$ represented by reflections through hyperplanes whose normal vector is denoted by $\vec{r}_i$. If the $\vec{r}_i$ span only a subspace of dimension $n_0 < n$, the group acts trivially on its orthogonal complement. This case is equivalent to a reflection group on a space of dimension n_0; from now on, we consider only the action on E_n of the "n-dimensional" reflection groups; the number n' of generators of such a group satisfies $n' \geq n$.

We will now prove that $n' = n$. The reflection hyperplanes of G partition E_n into $|G|$ convex cones; each one is the closure of a fundamental domain for the action of G on E_n. Choose one of these cones and orient its root vectors to the outside. The product of the reflections through two contiguous hyperplanes H_{r_i}, H_{r_j}, is a rotation in the 2-plane spanned by $\vec{r}_i$, $\vec{r}_j$ of order m given in (7.9). Then the scalar product of any pair of these normal vectors is ≤ 0; moreover the scalar product of each of them with a fixed vector $\vec{c}$

[4] See also his classical book [6].

in the interior of the cone is also < 0. It is easy to prove that these normal vectors are linearly independent. Indeed, assume the contrary: if $\vec{x}$ is any vector inside the cone, it can be written with two different decompositions $\vec{x} = \sum_i \alpha_i \vec{r}_i = \sum_j \beta_j \vec{r}_j$ with $\alpha_i < 0$, $\beta_j < 0$, the domains of i and j are two disjoint subsets of $1, 2, \ldots, n'$. Since $0 \leq (\vec{x}, \vec{x}) = \sum_{ij} \alpha_i \beta_j (\vec{r}_i, \vec{r}_j)$ is a sum of terms ≤ 0, it vanishes for $\vec{x} = \vec{0}$. So the r_i are linearly independent and $n' = n$.

The Gram matrix of these n normal vectors can be computed from the "Coxeter diagram" of the reflection group G. In it, each vertex corresponds to a generator r_i. Edges are drawn between the vertices r_i, r_j when the order m_{ij} of $\sigma_{r_i}\sigma_{r_j}$ is > 2 (see (7.13) and each edge carries as a label the value of m_{ij} (the tradition is to omit label 3 since (see below) at most one edge has a label > 3 in each connected part of the graph).

Proposition 37 *The irreducibility of a reflection group is equivalent to connectedness of its Coxeter graph.*

Proof: The reducible representation of a finite group G on the space E_n is a direct sum of p irreducible representations on the mutually orthogonal subspaces E_{n_α} with $1 \leq \alpha \leq p$, $\sum_\alpha n_\alpha = n$. If G is generated by reflections, a reflection $\sigma_{\vec{r}_i}$ maps a subspace onto itself only if $\vec{r}_i$ belongs to it. So the set of root vectors is partitioned into mutually orthogonal subsets labeled by α; each one generates a reflection group G_α acting on E_{n_α}. From the set of rules for building Coxeter graphs, the graph of G is disconnected into p pieces. Conversely, for a graph disconnected into p pieces labeled by α, β, $\cdots$, $(\vec{r}_i^{\alpha}, \vec{r}_j^{\beta}) = 0$ when $\alpha \neq \beta$, so the subspaces E_{n_α} are mutually orthogonal. Moreover, from (7.10), the subgroups $G_\alpha < G$ generated by each connected part commute between each other, and each G_α acts trivially on all E_{n_β} with $\beta \neq \alpha$; this shows that $G = \times_\alpha G_\alpha$ transforms each E_{n_α} into itself. □

The positivity of the Gram matrix implies that the Coxeter graph of an irreducible reflection group is a tree with at most one branching node, and gives some restriction on the nature of this node. So we have two cases to consider

i) *one branching node*: it has only three branches with the number of vertices $p, q, r \geq 2$ (one counts the vertex at the node) satisfying the relation:

$$\frac{1}{p} + \frac{1}{q} + \frac{1}{r} > 1, \tag{7.14}$$

and all edge labels are 3;

ii) *no branching nodes*: then all labels are 3 except possibly one of them.

A more refined study gives the list of finite irreducible reflection groups. As we have seen, every reflection group is a direct sum of irreducible reflection groups (including eventually the trivial reflection group $\{1\}$).

TAB. 7.1 – List of finite irreducible reflection groups. For each reflection group we give its traditional symbol with the dimension as a subscript, its order, and its Coxeter diagram. As is customary, we omit the label 3. It is clear from the diagrams that A_n is defined for all $n \geq 1$, B_n for $n \geq 2$, D_n for $n \geq 4$.

Group	Order	Coxeter diagram
A_n	$(n+1)!$	
B_n	$2^n n!$	 4
D_n	$2^{n-1}n!$	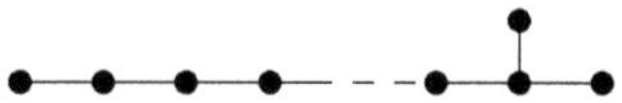
F_4	1152	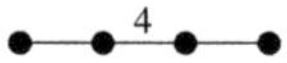 4
G_2	12	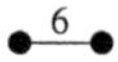 6
E_6	$72 \cdot 6!$	
E_7	$72 \cdot 8!$	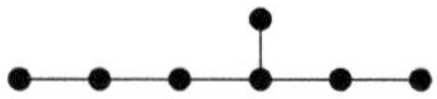
E_8	$192 \cdot 10!$	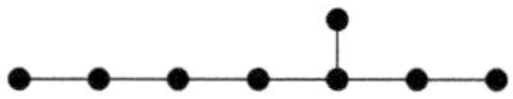

The following ones are not Weyl groups or symmetry groups of lattices:

Group	Order	Coxeter diagram	Group	Order	Coxeter diagram
H_2	10	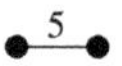 5	H_3	120	 5
H_4	14400	 5	$I_2^{(p)}$	$2p$	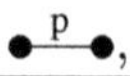 p, $p \geq 7$

Table 7.1 gives the list of finite irreducible reflection groups with the symbols used by most mathematicians; they were used first in the theory of Lie groups.

To give the abstract structure of most of these groups, we must introduce the following notation: $\mathcal{S}_n$ denotes the permutation group of n objects and $\mathcal{A}_n$ denotes its index 2 subgroup of even permutations[5]; $G\uparrow^n$ denotes the n-wreath product of G, i.e. the semi-direct product $G^n \rtimes \mathcal{S}_n$ where G^n is the direct product of n copies of G and $\mathcal{S}_n$ acts on it by permutations of these n factors. We have the isomorphisms:

$$A_n \sim \mathcal{S}_{n+1}, \quad B_n \sim O_n(\mathbb{Z}) \sim \mathbb{Z}_2\uparrow^n \sim A_1\uparrow^n, \quad D_n \sim \left(\mathbb{Z}_2^n/\mathbb{Z}_2^{\text{diag}}\right) \rtimes \mathcal{S}_n; \quad (7.15)$$

Indeed $A_1 \sim \mathbb{Z}_2$; its Coxeter diagram is a point. Notice that the diagram of D_5 could also be denoted E_5; similarly one could have defined D_3 as A_3 from the diagram shape (which justify the group isomorphism: $\mathbb{Z}_2^2 \rtimes \mathcal{S}_3 \sim \mathcal{S}_4$). We follow here the usual convention, but we shall use these remarks later. We have already seen that B_n is isomorphic to $O_n(\mathbb{Z})$, the group of orthogonal matrices with integer elements; it is generated by the diagonal matrices with diagonal elements ± 1 (they form the group $\mathbb{Z}_2^n$) and the group Π_n of $n \times n$ permutation matrices (all their elements are zero except one by line and

[5] The $\mathcal{A}$ of $\mathcal{A}_n$ is to remind the most usual name of this group: the "alternating group".

by column which is 1). To restrict B_n to its index 2 subgroup D_n, among the diagonal matrices defined above, one keeps only those of determinant 1. We denote by C_{nv} the dihedral group; its order is $|C_{nv}| = 2n$. Its definition by generators and relations is: $C_{nv}\,:\; x^2 = y^n = (xy)^2 = 1$. More isomorphisms of reflection groups are:

$$A_2 \sim C_{3v},\; B_2 \sim C_{4v},\; H_2 \sim C_{5v},\; G_2 \sim C_{6v},\; I_2^{(p)} \sim C_{pv},\; H_3 \sim \mathcal{A}_5 \times \mathbb{Z}_2. \tag{7.16}$$

7.1.2 Point symmetry groups of lattices invariant by a reflection group

As we have seen in (7.9), among the n-dimensional reflection groups, only the Weyl groups, i.e. those with $m_{ij} = 2,3,4,6$ are symmetry groups of n-dimensional lattices. While the others are not automorphisms of lattices in E_n, as *abstract* groups they can stabilize higher dimensional lattices[6]; indeed $H_2 < A_4,\quad H_3 < B_6,\quad H_4 < E_8,\quad I_2^{(p)} < B_p$.

From the knowledge of the Coxeter diagram of an irreducible Weyl group, we can write a quadratic form of the root lattice invariant by this group. Indeed, starting from (7.13) which defines a reflection group we have seen how to build a set of vectors $\vec{r}_i$ normal to the reflection hyperplanes of a fundamental cone and oriented outside the cone. These linearly independent vectors define a basis of a lattice invariant by the Weyl group; we denote by $Q := \{q_{ij}\}$ the corresponding Gram matrix. Equation (7.10) yields the following relations (depending on the integers m_{ij} which define the group) between non diagonal and corresponding diagonal elements of Q:

$$i \neq j, \quad \frac{q_{ij}}{\sqrt{q_{ii}q_{jj}}} = -\frac{1}{2}\sqrt{m_{ij} - 2 - (m_{ij} = 6)}. \tag{7.17}$$

This equation is independent of the length of the vectors $\vec{r}_i$; we verify case by case that we can require the elements of the Gram matrix Q to be relatively prime integers and this fixes the lengths of the root vectors $\vec{r}_i$. For instance in the groups of types A, D, E (with all non vanishing non-diagonal m's being 3), the reflections form a unique orbit; that must be also the case of the roots $\vec{r}_i$, so the non vanishing non-diagonal elements of the Gram matrix are -1 and the diagonal ones are $2 = N(\vec{r}_i)$. We give explicitly the

[6] Indeed H_2 is the symmetry group of the pentagon, $H_3 \sim \mathcal{A}_5 \times \mathbb{Z}_2$ that of the dodecahedron and the icosahedron (dual of each other), H_4 that of two dual regular polyhedra in 4 dimensions with respectively 120 and 600 faces, and the $I_2^{(p)}$ are the symmetry groups of the regular p-gons in the plane.

Gram matrices for $n = 8$:

$$Q(A_8) = \begin{pmatrix} 2 & -1 & 0 & 0 & 0 & 0 & 0 & 0 \\ -1 & 2 & -1 & 0 & 0 & 0 & 0 & 0 \\ 0 & -1 & 2 & -1 & 0 & 0 & 0 & 0 \\ 0 & 0 & -1 & 2 & -1 & 0 & 0 & 0 \\ 0 & 0 & 0 & -1 & 2 & -1 & 0 & 0 \\ 0 & 0 & 0 & 0 & -1 & 2 & -1 & 0 \\ 0 & 0 & 0 & 0 & 0 & -1 & 2 & -1 \\ 0 & 0 & 0 & 0 & 0 & 0 & -1 & 2 \end{pmatrix} \tag{7.18}$$

$$Q(D_8) = \begin{pmatrix} 2 & -1 & 0 & 0 & 0 & 0 & 0 & 0 \\ -1 & 2 & -1 & 0 & 0 & 0 & 0 & 0 \\ 0 & -1 & 2 & -1 & 0 & 0 & 0 & 0 \\ 0 & 0 & -1 & 2 & -1 & 0 & 0 & 0 \\ 0 & 0 & 0 & -1 & 2 & -1 & 0 & 0 \\ 0 & 0 & 0 & 0 & -1 & 2 & -1 & -1 \\ 0 & 0 & 0 & 0 & 0 & -1 & 2 & 0 \\ 0 & 0 & 0 & 0 & 0 & -1 & 0 & 2 \end{pmatrix} \tag{7.19}$$

The explicit form of $Q(E_n)$ depends on the way we label the nodes of the E_n diagram. Naturally the branching node has label $n-3$. Here we label $n-2$ the unique node of the short leg (above the line in the diagrams of Table 7.1) and $n-1, n$ those of the characteristic E_n leg.

$$Q(E_8) = \begin{pmatrix} 2 & -1 & 0 & 0 & 0 & 0 & 0 & 0 \\ -1 & 2 & -1 & 0 & 0 & 0 & 0 & 0 \\ 0 & -1 & 2 & -1 & 0 & 0 & 0 & 0 \\ 0 & 0 & -1 & 2 & -1 & 0 & 0 & 0 \\ 0 & 0 & 0 & -1 & 2 & -1 & -1 & 0 \\ 0 & 0 & 0 & 0 & -1 & 2 & 0 & 0 \\ 0 & 0 & 0 & 0 & -1 & 0 & 2 & -1 \\ 0 & 0 & 0 & 0 & 0 & 0 & -1 & 2 \end{pmatrix} \tag{7.20}$$

The quadratic forms $Q(A_n)$ and $Q(D_n)$ for $n > 8$ are obtained by the same modification which is obvious for $Q(A_n)$. One obtains the corresponding matrices for values of $n < 8$ by suppressing the first $8-n$ lines and columns. The matrix determinants are:

$$\det(Q(A_n)) = n+1, \quad \det(Q(D_n)) = 4, \quad \det(Q(E_n)) = 9-n. \tag{7.21}$$

These matrices invite us to define the quadratic forms for D and E below the conventional lower bound for n given in the caption of Table 7.1. For instance $Q(A_3)$ and $Q(D_3)$ are equivalent (by permuting the indices 1,2).[7] Similarly

[7] A_n, D_n are the Weyl groups of the Lie algebras SU_{n+1} and SO_{2n}. The algebras SU_4 and SO_6 are isomorphic. The Lie algebra of SO_4 is isomorphic to $SU_2 \times SU_2$.

$Q(E_5) \sim Q(D_5)$ (inverse the ordering of the matrix elements), $Q(E_4) \sim Q(A_4)$ (exchange the indices 1, 2).[8]

The Coxeter diagrams of the Weyl groups B_n, F_4, G_2 contain exponents m_{ij} with two different values: 3, 4 or 3, 6 for G_2. That corresponds to two conjugacy classes of reflections. From (7.17) one sees that the two orbits of roots can generate an integral lattice only if the two orbits of roots have different norms. That the matrix elements of the quadratic form of these root lattice be relatively prime impose the pair of values of the root norms to be 2,1 for B_n, 2, 4 for F_4, 2, 6 for G_2. So the quadratic forms defined by the Coxeter diagrams are

$$Q(B_8) = \begin{pmatrix} 2 & -1 & 0 & 0 & 0 & 0 & 0 & 0 \\ -1 & 2 & -1 & 0 & 0 & 0 & 0 & 0 \\ 0 & -1 & 2 & -1 & 0 & 0 & 0 & 0 \\ 0 & 0 & -1 & 2 & -1 & 0 & 0 & 0 \\ 0 & 0 & 0 & -1 & 2 & -1 & 0 & 0 \\ 0 & 0 & 0 & 0 & -1 & 2 & -1 & 0 \\ 0 & 0 & 0 & 0 & 0 & -1 & 2 & -1 \\ 0 & 0 & 0 & 0 & 0 & 0 & -1 & 1 \end{pmatrix}, \tag{7.22}$$

$$Q(F_4) = \begin{pmatrix} 2 & -1 & 0 & 0 \\ -1 & 2 & -2 & 0 \\ 0 & -2 & 4 & -2 \\ 0 & 0 & -2 & 4 \end{pmatrix}, \quad Q(G_2) = \begin{pmatrix} 2 & -3 \\ -3 & 6 \end{pmatrix} \tag{7.23}$$

The determinants of these matrices are:

$$\det(Q(B_n)) = 1, \quad \det(Q(F_4)) = 4, \quad \det(Q(G_2)) = 3. \tag{7.24}$$

So B_n is a self dual lattice. Let m be an $n \times n$ triangular lattice with 1's on the diagonal and above it and 0's below it. Then $mQ(B_n)m^\top = I_n$. Similarly we have the equivalence of quadratic forms:

$$m_F Q(F_4) m_F^\top = Q(D_4); \qquad m_G Q(G_2) m_G^\top = Q(A_2) \tag{7.25}$$

where m_F and m_G are:

$$m_F = \begin{pmatrix} 1 & 0 & 0 & 0 \\ 0 & 1 & 0 & 0 \\ -1 & -2 & -1 & 0 \\ -1 & -2 & -1 & -1 \end{pmatrix}; \qquad m_G = \begin{pmatrix} 1 & 0 \\ 1 & 1 \end{pmatrix}. \tag{7.26}$$

To summarize: the lattices with point symmetry groups B_n, F_4, G_2 are generated by their orbit of shortest roots and they are identical to the root lattices

[8] For the A, D, E systems, the matrices we have defined coincide with the Cartan matrices which play a great role in the theory of Lie algebras. It is worth recalling that the quadratic forms for E_6, E_7, E_8 have been first introduced in the study of perfect lattices by Korkin and Zolotarev [63]. That was more than fifteen years before the classification of the corresponding simple Lie algebras by Killing and Cartan.

I_n, D_4, A_2 respectively. These lattices have a second orbit of roots of norm $2, 4, 6$ respectively.

7.1.3 Orbit scalar products of a lattice; weights of a root lattice

For a lattice L we introduce the notation L_a for the set of vectors with norm a. For an integral lattice $L = \cup_{a \in \mathcal{N}} L_a$. We introduce the natural notation $\forall m \in \mathbb{Z},\ mL_a = \{m\vec{\ell},\ \vec{\ell} \in L_a\}$; hence $mL_a \subseteq L_{m^2 a}$.

When the lattice has a large symmetry group G (e.g. a maximal Bravais group) each L_a, for low values of a, is one, or the union of a few G-orbits. It is useful to define the following concept for non vanishing vectors:

$$\vec{v} \neq \vec{0} \neq \vec{\ell}, \quad \vec{v}, \vec{\ell} \in L, \quad |\vec{v}, \vec{\ell}\,|_G \stackrel{\text{def}}{=} \max_{\vec{x} \in G.\vec{v},\, \vec{y} \in G.\vec{\ell}} (\vec{x}, \vec{y}). \tag{7.27}$$

In words: $|\vec{v}, \vec{\ell}\,|_G$ is the maximum of the scalar product between these two vectors when they run through their respective G-orbits. We call this positive number the *scalar product of the two orbits*; we have defined it for any symmetry group of L. Equivalent definitions are:

$$\vec{v}, \vec{\ell} \in L, \ |\vec{v}, \vec{\ell}\,|_G = \max_{\vec{y} \in G.\vec{\ell}} (\vec{v}, \vec{y}) = \max_{\vec{x} \in G.\vec{v}} (\vec{x}, \vec{\ell}). \tag{7.28}$$

When the two vectors are in the same orbit, their orbit scalar product is equal to their norm. When G is the maximal symmetry group of the lattice, i.e. its Bravais group P_L^z, we will simply write $|\vec{v}, \vec{\ell}\,|$; since P_L^z contains $-I$, we have $|\vec{v}, \vec{\ell}\,| \geq 0$ or, when P_L^z is **R**-irreducible, $|\vec{v}, \vec{\ell}\,| > 0$.

The set of values of $|\vec{v}, \vec{\ell}\,|$ gives interesting information about the lattice. It has to satisfy some bounds: e.g. $N(\vec{v} - \vec{\ell}\,) \geq s(L)$ implies:

$$|\vec{v}, \vec{\ell}\,| \leq \frac{1}{2}((N(\vec{v}) + N(\vec{\ell}\,) - s(L)); \quad \forall \vec{s} \in \mathcal{S}(L), \ \ |\vec{s}, \vec{\ell}\,| \leq \frac{1}{2} N(\vec{\ell}\,). \tag{7.29}$$

A similar inequality will also be useful:

$$\vec{s}_i \in \mathcal{S}(L), \ \vec{s}_1 \neq \pm \vec{s}_2, \quad |(\vec{s}_1, \vec{s}_2)| \leq \frac{1}{2} s(L). \tag{7.30}$$

For lattices with high symmetry we will build their *tableau of orbit scalar products* with G as the symmetry group. *In order to avoid redundancy, we write in the tableau only the orbits of visible vectors.* This tableau is a symmetrical matrix whose rows and columns are labeled by L_a or L'_a, L''_a, $\cdots$ when several orbits have the same norms with the norm chosen in non-decreasing order. Here is the beginning of the tableau of the lattice called L_s in (7.11) and that we shall call from now on I_2; its Bravais group is $O_2(\mathbb{Z})$

(it is isomorphic to B_2):

		$\mathcal{R}_1$	$\mathcal{R}_2$	L_5	L_{10}	L_{13}	L_{17}	L_{25}	L_{26}	L_{29}	L_{34}
4	$\mathcal{R}_1$	1	1	2	3	3	4	4	5	5	5
4	$\mathcal{R}_2$	1	2	3	4	5	5	7	6	7	8
8	L_5	2	3	5	7	8	9	11	11	12	13
8	L_{10}	3	4	7	10	11	13	15	16	19	18
8	L_{13}	3	5	8	11	13	14	18	17	19	21
8	L_{17}	4	5	9	13	14	17	19	21	22	23
8	L_{25}	4	7	11	15	18	19	25	23	26	29
8	L_{26}	5	6	11	16	17	21	23	26	27	28
8	L_{29}	5	7	12	19	19	22	26	27	29	31
8	L_{34}	5	8	13	18	21	23	29	28	31	34

(7.31)

We have written $\mathcal{R}_1, \mathcal{R}_2$ instead of L_1, L_2 to emphasize that these are orbits of roots. The first column gives the number of lattice vectors in the orbit. Indeed the orbits of nonzero lattice vectors have either 8 or 4 vectors. The former case occurs when the coordinates of the orbit vectors are, with respect to the basis I_2, $(\varepsilon_1\mu_1, \varepsilon_2\mu_2)$ or $(\eta_2\mu_2, \eta_1\mu_1)$ with $\varepsilon_i^2 = 1 = \eta_i^2$ when the two positive integers μ_i are different. In the latter case the orbit is generated by a vector $(\mu, 0)$ or (μ, μ) (only the value $\mu = 1$ corresponds to visible vectors). Since $N(\vec{\ell}) = \mu_1^2 + \mu_2^2$, the only possible values of the norm are the sums of two squares. There can be two orbits with the same norm only for two different such decompositions. The smallest value for which that occurs is $N = 25 = 5^2 + 0 = 3^2 + 4^2$; then only the second orbit of the visible vectors is entered. The smallest norm value with two orbits of visible vectors is $170 = 7^2 + 11^2 = 1^2 + 13^2$. It is important to note that along a row or a column of such a tableau the value of elements may decrease locally; in (7.31) such examples are given by L_{13}, L_{25}, L_{34}.

We know (see section 3.4) that an integral lattice is a sublattice of its dual lattice. The dual of a root lattice L is also called the *weight lattice*[9] of L.

Definition: weights of an integral root lattice. The weights of an integral root lattice L are the vectors $\vec{w} \in L^*$ whose orbit scalar product with a root $\vec{r} \in L$ satisfies $|(\vec{w}, \vec{r})| = 1$.

For the lattices A_n; D_n, $n \geq 4$; E_n, $n = 6, 7, 8$, that we shall call for short, the *simple root lattices*, the Cartan quadratic forms defined above use a basis among the short vectors, i.e. the orbits of norm 2 roots. So the vectors of the dual basis are weights. The diagonal elements of the corresponding quadratic form show that for nearly all simple root lattices there are weights of different norms.

[9] This agrees with the theory of Lie algebra, but this is not the case for the definition of the weights; those defined here are akin to the "fundamental weights".

7.2 Lattices of the root systems

7.2.1 The lattice I_n

In order to give some examples of lattices and their duals, we have already introduced in section 3.4 some lattices we shall study in this section in relation with root systems.

In E_n, we choose an orthonormal basis $\{\vec{e}_i\}$. The Gram matrix of these vectors is I_n (the unit matrix) that we use as the symbol for the lattice they generate. This lattice is integral and self-dual. We have seen that its Bravais group is $O_n(\mathbb{Z})$ and that it is a maximal Bravais group in all dimensions. The $2n$ element set $\{\pm\vec{e}_i\}$ is $\mathcal{S}(I_n)$, the set of shortest vectors of the lattice I_n. It is an orbit of $O_n(\mathbb{Z})$. According to Proposition 35, $\mathcal{S}(I_n) = \mathcal{R}_1$ is an orbit of roots; so $|\mathcal{R}_1| = 2n$. The corresponding reflections $\sigma_{\vec{e}_i}$ are represented by diagonal matrices with all coefficients being 1 except for one entry which is -1. Since the roots of $\mathcal{R}_1$ generate I_n, it is a root lattice.

For $n > 1$ the lattice I_n has vectors of norm 2. Proposition 35 tells us also that these vectors are roots. They are, up to sign, $\vec{e}_j \pm \vec{e}_k$ with $1 \leq j < k \leq n$. So there are $4\binom{n}{2} = 2n(n-1)$ of them; it is easy to verify that they form an orbit of $O_n(\mathbb{Z})$ that we shall denote by $\mathcal{R}_2$. The corresponding reflections $\sigma_{\vec{e}_j \pm \vec{e}_k}$ have only n non-vanishing matrix elements. Since a reflection matrix has trace $n-2$, it must have at least $n-2$ elements of the diagonal equal to 1; so there are in $O_n(\mathbb{Z})$ only the two conjugacy classes of reflections that we have found and $\mathcal{R}(I_n) = \mathcal{R}_1 \cup \mathcal{R}_2$, $|\mathcal{R}(I_n)| = 2n^2$.

For $n > 1$ one verifies easily that the n vectors $\vec{b}_i$:

$$1 \leq i < n, \quad \vec{b}_i = \vec{e}_i - \vec{e}_{i+1}, \ \vec{b}_n = \vec{e}_n; \tag{7.32}$$

form a basis of the lattice I_n; the corresponding quadratic form is that of B_n (given in (7.22)). That shows the equivalence of quadratic forms: $Q(B_n) \sim I_n$.

7.2.2 The lattices D_n, $n \geq 4$ and F_4

In section 3.4, eq. (3.10) we defined D_n^r as a sublattice of index 2 of I_n:

$$D_n^r = \left\{ \sum_i \lambda_i \vec{e}_i, \ \sum_i \lambda_i \in 2\mathbb{Z} \right\}; \quad I_n/D_n^r = \mathbb{Z}_2; \quad \mathrm{vol}(D_n^r) = 2. \tag{7.33}$$

and noticed that D_n^r is an even integral lattice. Its shortest vectors are of norm 2 and by Proposition 35 they are roots. Obviously they generate D_n^r; thus it is a root lattice (our notation is justified!); but to follow the commonly used notation, from now on we simply denote it by D_n. From the definition of the lattice given in equation (3.10), the point symmetry group B_n acting on I_n transforms the index 2 sublattice D_n into itself. So B_n is a group of isomorphisms of D_n, and we have already shown that $\mathcal{R}_2$ is an orbit.

We expect another orbit of roots whose reflections and those corresponding to $\mathcal{R}_2$ will generate B_n. Beware that the tableau of D_n is not a subset of that of I_n. Indeed the double of the roots in $\mathcal{R}_1 \subset I_n$ are not visible vectors in I_1 but are visible in D_n^r; and they are roots of it. The set of these roots, $\{(\pm 2, 0^{n-1})\}$, form a B_n orbit of roots that has $2n$ elements. We denote it by $\mathcal{R}_4$; with $\mathcal{R}_2$ it defines the $2n^2$ reflections of B_n.

We can extract the following basis from $\mathcal{R}_2$:

$$1 \leq i < n, \quad \vec{b}_i = \vec{e}_i - \vec{e}_{i+1}, \; \vec{b}_n = \vec{e}_{n-1} + \vec{e}_n. \tag{7.34}$$

The corresponding Gram matrix is exactly the quadratic form $Q(D_n)$ defined by (7.19).

One can prove that, by a change of basis if necessary, one can always transform the quadratic form of an integral lattice into one represented by a tridiagonal matrix[10]. For D_n such a change of basis is obtained by replacing the root $\vec{b}_n$ in (7.34) by the root $\vec{b}'_n = 2\vec{e}_n$. The Gram matrix for $n = 8$ is

$$Q'(D_8) = \begin{pmatrix} 2 & -1 & 0 & 0 & 0 & 0 & 0 & 0 \\ -1 & 2 & -1 & 0 & 0 & 0 & 0 & 0 \\ 0 & -1 & 2 & -1 & 0 & 0 & 0 & 0 \\ 0 & 0 & -1 & 2 & -1 & 0 & 0 & 0 \\ 0 & 0 & 0 & -1 & 2 & -1 & 0 & 0 \\ 0 & 0 & 0 & 0 & -1 & 2 & -1 & 0 \\ 0 & 0 & 0 & 0 & 0 & -1 & 2 & -2 \\ 0 & 0 & 0 & 0 & 0 & 0 & -2 & 4 \end{pmatrix}. \tag{7.35}$$

For $n > 4$, the tableau of orbit scalar products of D_n is therefore the union of the even norm vector orbits of I_n and the orbit $\mathcal{R}_4$ of roots $\{(\pm 2, 0^{n-1})\}$. It is trivial to check that the orbit $L'_4 = \{([\pm 1]^4, 0^{n-4})\}$ and the one (when $n = 5$) or two (when $n \geq 6$) orbits of L_6 are not root orbits; so according to Proposition 34, the lattice has no roots outside $\mathcal{R}_2$ and $\mathcal{R}_4$. As we have seen the reflections corresponding to these two orbits generate the holohedry B_n, but in a Bravais class different than $O_n(\mathbb{Z})$, since the tableau of B_n and D_n differ by more than a dilation.

For $n = 4$, one verifies easily that both B_4-orbits of norm 4 (i.e. in L_4) $\{(\pm 2, 0^3)\}$ and $\{([\pm 1]^4)\}$ are root orbits. That is exceptional and shows that the holohedry is larger than B_4. To verify that a given set of 4 lattice vectors of D_4 forms a basis for this lattice, we need only to verify that the determinant of its Gram matrix is 4. That is the case of the four vectors:

$$D_4 \text{ basis }: \vec{b}_1 = \vec{e}_1 - \vec{e}_2, \; \vec{b}_2 = \vec{e}_2 - \vec{e}_3, \; \vec{b}_3 = 2\vec{e}_3, \; \vec{b}_4 = -\vec{e}_1 - \vec{e}_2 - \vec{e}_3 - \vec{e}_4. \tag{7.36}$$

Their Gram matrix is $Q(F_4)$, given in (7.23). Since F_4 is a maximal finite subgroup of $GL_4(\mathbb{Z})$ it is the Bravais group of D_4. Sometimes F_4 is used as a label of the lattice D_4.

[10] That is a matrix whose non-vanishing elements q_{ij} satisfy the condition $|i - j| \leq 1$.

We give now the beginning of the tableau of orbit scalar products for $F_4 \equiv D_4$. The first column gives the cardinal of the F_4-orbit, the second one gives the stabilizer, the third one gives the nature of the lattice vectors (i.e. their components in the orthonormal basis of the space). We recall that B_4 is a subgroup of index 3 of F_4 and $|F_4| = 1152$.

$$\begin{array}{cccccccc} & & & & \mathcal{R}_2 & \mathcal{R}_4 & L_6 & L_{10} \\ 24 & B_3 & \{([\pm1]^2, 0^2)\} & \mathcal{R}_2 & 2 & 2 & 3 & 4 \\ 24 & B_3 & \{([\pm1]^4) \cup (\pm2, 0^3)\} & \mathcal{R}_4 & 2 & 4 & 4 & 6 \\ 96 & A_1 \times A_2 & \{(\pm2, [\pm1]^2, 0)\} & L_6 & 3 & 4 & 6 & 7 \\ 144 & B_2 & \{([\pm2]^2, [\pm1]^2)\} \cup \{(\pm3, \pm1, 0^2)\} & L_{10} & 4 & 6 & 7 & 10 \end{array} . \tag{7.37}$$

7.2.3 The lattices D_n^*, $n \geq 4$

We already defined in section 3.4 the dual lattice of D_n; there we wrote it D_n^w, where w is the initial letter of the word *weight*. Indeed D_n^* is the lattice generated by the weights of the root lattice D_n.

As in section 3.4, from the orthonormal vectors $\vec{e}_i$ we define the vectors:

$$\vec{w}_n \equiv \vec{w}_n^+ = \frac{1}{2}\sum_{i=1}^{n} \vec{e}_i, \quad \vec{w}_n^- = \vec{w}_n^+ - \vec{e}_n, \quad N(\vec{w}^\pm) = \frac{n}{4}. \tag{7.38}$$

Then we can use either vector for the decomposition of D_n^* into two cosets of the lattice I_n:

$$D_n^* = I_n \cup (\vec{w}_n + I_n), \quad \mathrm{vol}(D_n^*) = \frac{1}{2}. \tag{7.39}$$

Dual lattices have the same point symmetry group and their Bravais groups are contragredient. (See the definition of the contragredient representation in 2.6.)The only problem is to know whether these Bravais groups are identical (i.e. conjugate in $GL_n(\mathbb{Z})$). We have two cases to consider: $n = 4$, holohedry F_4, and $n > 4$, holohedry B_n.

i) $n = 4$. Then $N(\vec{w}_n^\pm) = 1$, so their B_4 orbit $\{\frac{1}{2}([\pm1]^4)\}$ and that of the vectors $\{\pm\vec{e}_i\} \equiv \{(\pm1, 0^3)\}$ form the 24 element set $S(D_4^*)$ of shortest vectors. They form a F_4 orbit of roots (identical to $\frac{1}{2}\mathcal{R}_4$ where $\mathcal{R}_4$ is one of the root orbits of D_4 (see (7.37)) and they generate the lattice. The other orbit $\mathcal{R}_2$ of 24 roots is exactly that of the shortest vectors of D_4. It is straightforward to prove that the tableau of D_4^* is obtained from that of D_4^r by multiplication by $\frac{1}{2}$. It is also easy to verify that the four vectors $\vec{w}_n, -\vec{e}_1, \vec{e}_1 - \vec{e}_2, -\vec{e}_2 - \vec{e}_3$ form a basis of D_4^* and the corresponding Gram matrix is $\frac{1}{2}q(F_4)$. That proves that the lattice $\frac{1}{\sqrt{2}}F_4$ is isodual and that F_4 has only one Bravais class.

ii) $n > 4$. Then $N(\vec{w}_n^\pm) = \frac{n}{4}$, so the set of shortest vectors $\mathcal{S}(D_n^*)$ consists of the $2n$ vectors $\pm\vec{e}_i$. $\mathcal{S}(D_n)$ does not generate the lattice[11]. Since $|\mathcal{S}(D_n)| = (n-1)|\mathcal{S}(D_n^*)|$ the two tableaus cannot be proportional, so for $n > 4$ the holohedry B_n has three different Bravais classes, corresponding to the Bravais groups of the three lattices $D_n < I_n < D_n^*$.

We recall here that the weights of the root lattice D_n are the $2n$ vectors $\pm\vec{e}_i$ of norm 1 and the 2^n vectors $\{([\pm\frac{1}{2}]^n)\}$ of norm $\frac{n}{4}$. They form two orbits of B_n.

7.2.4 The lattices D_n^+ for even $n \geq 6$

Using the fact that the sum of the coordinates of the vectors of D_n is even, we already verified that:

$$n > 2, \quad 2\vec{w}_n^\pm \begin{cases} \notin D_n & \text{when } n \text{ is odd} \\ \in D_n & \text{when } n \text{ is even.} \end{cases} \tag{7.40}$$

So the four cosets of D_n in D_n^*,

$$D_n, \quad D_n + \vec{w}_n^+, \qquad D_n + \vec{e}_n, \quad D_n + \vec{w}_n^- \tag{7.41}$$

form a group $\mathbb{Z}_4$ when n is odd, $\mathbb{Z}_2 \times \mathbb{Z}_2$ when n is even (this was already proved in (3.12)). So when n is odd and ≥ 3, we have studied the three lattices invariant under B_n. When n is even, each of the three non-trivial cosets of D_n generates with that lattice a sublattice of volume 1, having index 2 in D_n^*. One of them is I_n. The others are

$$n \text{ even}: \; D_n^\pm = D_n \cup (\vec{w}_n^\pm + D_n); \quad \det(D_n^\pm) = 1. \tag{7.42}$$

Since the vectors $\vec{w}^\pm$ are exchanged by the reflection $\sigma_{\vec{e}_n} \in B_n$ which transforms D_n into itself, $\sigma_{\vec{e}_n}$ exchanges the two lattices $D_n^\pm$. That proves that they have the same symmetry, i.e. the same Bravais class.

Without going into details we just formulate here the results of the description of D_n^+ lattices in a theorem

Theorem 11 *Let $0 < k \in \mathbb{N}$. For $n = 4k$, the lattices D_n^+ are integer self-dual; their Bravais group is a maximal subgroup of $GL_{4k}(\mathbb{Z})$ and one of the 4 arithmetic classes of D_n except for $k = 1$, $D_4^+ \sim I_4$ and $k = 2$, $D_8^+ = E_8$. For $n = 4k+2$, the lattices D_n^+ are isodual with the Bravais group D_n and $\sqrt{2}D_n^+$ are integral lattices.*

7.2.5 The lattices A_n

For small dimensions, $n \leq 4$, it is easy to study directly the lattices invariant under A_n. But the easiest method which can be generalized to arbitrary

[11] Historically, D_5^* was the first known lattice not generated by a set of successively linearly independent shortest vectors (Dirichlet's remark).

n is to study these lattices as sublattices of I_{n+1}. In the space E_{n+1} we choose an orthonormal basis $\alpha, \beta \in \mathcal{N}_n, (\vec{e}_\alpha, \vec{e}_\beta) = \delta_{\alpha\beta}$, where $\mathcal{N}_n$ denotes the set $\{0, 1, \ldots, n\}$ of the first $n+1$ non negative integers, and we define

$$\vec{e} = \frac{1}{n+1} \sum_{\alpha \in \mathcal{N}_n} \vec{e}_\alpha, \quad N(\vec{e}) = \frac{1}{n+1}. \tag{7.43}$$

We denote by H_n the subspace orthogonal to $\vec{e}$. The sums of the coordinates of the points in H_n are equal to zero. It will be useful to consider the vectors in H_n:

$$\vec{u}_\alpha = \vec{e}_\alpha - \vec{e} \tag{7.44}$$

and also the set of norm 2 vectors in H_n,

$$\mathcal{R}_2 = \{\vec{r}_{\alpha\beta} = \vec{e}_\alpha - \vec{e}_\beta = \vec{u}_\alpha - \vec{u}_\beta\}, \quad |\mathcal{R}_2| = n(n+1). \tag{7.45}$$

Then we verify that each reflection $\sigma_{\vec{r}_{\alpha\beta}}$ exchanges the basis vectors $\vec{e}_\alpha, \vec{e}_\beta$ and leaves the others fixed. That proves that the reflections associated with the vectors of $\mathcal{R}_2$ generate the group $\mathcal{S}_{n+1}$ of permutations of the basis vectors of E_{n+1}. This group leaves fixed the vector $\vec{e}$; it acts linearly and irreducibly on the subspace H_n. In this section, the n-dimensional lattices that we study are in H_n.

We first prove that the lattice A_n is the intersection of the lattice I_{n+1} with H_n; explicitly:

$$A_n = I_{n+1} \cap H_n. \tag{7.46}$$

It is easy to verify from this definition that this lattice is an even integral lattice; its shortest vectors are of norm 2. They form the set $\mathcal{R}_2$ and generate the lattice. We can take as a basis the set of n vectors:

$$\{\vec{r}_i = \vec{r}_{i-1,i} = \vec{e}_{i-1} - \vec{e}_i = \vec{u}_{i-1} - \vec{u}_i\} \subset \mathcal{R}_2, \quad i \in \mathcal{N}_n^+, \tag{7.47}$$

where $\mathcal{N}_n^+ = \{1, 2, \ldots, n\}$ denotes the set of the first n positive integers. The Gram matrix of the $\vec{r}_i$'s, $q_{ij} = 2\delta_{ij} - (|i-j| = 1)$ (for a definition of a Boolean function, see (7.10)) is usually called the Cartan matrix of (the Lie algebra) A_n. Moreover we have shown again explicitly that $\mathcal{S}(A_n) = \mathcal{R}_2$, is a set of roots, so A_n is a root lattice invariant by the symmetric group on $n+1$ letters (see (7.15)). Since the symmetry through the origin, $-I_{n+1}$, is a symmetry of any lattice, we use the notation $\overline{A}_n$ for the $2(n+1)!$ element group generated by A_n and $-I_{n+1}$. The Bravais group $\overline{A}_n$ is defined by its linear representation on H_n; it is a maximal irreducible subgroup of $GL_n(\mathbb{Z})$. The set of roots $\mathcal{R}_2$ forms an orbit of $\overline{A}_n$. Since the permutation of α, β in (7.45) has the same effect as a change of sign, $\mathcal{R}_2$ is also the orbit $A_n : A_{n-2}$ of A_n.

From $\mathcal{R}_2$ we can extract another interesting basis:

$$i \in \mathcal{N}_n^+, \quad \vec{b}_i = \vec{e}_i - \vec{e}_0. \tag{7.48}$$

If we denote by J_n the $n \times n$ matrix all of whose elements are 1, the Gram matrix corresponding to the basis (7.48) is:

$$Q(A_n) = I_n + J_n, \quad \text{with } (J_n)_{ij} = 1. \tag{7.49}$$

7.2.6 The lattices A_n^*

It is easy to compute the dual basis of (7.48) and the corresponding quadratic form:

$$\vec{b}_i^* = \vec{e}_i - \vec{e}, \quad Q(A_n^*) = I_n - \frac{1}{n+1} J_n \quad \det(Q(A_n^*)) = \frac{1}{n+1}. \tag{7.50}$$

To find the set $\mathcal{W}$ of weights of the root lattice A_n one has to look for the elements of A_n^* whose scalar product with the roots $\vec{r}_{\alpha\beta}$ are $\pm 1,\ 0$. An easy computation leads to (we recall that $\subset$ is a strict inclusion):

$$\emptyset \neq \mathcal{A} \subset \mathcal{N}_n, \quad \mathcal{W} = \{\vec{w}_{\mathcal{A}} = \sum_{\alpha \in \mathcal{A}} \vec{u}_\alpha\}, \quad \vec{w}_{\mathcal{A}} + \vec{w}_{\overline{\mathcal{A}}} = 0; \quad |\mathcal{W}| = 2(2^n - 1), \tag{7.51}$$

where $\overline{\mathcal{A}}$ is the complement in $\mathcal{N}_n$ of the subset $\mathcal{A}$.

The set $\mathcal{W}$ splits into n orbits of the Weyl group A_n, each orbit containing all $\vec{w}_{\mathcal{A}}$ whose defining subsets $\mathcal{A}$ have the same cardinal $|\mathcal{A}| = k$; we denote these orbits by $\mathcal{W}_k$ and note that $|\mathcal{W}_k| = \binom{n+1}{k}$. We can choose as a representative of these n orbits:

$$\vec{w}_k = \sum_{i=1}^{k} \vec{u}_{i-1}, \quad 1 \leq k \leq n. \tag{7.52}$$

These n vectors form the dual basis of that of A_n defined in (7.47) (indeed, $(\vec{w}_i, \vec{r}_j) = \delta_{ij}$); so the $\vec{w}_i$'s define an A_n^* basis.

7.3 Low dimensional root lattices

To conclude the chapter on root lattices we return to the examples of three and four dimensional lattices which are at the same time root lattices. Table 7.2 gives the group, its order, combinatorial type, and graphical visualization.

TAB. 7.2 – Three and four dimensional lattices associated with root systems of reflection groups.

Group	Order	Type	Graph
		n=3	
B_3	48	**6.8**-0	
A_3	48	**12.14**-0	
A_3^*	48	**14.24**-8	
		n=4	
B_4	384	**8.16**-0	
F_4	1152	**24.24**-0	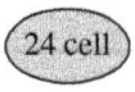 24 cell
$G_2 \uparrow G_2$	288	**12.36**-12	
$G_2 \otimes G_2$	144	**30.102**-36	
A_4	240	**20.30**-0	
A_4^*	240	**30.120**-60	

Chapter 8

Comparison of lattice classifications

In previous chapters we have discussed translation lattices from the point of view of their symmetry, their Voronoï cells and associated quadratic forms. In this chapter we analyze the applications of these different approaches to the most evident and straightforward physical example, the description and classification of periodic crystal structures, and compare the advantages and disadvantages of alternative approaches and possibilities of their generalizations to arbitrary dimension.

We follow in this analysis the works by Michel and Mozrzymas [76] and Michel [75, 73, 74].

We remember that one-to-one correspondence exists between the set $\mathcal{B}_n$ of translation lattice bases defined in the Euclidean space R^n with a fixed orthonormal basis $\{e_i\}$, $e_i e_j = \delta_{ij}$, and elements of $GL_n(\mathbb{R})$. Every basis $b \in \mathcal{B}_n$ defines a lattice L_n

$$L_n = \left\{ \sum_{i=1}^{n} n_i b_i, \quad n_i \in Z \right\}; \tag{8.1}$$

with all other possible bases being of the form

$$b_i' = \sum_{j=1}^{n} m_{ij} b_j, \quad m_{ij} \in GL_n(\mathbb{Z}). \tag{8.2}$$

The relation (8.2) shows that $\mathcal{L}_n$, the set of lattices of dimension n is the variety:

$$\mathcal{L}_n = GL_n(\mathbb{R}) : GL_n(\mathbb{Z}) \equiv \mathcal{B}_n | GL_n(\mathbb{Z}). \tag{8.3}$$

Let us denote L_n^0 the set of lattices obtained from L_n by an orthogonal transformation, i.e. the orbit of the $O(n)$ group action on $\mathcal{L}_n$. The corresponding set of orbits we denote by $\mathcal{L}_n^0$

$$\mathcal{L}_n^0 = \mathcal{L}_n | O(n). \tag{8.4}$$

The simplest initial classification of lattices by symmetry is given by the stabilizers of orbits of the $O(n)$ action. The system of strata, $\mathcal{L}_n||O(n)$ of the $O(n)$ action on the set $\mathcal{L}_n$ of lattices defines crystal systems. Note that a relatively small number of point symmetry groups (subgroups of $O(n)$) can appear as stabilizers of $O(n)$ orbits on the set of lattices.

From equations (8.2), (8.3) we deduce that the set of $O(n)$ orbits $\mathcal{L}_n^0$ is the set of double cosets: $GL_n(\mathbb{Z})\backslash GL_n(\mathbb{R})/O(n)$. This means that if b is a basis of L_n, the set of bases of lattices L_n^0 is the double coset

$$GL_n(\mathbb{Z})bO(n) \equiv \{mbr^{-1}, \forall m \in GL_n(\mathbb{Z}), \forall r \in O(n)\}. \tag{8.5}$$

Alternative interpretation of this double coset is an orbit of the direct product $GL_n(\mathbb{Z}) \times O(n)$ acting on $GL_n(\mathbb{R})$ through $b \mapsto mbr^{-1}$. The stabilizer H of b is the subgroup $\{(m,r) \in GL_n(\mathbb{Z}) \times O(n), mbr^{-1} = b\}$.

Let π_z and π_o be the canonical projections of $GL_n(\mathbb{Z}) \times O(n)$ on its factors: $\pi_z(m,r) = m$, $\pi_o(m,r) = r$. The geometrical interpretation of H is as follows: $\pi_o(H)$ is the group of orthogonal transformations r which transform the lattice L_n into itself because any basis $m'b$ of L_n transforms into the basis $m'br^{-1} = m'mb$. The stabilizer H is the point symmetry group of the lattice (holohedry of the lattice is the historical terminology). The stabilizer is defined up to conjugation in $GL_n(\mathbb{Z}) \times O(n)$; moreover, there are the isomorphisms: $\pi_z(H) \sim H \sim \pi_o(H)$. We have noted that the conjugation class $[\pi_o(H)]_{O(n)}$ defines the stratum named the crystal system. There are four crystal systems in dimension two, seven crystal systems in dimension 3, 33 (+7 taking into account enantiomorphic groups) crystal systems in dimension 4. In dimensions 5 and 6 there are respectively 59 and 251 crystal systems.

The classification of stabilizers $[\pi_z(H)]_{GL_n(\mathbb{Z})}$ up to conjugation in $GL_n(\mathbb{Z})$ defines the Bravais class of the lattice L_n. We can define Bravais classes of lattices also as strata $\mathcal{Q}_n||GL_n(\mathbb{Z})$ of $GL_n(\mathbb{Z})$ action on the set of quadratic forms, $\mathcal{Q}_n$, associated to all lattices L_n^0.

Let b be the basis of lattice L_n and $bb^\top$ a symmetric positive definite matrix (=quadratic form) with elements $(bb^\top)_{ij} = b_i b_j$. We denote $\mathcal{Q}_n$ the set of positive quadratic forms which is a convex cone. The polar decomposition of invertible matrices: $b = \sqrt{bb^\top}s = s\sqrt{b^\top b}$, $s \in O(n)$ shows that

$$\mathcal{Q}_n = GL_n(\mathbb{R}) : O(n) = GL_n(\mathbb{R})|O(n). \tag{8.6}$$

This means that $\mathcal{Q}_n$ can be identified with left cosets of $O(n)$ in $GL_n(\mathbb{R})$, or else as a space of orbits of $O(n)$ acting by left multiplication on $GL_n(\mathbb{R})$. The action of $GL_n(\mathbb{Z}) \times O(n)$ on $b \in GL_n(\mathbb{R})$ can be transported to the action on $bb^\top \in \mathcal{Q}_n$. The group $O(n)$ acts trivially and $GL_n(\mathbb{Z})$ acts through:

$$\forall m \in GL_n(\mathbb{Z}), \quad bb^\top \mapsto mbb^\top m^\top. \tag{8.7}$$

The orbit of $GL_n(\mathbb{Z})$ is the set of quadratic forms associated to all bases of all lattices of L_n^0. This allows us to give an alternative definition: Bravais classes

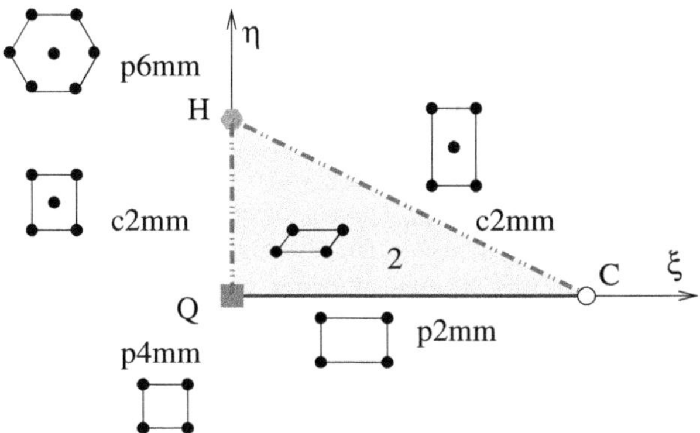

FIG. 8.1 – A fundamental domain of the action of $GL_2(\mathbb{Z})$ on the cone of positive quadratic forms q or, equivalently, on intrinsic lattices. Positive quadratic forms are parameterized by $\xi = (q_{11} - q_{22})/(\mathrm{Tr}\ q)^{-1}$, $\eta = 2q_{12}(\mathrm{Tr}\ \mathrm{q})^{-1}$, where $\mathrm{Tr}\ q = q_{11} + q_{22} > 0$. With this parameterization the quadratic form becomes $q = (1/2)(\mathrm{Tr}\ q)(I_2 + \xi\sigma_3 + \eta\sigma_1)$ with σ_1, σ_3 being usual Pauli matrices. The positivity implies $\xi^2 + \eta^2 < 1$. The fundamental domain is the triangle HQC minus the vertex $C(1, 0)$ which belongs to the surface of the cone. $H(0, 1/2)$ represents the $p6mm$ lattices, $Q(0, 0)$ the $p4mm$ lattices, the side QC the $p2mm$ lattices. The two sides QH and HC represent cmm lattices with, respectively, four shortest vectors (half of the diagonal is shorter than the sides of rectangle) and two shortest vectors (half of the diagonal is longer than one of the sides).

of lattices are the strata $\mathcal{Q}_n||GL_n(\mathbb{Z})$ of the action (8.7), with the quadratic form $bb^\top$ being associated to the base b of L_n.

The fundamental domain of the stratification of the cone of positive quadratic forms by $GL_n(\mathbb{Z})$ action is shown in figure 8.1. Each stratum corresponds to the Bravais class of two-dimensional lattices. The numbers of Bravais classes in dimensions $d = 1, 2, 3, 4, 5, 6$ are respectively 1, 5, 14, 64[1], 189, and 841 [81, 89].

In order to demonstrate the relation between Bravais classes and point symmetry groups of lattices (i.e. crystal systems) we note first that there exists a natural mapping ϕ from the set of conjugation classes of finite subgroups of $GL_n(\mathbb{Z}) < GL_n(\mathbb{R})$ into subgroups of $O(n)$.

This statement follows from the well known fact that all finite subgroups of $GL_n(\mathbb{R})$ are conjugate to a subgroup of $O(n) < GL_n(\mathbb{R})$, and the existence of natural mapping from the conjugation classes of subgroups of $GL_n(\mathbb{Z}) < GL_n(\mathbb{R})$ into subgroups of $GL_n(\mathbb{R})$. If two finite subgroups of $O(n)$ are conjugate in $GL_n(\mathbb{R})$, they are conjugate in $O(n)$ as well. This gives

[1] 10 Bravais classes are split into enantiomorphic pairs and if one counts enantiomorphic forms as different, there are 74 Bravais classes in dimension 4.

the correspondence

$$\phi\left([\pi_z(H)]_{GL_n(\mathbb{Z})}\right) = [\pi_o(H)]_{O_n}. \tag{8.8}$$

The restriction of ϕ to Bravais classes gives a mapping $\tilde{\phi}$ of $\{BC\}_n$, from the set of Bravais classes, on $\{BCS\}_n$ the set of crystal systems in n-dimensional space. (Louis Michel [76, 75] uses (BCS)=Bravais Crystallographic Systems instead of simply Crystal Systems in order to stress the difference with the "crystal family" notion widely used in crystallography.) Note, however, that not all conjugation classes which are inverse images ϕ^{-1} of crystal systems are Bravais classes.

8.1 Geometric and arithmetic classes

We have seen that very small number of finite subgroups of $O(n)$ could be realized as symmetry groups of a translation lattice, i.e. to be a point group defining the crystal system (a holohedry). At the same time any subgroup of a holohedry group can be a point symmetry group of a multiregular system of points or, in more physical words of a crystal formed of several types of atoms. Point symmetry group of a n-dimensional crystal defined up to conjugation in $O(n)$ is named a geometric class. Geometric classes form a partially ordered set which includes all the holohedries. Partially ordered set of three-dimensional geometric classes is represented in Figure 8.2. It includes, in particular seven groups which are the holohedries and characterize the crystal systems. In 3-dimensional space there are 32 geometric classes or 32 crystallographic point groups. The adjective "crystallographic" is used to stress that the existence of a translation lattice imposes certain restrictions on subgroups of $O(n)$ to be a point symmetry group of a lattice. In dimensions 4, 5, and 6 there are respectively 227, 955, and 7103[2] geometric classes. If one counts enantiomorphic pairs as different, then in dimension 4 there are 227+44 different geometric classes.

It should be noted that different geometric classes can be isomorph, i.e. they correspond to the same abstract group. Among the 32 geometric classes for three-dimensional crystals there are only 18 non-isomorph abstract groups. The isomorphism relation between geometric classes is illustrated in Table 8.1. In dimension two among 10 geometric classes there is only one pair of isomorph groups, namely $C_2 \sim C_s$. In dimensions 4, 5, and 6 the numbers of abstractly non-isomorph geometric classes are respectively 118, 239, and 1594.

We have described all possible geometric classes by looking for all subgroups of point groups characterizing crystal systems (symmetry of translation lattices). We can also study conjugacy classes of finite subgroups of $GL_n(\mathbb{Z})$, i.e. all subgroups of Bravais groups. The conjugacy classes of finite

[2] For dimension 6 the number of geometric classes given in [89] is 7104.

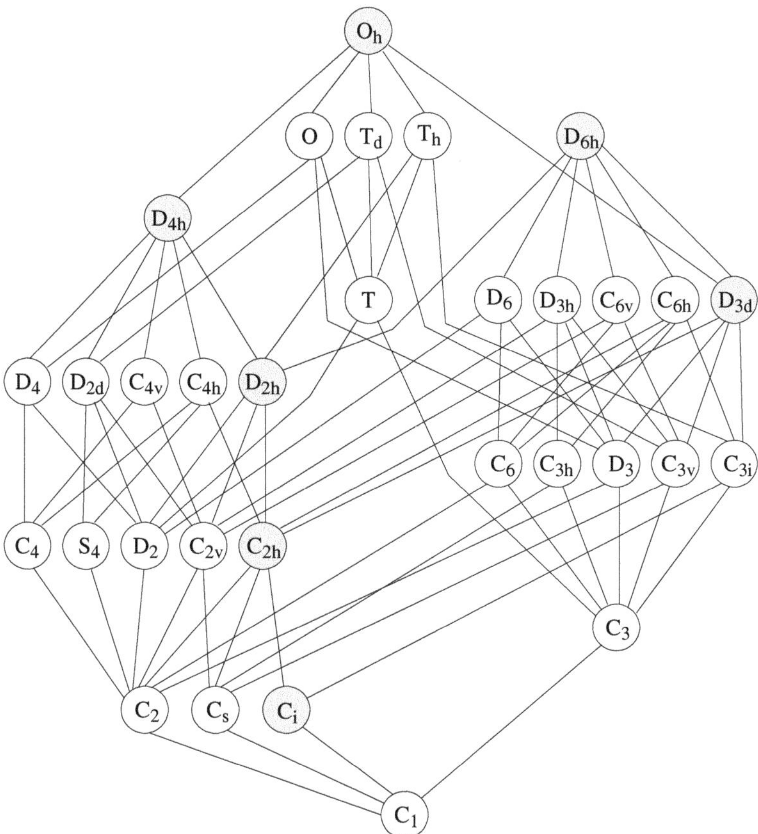

FIG. 8.2 – A partially ordered set of 3D-geometric classes up to their equivalence in $O(3)$. Seven point groups corresponding to crystal systems are shown by shading.

subgroups of $GL_n(\mathbb{Z})$ are named arithmetic classes. Arithmetic classes form a partially ordered set which includes, in particular, all Bravais groups.

It is known that the number of conjugacy classes of finite subgroups of $GL_n(\mathbb{Z})$ is finite. For $n = 1, 2, 3, 4, 5, 6$ this number is 2, 13, 73, 710(+70), 6079, and 85311 (+30) [89]. In parenthesis the number of enantiomorphic pairs is indicated.

The partially ordered set of arithmetic classes for each dimension can have several maximal elements. These maximal arithmetic classes are always the Bravais groups. All arithmetic classes can be described as a subgroups of maximal ones. The same is naturally valid for Bravais groups. Thus it is important to know the complete list of maximal arithmetic classes

TAB. 8.1 – The 32 crystallographic geometric classes and their 18 isomorphy classes. The isomorphy classes are listed in column 1 and are defined as direct products of cyclic groups Z_n, dihedral groups c_{nv}, permutation group of four objects $\mathcal{S}_4$, and its subgroup of even permutations $\mathcal{A}_4$. In column 2 the corresponding geometric classes are listed in ITC and Schönflies notations.

Isomorphic	Geometric
1	$1 = C_1$
Z_2	$\bar{1} = C_i$, $m = C_s$, $2 = C_2$
$Z_2 \times Z_2$	$2/m = C_{2h}$, $mm2 = C_{2v}$, $222 = D_2$
$Z_2 \times Z_2 \times Z_2$	$mmm = D_{2h}$
Z_3	$3 = C_3$
$Z_2 \times Z_3$	$6 = C_6$, $\bar{3} = C_{3i} \equiv S_6$, $\bar{6} = C_{3h}$
Z_4	$4 = C_4$, $\bar{4} = S_4$
$Z_2 \times Z_4$	$4/m = C_{4h}$
c_{3v}	$3m = C_{3v}$, $32 = D_3$
$Z_2 \times Z_2 \times Z_3$	$6/m = C_{6h}$
c_{4v}	$4mm = C_{4v}$, $422 = D_4$, $\bar{4}m2 = D_{2d}$
$c_{4v} \times Z_2$	$4/mmm = D_{4h}$
$c_{3v} \times Z_2$	$6mm = C_{6v}$, $622 = D_6$, $\bar{3}m = D_{3d}$, $\bar{6}m2 = D_{3h}$
$c_{3v} \times Z_2 \times Z_2$	$6/mmm = D_{6h}$
$\mathcal{A}_4$	$23 = T$
$\mathcal{A}_4 \times Z_2$	$m\bar{3} = T_h$
$\mathcal{S}_4$	$\bar{4}3m = T_d$, $432 = O$
$\mathcal{S}_4 \times Z_2$	$m\bar{3}m = O_h$

(i.e. maximal finite subgroups) of $GL_n(\mathbb{Z})$. The number of maximal arithmetic classes for $n = 1, 2, 3, 4, 5$ is 1, 2, 4, 9, and 17.

There exists a natural map between arithmetic and geometric classes in the d-dimension. Figure 8.3 illustrates this map in dimension two.

For three-dimensional lattices the correspondence between arithmetic and geometric classes is represented in Table 8.2.

8.2 Crystallographic classes

Geometric and arithmetic classes characterize only partially the symmetry of a multiregular system of points. The complete infinite discrete symmetry group which includes all translations as well is named the crystallographic space group. The crystallographic space groups are the subgroups of the Euclidean group Eu_d which contain a d-dimensional lattice of translations.

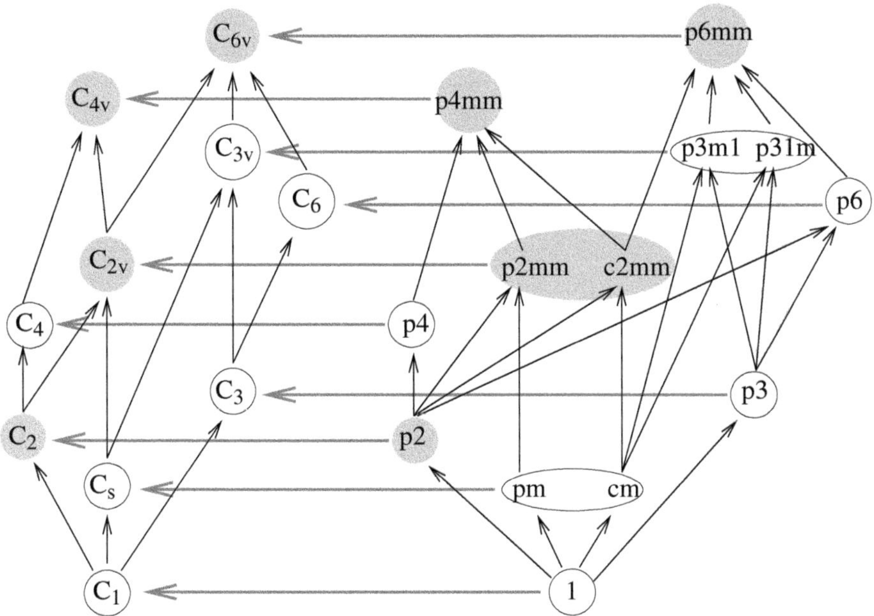

FIG. 8.3 – A surjective map $\{AC\}_2 \to \{GC\}_2$ from the partially ordered set of Arithmetic Classes, i.e. conjugacy classes of finite subgroups of $GL_2(\mathbb{Z})$ (the right part of the diagram), to partially ordered set of Geometric Classes, i.e. conjugacy classes of finite subgroups of $O(2)$ (the left part of the diagram), for two-dimensional lattices. Bravais classes form a subset of the arithmetic classes; they are indicated by shading on the right part. Crystallographic Bravais systems (crystal classes = holohedry) form a subset of the geometric classes; they are indicated by shading on the left part.

The Euclidean group Eu_d is the semi-direct product of the orthogonal group by the translations. Applying the construction of a semi-direct product to an arithmetic class (P^z finite subgroup of $GL_d(\mathbb{Z})$) and translation lattice we can define a space group. The so obtained space group does not depend on the choice of the group from a given arithmetic class. Any conjugated group mP^zm^{-1}, with $m \in GL_d(\mathbb{Z})$, results in the same space group. But such construction of space groups gives only a part of all possible space groups.

The space groups obtained as semi-direct products are named symmorphic in crystallography. Their number (and notation) coincides with the number (and notation) of arithmetic classes. In dimension 2 and 3 there are respectively 13 and 73 symmorphic space groups. In general, each arithmetic class $[P]_{GL_n(\mathbb{Z})}$, i.e. conjugacy class in $GL_n(\mathbb{Z})$ of a point symmetry group P allows us to construct a set of crystallographic groups by including lattice translations. This procedure is named group extension. Equivalence classes of

TAB. 8.2 – Correspondence between 3D geometric and arithmetic classes. Bravais crystallographic systems (holohedry) are shown among geometric classes by light shading. Bravais classes are differentiated among arithmetic classes by shading. Maximal geometric and arithmetic classes are underlined.

Group order	Geometric classes	Arithmetic classes
48	$\underline{O_h}$	$\underline{Pm\bar{3}m,\ Fm\bar{3}m,\ Im\bar{3}m}$
24	O	$P432,\ F432,\ I432$
	T_h	$Pm\bar{3},\ Fm\bar{3},\ Im\bar{3}$
	T_d	$P\bar{4}3m,\ F\bar{4}3m,\ I\bar{4}3m$
	$\underline{D_{6h}}$	$\underline{P6/mmm}$
16	D_{4h}	$P4/mmm,\ I4/mmm$
12	T	$P23,\ F23,\ I23$
	D_{3d}	$R\bar{3}m$, $P\bar{3}m1,\ P\bar{3}1m$
	C_{6v}	$P6mm$
	C_{6h}	$P6/m$
	D_{3h}	$P\bar{6}2m,\ P\bar{6}m2$
	D_6	$P622$
8	C_{4v}	$P4mm,\ I4mm$
	C_{4h}	$P4/m,\ I/mm$
	D_{2d}	$P\bar{4}2m,\ P\bar{4}m2,\ I\bar{4}2m,\ I\bar{4}m2$
	D_4	$P422,\ I422$
	D_{2h}	$Pmmm,\ Cmmm,\ Fmmm,\ Immm$
6	C_{3v}	$R3m,\ P3m1,\ P31m$
	D_3	$R32,\ P321,\ P312$
	S_6	$R\bar{3},\ P\bar{3}$
	C_{3h}	$P\bar{6}$
	C_6	$P6$
4	C_4	$P4,\ I4$
	S_4	$P\bar{4},\ I\bar{4}$
	C_{2v}	$Pmm2,\ Cmm2,\ Amm2,\ Fmm2,\ Imm2$
	D_2	$P222,\ C222,\ F222,\ I222$
	C_{2h}	$P2/m,\ C2/m$
3	C_3	$R3,\ P3$
2	C_2	$P2,\ C2$
	C_s	$Pm,\ Cm$
	C_i	$P\bar{1}$
1	C_1	$P1$

extensions form second cohomology group $H^2(P, L_n)$. A formal mathematical description of group extensions can be found in [69]. (Note that the first set of lectures on applications of cohomology of groups in physics was given in a Physics summer school by Louis Michel in 1964.) Explicit construction of group extensions for two-dimensional and three-dimensional space groups via intermediate definition of non-symmorphic elements is discussed in [75]. Note however, that if the space group contains nonsymmorphic element, the group is non-symmorphic, but the contrary is not right. In dimension three there are two non-symmorphic groups, namely $I2_12_12_1$ and $I2_13$ (given in ITC notation) which do not have non-symmorphic elements.

We mention here that the total number of crystallographic space groups in dimensions 2, 3, 4, 5, and 6 is respectively : 17, 219(+11), 4783(+111), 222018(+79), and 28927922(+7052) [89]. In parenthesis the number of enantiomorphic pairs of space groups is given.

We can now summarize the relations between different symmetry classes introduced in this section for multiregular system of points.

The diagram below uses the notations :
$\{CC\}$ - crystallographic classes;
$\{AC\}$ - arithmetic classes;
$\{GC\}$ - geometric classes;
$\{BC\}$ - Bravais classes;
$\{BCS\}$ - Bravais crystallographic systems.

$$\begin{array}{ccccc} \{CC\} & \xrightarrow{\alpha} & \{AC\} & \xrightarrow{\phi} & \{GC\} \\ & \searrow \beta & \downarrow \gamma & & \\ & & \{BC\} & \xrightarrow{\bar{\phi}} & \{BCS\} \end{array}$$

8.3 Enantiomorphism

We have noted on several occasions that numbers of different objects given in mainly physical and mainly mathematical literature turn out to be different. One of the sources of such difference is the different treatment of enantiomorphic objects [42, 89]. The best known example of that kind is the following "mathematical" and "physical" statements.

i) There exist 219 abstractly non-isomorph three-dimensional crystallographic (Fedorov or space) groups (typically mathematical statement).

ii) There exist 230 crystallographic (Fedorov or space) groups (typical statement in crystallography or in physics).

The difference between these two statements is due to fact that two groups written by the same set of matrices but in frames of different orientation can be considered as equivalent or as different. Such two groups form an enantiomorphic pair. In particular, there are 11 enantiomorphic pairs of crystallographic three-dimensional groups and this gives the explanation of reference to 219 or to 230 3D-groups.

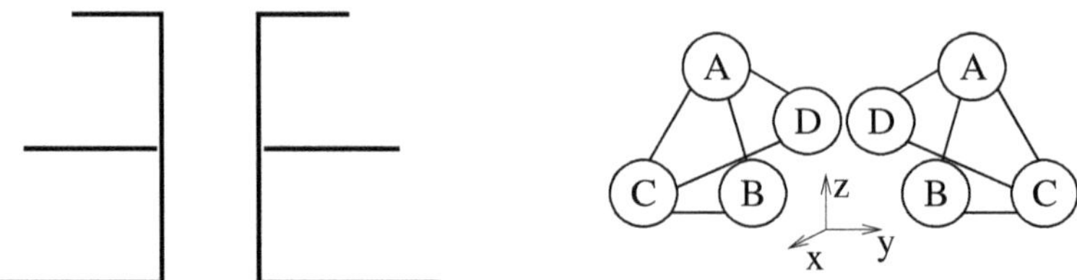

FIG. 8.4 – Enantiomorphic pair or objects in two-dimensional space (left) and in three-dimensional space (right).

We have also mentioned that in the general n-dimensional case the enantiomorphic pairs of Bravais lattices, of arithmetic and geometric classes, of Bravais crystal systems exist.

Let us discuss this subject briefly by starting with the definition of enantiomorphic or chiral objects which is not very precise but allows us to unify the treatment of very different objects and constructions from the point of view of enantiomorphism.

Two objects that are equivalent by an affine transformation but not by an orientation preserving transformation are called an enantiomorphic pair, each member of an enantiomorphic pair is said to be enantiomorphic or chiral.

Figure 8.4 shows two- and three-dimensional examples of a pair of objects which can be easily transformed one into another by applying reflection which is an improper symmetry transformation. At the same time there are no two-dimensional or three-dimensional orientation preserving transformations between members of each pair. Note, however, that if two dimensional objects are considered as situated (immersed) in three-dimensional space a two-dimensional reflection can be realized as a pure three-dimensional rotation.

Now, before turning to a discussion of enantiomorphic symmetry classes we need first to be precise about what we mean by equivalence under affine transformations or under orientation preserving affine transformations. Equivalence between symmetry groups or classes means that two objects belong to the same conjugacy class of the group $\mathcal{G}$ used for the classification. In the case of arithmetic classes the group $\mathcal{G}$ is taken to be $GL_n(\mathbb{Z})$. For geometric classes we look for equivalence within the $GL_n(\mathbb{Q})$ group. The most fine classification into space group types (crystallographic classes) is done within the affine group $\mathcal{A}(n, \mathbb{R})$.

An equivalence class is given as the orbit of a member H of the class under a chosen group $\mathcal{G}$ of transformations. If group $\mathcal{G}$ contains a transformation σ that does not preserve the orientation the group $\mathcal{G}$ can be split into a disjointed union of the two cosets with respect to the subgroup $\mathcal{G}^+$ of orientation preserving transformations:

$$\mathcal{G} = \mathcal{G}^+ \cup \sigma \cdot \mathcal{G}^+. \tag{8.9}$$

The group H and its orientation-reversed transform $H' := \sigma^{-1}H\sigma$ form an enantiomorphic pair if and only if H' is not contained in the orbit of H under $\mathcal{G}^+$.

Equivalent more formal formulations are:

Proposition 38 *A group H is enantiomorphic if and only if the normalizer (stabilizer) $N(H)$ in $\mathcal{G}$ is contained in $\mathcal{G}^+$.*
A group H is not enantiomorphic (achiral) if and only if the normalizer $N(H)$ in $\mathcal{G}$ contains an orientation-reversed transformation.

Proof Assume that H and its transform H' do not form an enantiometric pair, then H' is contained in the orbit of H under $\mathcal{G}^+$ and thus there exists $g_0 \in \mathcal{G}^+$ such that $g_0^{-1}Hg_0 = H' = \sigma^{-1}H\sigma$. This shows that $\sigma \cdot g_0^{-1} \in N(H)$ and since $\sigma \cdot g_0^{-1} \notin \mathcal{G}^+$, $N(H) \not\subseteq \mathcal{G}^+$. On the other hand, if $N(H) \not\subseteq \mathcal{G}^+$, there exists $g_1 \in N(H)$ with $g_1 \notin \mathcal{G}^+$. We then have $g_1 \cdot \sigma \in \mathcal{G}^+$ and $(g_1 \cdot \sigma)^{-1}H(g_1 \cdot \sigma) = \sigma^{-1}(g_1^{-1}Hg_1)\sigma = \sigma^{-1}H\sigma = H'$, thus H' is contained in the orbit of H under $\mathcal{G}^+$. □

Let H be an arbitrary point group. If the normalizer $N(H)$ of the group H in the group of all symmetry transformations $\mathcal{G}$ includes an improper rotation (with determinant -1), then the rotations of group H can be represented by the same matrices in frames with different orientation and vice versa. In fact, suppose that in some frame $\mathcal{F}$ the rotations of group H are described by the matrices $\{E, A, B, \ldots\}$ and that C is the transformation $C \in N(H)$ with $\det(C) = -1$. Such transformation C takes frame $\mathcal{F}$ into the frame $\mathcal{F}'$ with opposite orientation. In $\mathcal{F}'$ the rotations of the group H are described by the matrices $\{E, C^{-1}AC, C^{-1}BC, \ldots\}$, and since $C \in N(H)$ the transformed set of matrices coincides with the initial one. Consequently, the rotations of group H in frames of different orientation are described by the same matrices. The converse is also true.

In an odd-dimensional space the normalizer of any point group contains an improper rotation (for example, one can take as such a transformation the reflection in a point). Therefore, enantiomorphic pairs of point groups do not exist in odd-dimensional spaces. One can easily verify that there are no enantiomorphic point groups in two-dimensional space as well because the normalizer of any two-dimensional symmetry group C_n always contains a reflection. The conjugation by reflection simply leads to reversing the direction of rotation.

8.4 Time reversal invariance

Depending on physical properties we are interested in one or another classification of lattices and crystals and it is important to find the most appropriate classification for a concrete subject under study. Many experiments deals with functions defined on the Brillouin zone and consequently it is important to know, in particular the symmetry properties of functions defined on the Brillouin zone. These properties are strongly related to the action of the space symmetry group on the Brillouin zone.

Let us recall briefly the definition of the Brillouin zone and the space group action on it.

We have defined L^*, the dual lattice of the lattice L as the set of vectors whose scalar products with all $\ell \in L$ are integers. In physics one prefers to consider the reciprocal lattice, which is $2\pi L^*$. This lattice is relevant to diffraction experiments (with X-rays, neutrons, electrons) with crystals possessing translation lattice L. It corresponds to the Fourier transform; the momentum variable is usually denoted by $\mathbf{k}$ and the vector space of $\mathbf{k}$'s is called the momentum or the reciprocal space. A unitary irreducible representation (unirrep) of the translation group is given by $\mathbf{k}(x) = \exp[i(\mathbf{k}\cdot\mathbf{x}]$. Here we are interested in the subgroup of the translation group R^d defined by the lattice of translations L. By restriction to L, two unirreps $\mathbf{k}$ and $\mathbf{k}'$ of R^d such that $\mathbf{k}' - \mathbf{k} \in 2\pi L^*$, yield the same unirrep of L. So the set $\hat{L}$ of inequivalent unirreps is

$$L \ni \ell \mapsto \mathbf{k}(\ell) = e^{i\mathbf{k}\cdot\ell}, \quad \hat{L} = \{\mathbf{k} \bmod 2\pi L^*\}. \tag{8.10}$$

Equivalently, with a choice of dual bases (see section 3.4)

$$\ell = \sum_j \mu_j \mathbf{b}_j. \quad \mathbf{k} = \sum_j \mathbf{b}_j^*, \quad \ell \mapsto e^{i\sum_j \mathbf{k}_j\mu_j}, \quad \mu_j \in Z, \quad k_j \bmod 2\pi. \tag{8.11}$$

The set $\hat{L}$ of the unirreps has the structure of a group, with the group law

$$\mathbf{k} \equiv \left(\mathbf{k}^{(1)} + \mathbf{k}^{(2)}\right) \bmod 2\pi L^* \Leftrightarrow k_j \equiv \left(k_j^{(1)} + k_j^{(2)}\right) \bmod 2\pi. \tag{8.12}$$

This group is called the *dual group* of L by mathematicians and the Brillouin zone ($=BZ$) by physicists. It is isomorphic to the group

$$\hat{l} = BZ \sim U_1^d. \tag{8.13}$$

We denote by $\hat{k}$ the elements of BZ in order to distinguish clearly between $\mathbf{k}$ and $\hat{k} =: \mathbf{k} \bmod 2\pi L^*$. The Bravais group P_L^z of L acts on BZ through its contragredient representation $\hat{P}_L^z =: (P_L^z)^{-1}$. More generally, since by definition of BZ the translation group acts trivially, a space group G acts through its quotient

$$G \xrightarrow{\rho} G/L = P^z. \tag{8.14}$$

So the space groups belonging to the same arithmetic class P^z have the same action. As usual, we denote by $G_{\hat{k}}$ the stabilizer in G of $\hat{k} \in BZ$ and $P_{\hat{k}}^z$ the stabilizer in P^z. The latter stabilizer depends only on the arithmetic class;

beware that for a given $\hat{k}$ the stabilizers $G_k = \theta^{-1}(P^z_{\hat{k}})$ for[3] the different space groups of the same arithmetic class P^z are, in general, non-isomorphic. Notice that the G_k's are also space groups.

Detailed analysis of the group action on the BZ is done in [75], chapters 4, 5. Here we discuss only the effect of the time reversal operation, $\mathcal{T}$. In classical Hamiltonian mechanics if, at a given instant one reverses the momenta, the trajectories are unchanged but they are followed in the reverse direction. That symmetry has been called *time reversal*, we denote it by $\mathcal{T}$. The fundamental contribution to time reversal representation in quantum mechanics is done by Wigner [98] who showed that $\mathcal{T}$ is represented by an anti-unitary operator. To see the effect of the time reversal on the space group action on BZ it is necessary to note that the change of sign of momenta transforms a unirrep of the group L into its complex conjugate. Taking into account that BZ is the set of inequivalent unitary irreducible representations of L we conclude that the change of sign of momenta corresponds on BZ to the transformation $\hat{k} \leftrightarrow -\hat{k}$. For simplicity we study here $\mathcal{T}$ only when the spin coordinates do not intervene explicitly. Time reversal invariance is a symmetry of many equilibrium states. As a consequence of that the real functions on BZ describing their physical properties, e.g. the energy function, must satisfy the relation $E(\hat{k}) = E(-\hat{k})$. The effect of this symmetry can be obtained by enlarging P^z, the group acting effectively on BZ with $-I_d$, when P_z does not already contain the symmetry through the origin. We denote this enlarged group by $\check{P}^z$, this is simply P^z for the 7, 24 arithmetic classes (for $d = 2, 3$) which contain the symmetry through the origin.

For two-dimensional systems adding time reversal decreases the number of arithmetic classes to study from 13 till seven (see Table 8.3). For three-dimensional systems the number of different arithmetic classes decreases from 73 till 24 (see Table 8.4).

8.5 Combining combinatorial and symmetry classification

We have seen in Chapter 6 that translation lattices can be characterized by the combinatorial type of their Voronoï parallelohedron. In its turn each combinatorial type of Voronoï cells can be additionally split into different symmetry classes (Bravais classes). Voronoï cells of the same combinatorial type can have different symmetry groups and moreover the same symmetry group can act differently on the face lattice of a given Voronoï cell.

The classification of Voronoï cells into combinatorial types gives for $d = 2$ only two combinatorially different polygons, a hexagon (which is generic or

[3] The map θ is not invertible, so θ^{-1} alone has no meaning; but it is an accepted tradition to denote $\theta^{-1}(P_k)$ the counter image of P_k by θ, i.e. the unique subgroup of G such that $\theta(G_k) = P^z_{\hat{k}}$.

TAB. 8.3 – Correspondence between $2d$-arithmetic classes before and after inclusion of time reversal invariance. Five arithmetic classes of the Bravais groups are indicated between ().

Class with inversion	Class without inversion
$(p2)$	$p1$
$(p2mm)$	pm
$(c2mm)$	cm
$p4$	
$(p4mm)$	
$p6$	$p3$
$(p6mm)$	$p3m1$, $p31m$

TAB. 8.4 – Arithmetic classes (in dimension 3) obtained by adding $-I_3$, corresponding to inclusion of time reversal invariance. The numbers at the left of arithmetic class show the number of space groups belonging to each of the 24 arithmetic classes in the case of time reversal symmetry. The 14 arithmetic classes of the Bravais groups are given between ().

2	$(P\bar{1})$	$P1$	5	$(R\bar{3}m)$	$R32, R3m$
8	$(P2/m)$	$P2, Pm$	4	$P\bar{3}$	$P3$
5	$(C2/m)$	$C2, Cm$	7	$P\bar{3}1m$	$P312$, $P31m$
30	$(Pmmm)$	$P222, Pmm2$	7	$P\bar{3}m1$	$P321, P3m1$
15	$(Cmmm)$	$C222, Cmm2, Amm2$	9	$P6/m$	$P6, P\bar{6}$
5	$(Fmmm)$	$F222, Fmm2$	18	$(P6/mmm)$	$P622, P6mm,$ $P\bar{6}m2, P\bar{6}2m$
9	$(Immm)$	$I222, Imm2$	5	$Pm\bar{3}$	$P23$
9	$P4/m$	$P4, P\bar{4}$	10	$(Pm\bar{3}m)$	$P432, P\bar{4}32$
40	$(P4/mmm)$	$P422, P4mm,$ $P\bar{4}2m, P\bar{4}m2$	3	$Fm\bar{3}$	$F23$
5	$I4/m$	$I4, I\bar{4}$	8	$(Fm\bar{3}m)$	$F432, F\bar{4}3m$
14	$(I4/mmm)$	$I422, P4mm,$ $I\bar{4}m2, I\bar{4}2m$	4	$Im\bar{3}$	$I23$
2	$R\bar{3}$	$R3$	6	$(Im\bar{3}m)$	$I432, I\bar{4}3m$

primitive) and a rectangle. For $d = 3$ there are five combinatorially different polytopes (see Chapter 6).

The splitting of a combinatorial type of 2D-lattice into Bravais classes is shown in Table 8.5 (see also Figures 8.1 and 6.13). It should be noted that Table 8.5 counts only those regions of the cone of positive quadratic forms which belong to a fundamental domain with respect to $GL_2(\mathbb{Z})$ action, or in other words to reduced quadratic forms. The $c2mm$ Bravais group appears twice in Table 8.5 because the fundamental domain of the $GL_2(\mathbb{Z})$ action

TAB. 8.5 – Splitting of combinatorial types of 2D-lattices into Bravais classes. The dimension of the region of the cone of positive quadratic forms is given in the last column. For the *c2mm* Bravais group two connected components are shown.

Bravais group	Hexagonal cell	Rectangular cell	Dimension
p6mm			1
p4mm		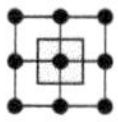	1
c2mm[a]	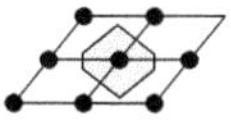		2
c2mm[b]	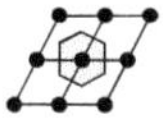		2
p2mm		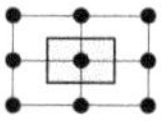	2
p2	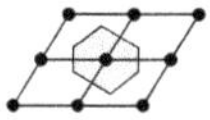		3

[a]Lattices with four shortest vectors (half of the diagonal is shorter than the sides of the rectangle).
[b]Lattices with two shortest vectors (half of the diagonal is longer than two sides of the rectangle).

on the cone of positive quadratic forms includes two connected components formed by hexagonal cells with *c2mm* symmetry (see Figure 8.1). In order to deform continuously the *c2mm* cell from one connected component into the *c2mm* cell belonging to another connected component, we need to construct a path which crosses at least *p6mm* stratum, or goes through a *p2* generic stratum. In dimension 2 the correspondence between combinatorial and symmetry classifications is rather simple. Each Bravais group is compatible with only one combinatorial type of the Voronoï cell.

Five combinatorial types of three-dimensional Voronoï cells are described in Table 8.6 (see also Figure 6.13). Note that to see the correspondence between the Delone notation used in Table 8.6 and the graphical representation used in Figure 6.13 it is sufficient to simply remove edges with black points from the Delone representation.

The systematic procedure of simultaneous analysis of combinatorial type and the Bravais symmetry type of the lattices in dimension 3 was realized by Delone on the basis of initial Voronoï studies (see also [32, 35]).

To characterize the three-dimensional lattice given by three translation vectors $\mathbf{a}, \mathbf{b}, \mathbf{c}$, Delone uses instead of the six standard parameters a^2, b^2, c^2, $g = (\mathbf{a} \cdot \mathbf{b})$, $h = (\mathbf{a} \cdot \mathbf{c})$, $k = (\mathbf{b} \cdot \mathbf{c})$, the ten parameters associated with vectors $\mathbf{a}, \mathbf{b}, \mathbf{c}$ and $\mathbf{d} = -(\mathbf{a} + \mathbf{b} + \mathbf{c})$. These parameters, are : the squares of the lengths of vectors $\mathbf{a}, \mathbf{b}, \mathbf{c}, \mathbf{d}$, denoted by a^2, b^2, c^2, d^2, and their scalar products

TAB. 8.6 – Three dimensional Voronoï cells. Column 1 gives the Delone symbol (see text). Column 2: the dimension of their domain in the cone of positive quadratic ternary forms, $\mathcal{C}_+(\mathcal{Q})_3$. Column 3: the number of hexagonal and quadrilateral faces. Columns 4, 5, 6: $|F|$ the total number of faces, $|E|$ the number of edges, $|V|$ the total number of vertices. Column 7: the number of vertices of valence 3 and 4. Column 8: $|C|$ the number of corona vectors. Column 9: the number of shortest vectors in each of the 7 non trivial $L/2L$ cosets.

Delone	dim	6 - 4	$\lvert F\rvert$	$\lvert E\rvert$	$\lvert V\rvert$	3 - 4	$\lvert C\rvert$	$L/2L$
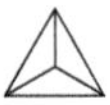	6	8 - 6	14	36	24	24 - 0	14	2 2 2 2 2 2 2
	5	4 - 8	12	28	18	16 - 2	16	2 2 2 2 2 2 4
	4	0 - 12	12	24	14	8 - 6	18	2 2 2 2 2 2 6
	4	2 - 6	8	18	12	12 - 0	20	2 2 2 2 4 4 4
	3	0 - 6	6	12	8	8 - 0	26	2 2 2 4 4 4 8

g, h, k, l, m, n introduced earlier by Selling and used to describe reduced forms. These 10 parameters are naturally linearly dependent. The sum of numbers in one line of the following table

	a	**b**	**c**	**d**
a	a^2	k	h	l
b	k	b^2	g	m
c	h	g	c^2	n
d	l	m	n	d^2

is zero. The advantage of using these ten parameters is the clear visualization of different combinatorial and symmetry types of Voronoï cells.

A general Delone symbol 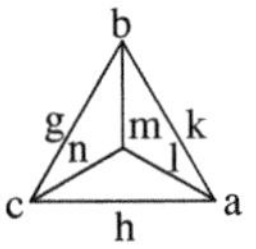 could be imagined as a projection of a tetrahedron $ABCD$ with vertices corresponding to ends of the vectors $\mathbf{a}, \mathbf{b}, \mathbf{c}, \mathbf{d}$. The edges are labeled by numbers g, h, k, l, m, n. If one thinks of this symbol as a three-dimensional model of a tetrahedron its vertices and edges turn out to be equivalent.

Delone has shown that 24 sorts of lattices exist. They are nowadays referenced as 24 Delone sorts of lattices. Without going into details of mathematical justifications (see [42, 73]) we summarize here just the main results explaining the graphical representation of Delone symbols for lattices of different combinatorial type and of different symmetry.

Different combinatorial types of Voronoï cells are described by Delone symbols with 0, 1, 2, or 3 zeros on the edges of the Delone symbol. It is not possible to put 4 zeros because any quadratic form in $\mathcal{Q}_3$ with only two λ's has a zero determinant. For the same reason it is not possible to put 3 zeros on 3 edges with a common vertex. To have only five different possibilities for the distribution of zeros on the edges (as table 8.6 shows where zero on the edge is symbolized by a black dot) it is sufficient to check that two possible distributions of three zeros give the same combinatorial type, namely a rectangular parallelepiped.

In order to characterize Bravais symmetry it is sufficient to indicate the Delone symbol edges equivalent by symmetry. This is typically done by putting the same number of dashes on equivalent edges. Table 8.7 gives the complete description of 24 sorts of Delone lattices through their Delone symbols and the distribution of Delone sorts into combinatorial and symmetry (Bravais) types.

From Table 8.7 it follows that nine Bravais classes are compatible each with only one combinatorial type of Voronoï cell. On the other hand, for the $C2/m$ Bravais group there are two different Delone sorts of the combinatorial 14-24 type and two Delone sorts of the combinatorial 12-14 type. This illustrates the existence of two alternative C_{2h} group actions on the same combinatorial type of the Voronoï cell.

With increasing dimension the number of combinatorial types of Voronoï cells increases rapidly, as well as the number of Bravais classes. Thus the detailed classification performed by Delone for three dimensional lattices and even more simpler classifications into individual combinatorial types or into symmetry types can become unrealizable or even unutilizable because of extremely large number of members. The reasonable classification should be based on new more crude invariants and types or on statistical distributions over different types of lattices from one side and on the description of some extremal types of lattices (for example with maximal symmetry).

TAB. 8.7 – A list of the Delone symbols describing the Voronoï cells of the lattices belonging to a Bravais class. The combinatorial description of the Voronoï cell is given by the symbol $|F| - |V|$ indicating the number of facets $|F|$ and the number of vertices $|V|$. The first column lists the Bravais classes. The last column gives the dimensions up to dilation, of the different domains of cells in the cone of positive quadratic forms $\mathcal{C}_+(\mathcal{Q}_3)$.

Voronoï	14-24	12-18	12-14	8-12	6-8	dim.
Cubic P						0
Cubic F						0
Cubic I						0
Hexa P						1
Trigo R						1, 1
Tetra P						1
Tetra I						1, 1
Ortho P						2
Ortho C				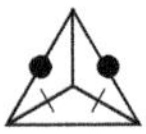		2
Ortho F	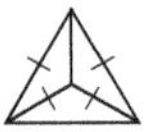					2
Ortho I						2, 2, 1
Mono P				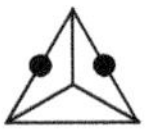		3
Mono C	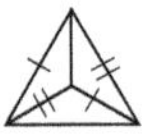					3, 3, 2
Mono C	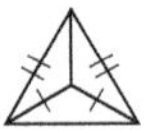					3, 2
Tricli P	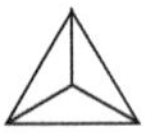					5, 4, 3

Chapter 9

Applications

Lattices appear naturally in rather different domains of natural science and formal mathematics. The goal of the present chapter is to discuss briefly several examples of problems which are tightly related with the lattice constructions, lattice classifications, and use lattices as an initial point for more elaborated mathematical and physical models and processes.

9.1 Sphere packing, covering, and tiling

One of the most simply formulated practical problem leading to the study of lattices is the classical problem of packing spheres (or balls). We can think about canon balls or about oranges of the same dimension and try to find the packing that maximizes the density assuming that the dimension of the box to pack the balls is infinitely bigger than the ball dimension. To make this "practical" problem more mathematically sound we can generalize it to an arbitrary dimension and to look for solutions for more restricted problem by imposing the periodicity condition on packing (lattice packing) and more general packing without periodicity.

The solution of this problem is trivial in dimension 1 (see Figure 9.1). One-dimensional spheres ($\sim$ intervals) fill completely one-dimensional space (line).

The solution for the dimension two is also simple (we need to pack disks on the plane, see Figure 9.2). Each disk can be surrounded by six neighboring discs. Continuing this local packing we get the hexagonal lattice which is the densest packing of the 2-D discs.

The density of hexagonal packing can be easily calculated by noting that for discs of radius R, each elementary cell is a rhomb with diagonals equal 2R and $2\sqrt{3}R$. The area occupied by the disk in each elementary cell is equal exactly to the area of one disk, πR^2 (two sectors of $2\pi/6$ and two sectors of $2\pi/3$). The area of the elementary cell is $2\sqrt{3}R^2$. Thus the density is $\pi/(2\sqrt{3}) \approx 0.9069$.

FIG. 9.1 – Densest packing of 1-dimensional spheres on a line.

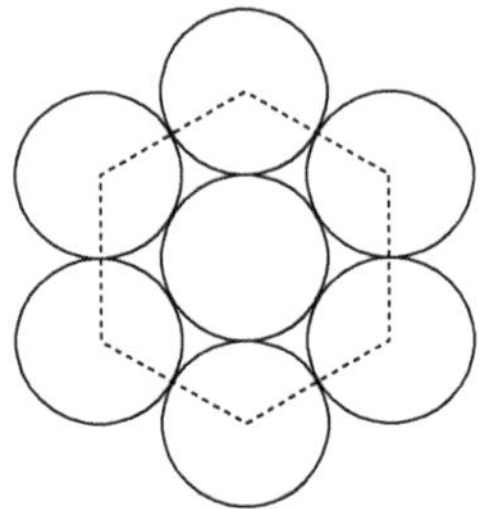

FIG. 9.2 – The most dense packing of two-dimensional spheres (discs) on a plane is a hexagonal lattice packing.

In dimension three the problem of sphere packing is less trivial. The origin of the difficulty can be easily understood if we take one ball and try to put around it the maximal number of identical balls touching it. It is easy to check that it is possible to put 12 balls in contact with one ball but there is still some free space between 12 balls and they can move rather freely being always in contact with the central ball. It is not easy to prove that it is impossible to put the thirteenth ball in contact with the central one. The contact number (i.e. the maximal number of balls which can be put in contact with one ball) is not known for the majority of dimensions $d \geq 4$. It is known that contact numbers in dimensions 8 and 24 are respectively 240 and 196560. These solutions are known because in dimensions 8 and 24 the arrangement of balls around one central ball is unique. These arrangements correspond to the lattice E_8 and to one of the forms of the 24-dimensional Leech lattice.

At the same time it is easy to suggest the packing for 3-D balls (in fact even infinity different versions) which can be thought to be the densest packing. We can start with one layer of balls forming a hexagonal 2D-lattice. Then the next layer can be posed in such a way as to put balls in cavities of the second layer, and so on .... As soon as the number of cavities is twice the number of balls there are two ways to position the next layer. The periodic structure with the shortest period corresponds to the sequence of layers $ABAB\ldots$ This packing is named the hexagonal close packing. The periodic packing of the form $ABCABC\ldots$ corresponds to the structure named face-centered cubic lattice (see Figure 9.3). The density of all packings corresponding to any sequence (periodic or not) of hexagonal layers is $\pi/\sqrt{18}$. Each ball in these packings has 12 neighbors. Although there exist a number of different proofs that the mentioned above packings are the densest ones among lattice packings, only

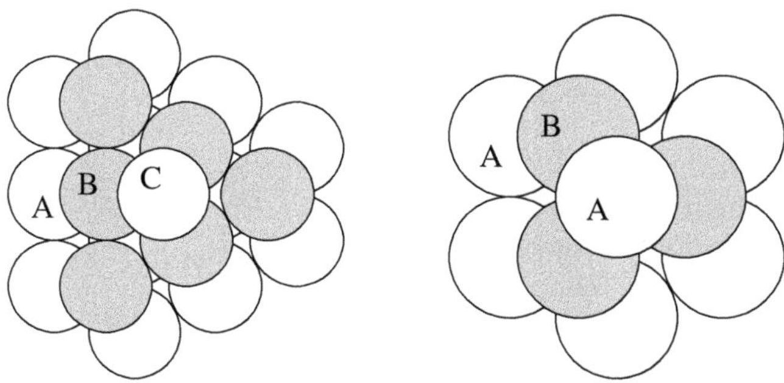

FIG. 9.3 – ABC and ABA packing of hexagonal layers of balls.

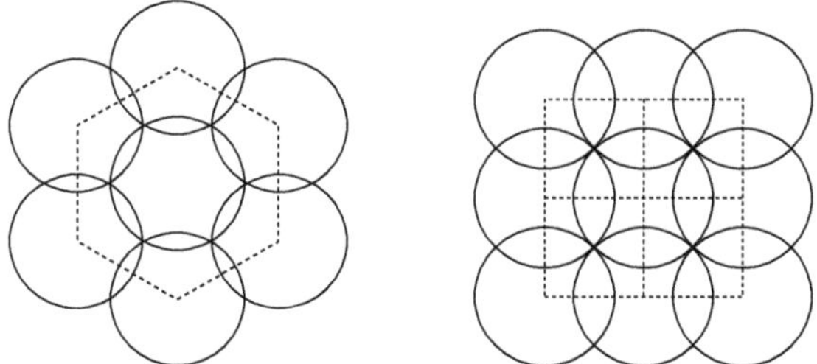

FIG. 9.4 – Covering plane by discs. Center of discs form hexagonal (left) and square (right) lattice. Hexagonal covering is less dense than the square lattice covering.

recently has a computer assisted proof appeared that this statement remains valid for arbitrary non-lattice packings in three-dimensional space.

Nowadays, the solution for the densest packing of spheres is known in many dimensions. The density of the known densest packing varies with dimension in rather irregular fashion. It is also not clear in advance what kind of lattice corresponds to the densest packing for a given dimension.

A problem tightly related to packing is the covering by spheres. Now it is necessary to find the arrangement of overlapping spheres covering the whole space and having the lowest density. The answer is again trivial for the one-dimensional problem. For the two-dimensional problem the hexagonal lattice gives again the best solution (the lowest density) for the covering problem. Figure 9.4 shows the comparison of the coverings obtained for the square lattice and for the hexagonal lattice. For the hexagonal lattice the overlapping of spheres is smaller and the density of covering is lower, namely $2\pi/(3\sqrt{3}) \approx 1.2092$ whereas for the square lattice the density of covering is $\pi/2 \approx 1.5708$.

For three-dimensional lattices the lowest density covering is given by a body-centered lattice, in spite of the fact that the densest sphere packing is associated with another, face-centered cubic lattice.

TAB. 9.1 – Classes of the symmetry groups H of mesomorphic phases of matter: $[E(3) : H]$ compact, H_0 = largest connected subgroup of H, $T_H = H \cap T$, where T is the translation subgroup of $E(3)$.

Class	T_H	H_0
Ordinary nematics	R^3	$R^3 \times U(1)$
Exceptional nematics	R^3	R^3
Cholesterics (chiral)	$R^2 \times Z$	R^3
Smectics A	$R^2 \times Z$	$R^2 \times U(1)$
Smectics C	$R^2 \times Z$	R^2
Chiral smectics C	R^2	R^2
Rod lattices (e.g., lyotropics)	$R \times Z^2$	R
Crystals	Z^3	$\{1\}$

The problem of ball packing can be formulated in a much wider sense than simply as a problem of the densest sphere packing. From the point of view of the description of packing of atoms or molecules in crystals it is natural to ask about regular or lattice packing of balls which are stable in a certain sense (see [11, 5]).

9.2 Regular phases of matter

We want to discuss here briefly the relation of lattices to the classification of different phases of matter, which is more general than just the classification of crystals. In fact, a simultaneous discussion of different mesomorphic phases of matter was suggested by G. Friedel in 1922 [55]. He suggested to treat both crystals and liquid crystals on the basis of symmetry arguments. We follow here the description of the mesomorphic phases of matter done by Louis Michel in [71] on the basis of the symmetry breaking scheme applied to $E(3)$, the three-dimensional Euclidean group. The idea of this classification is to describe the possible stabilizers (little groups in physical terminology) of transitive states. The equilibrium states of matter are associated with the symmetry group which is a subgroup H of $E(3)$. The classes of the symmetry groups H of mesomorphic states of matter are listed in Table 9.1.

Symmetry groups H are defined up to conjugation. When H is discrete, the phase is a crystal. The characteristic lengths of the crystal is not of importance for physical applications, but the difference between left-handed and right-handed crystals can be eventually important for certain physical properties. This is the reason to classify crystals up to a conjugation in the connected affine group (see section 8.3). This yields 230 crystal symmetries. The same classification principle leads to an infinity of other H subgroups. They can be put in families according to the topology of their largest connected subgroup H_0 and their intersection $H \cap T = T_H$ with the translation subgroup of $E(3)$. These broad classes are listed in Table 9.1.

Short description of the most important other mesomorphic phases is as follows.

In nematics, the molecules are aspherical; their positions are distributed at random as in a liquid, but they are aligned. In ordinary nematics H is the semi-direct product $R^3 \wedge D_{\infty h}$. This means that the orientation of the molecules causes them to yield only axially symmetric quadrupole effects even when the molecules have no axial symmetry. Near the solidification temperature, molecules with strong deviation from axial symmetry may rotate less easily and exceptional nematics can be observed (e.g. birefringent quadrupoles with three unequal axes).

Cholesterics are constructed of polar molecules; their symmetry group H contains all the translations in a plane and, with a perpendicular axis, a continuous helicoidal group. They appear frequently in biological tissues.

In smectics the molecules are distributed in parallel monomolecular or bimolecular layers, and they are aligned either perpendicularly (smectics A) or obliquely (smectics C) to the layers. In chiral smectics C inside each layer the polar molecules are oriented with a constant oblique angle, but the azimuth of this orientation turns by a constant angle θ from one layer to the next and two different subclasses are possible depending on whether θ/π is rational or not.

The classification of mesomorphic phases of matter described above is based on the spatial distribution of atomic positions with each atom being represented as a point in real physical space. Naturally, the points representing the localized atoms in space are associated with heavy atomic nuclei (eventually together with some internal electrons), whereas (outer) electrons are distributed in space in the presence of the lattice formed by localized atomic cores.

From the physical point of view it is quite interesting and important to find if there are some more general restrictions which allows us to introduce some universality classes of matter which persist even if periodicity is broken. It is possible to look for such criteria which are due to global topological effects (invariants) which cannot be removed under small deformation breaking symmetry. Classification of universal classes of topological states of matter takes into account the global symmetries like time reversal, charge conjugation, and their combination. The origin of particles themselves (fermions or bosons) is equally important. This subject has become very popular now due to the discovery of such new topological phases of matter as topological insulators or topological superconductors [27].

9.3 Quasicrystals

We cannot avoid to mention the application of lattice geometry to study quasicrystals, or aperiodic regular structures. This is mainly due to the fact that aperiodic crystals can be naturally described as projections of a higher dimensional periodic structure to a subspace of lower dimension. In order to obtain an aperiodic structure the subspace on which such a projection is realized should be irrational with respect to the lattice vectors of the initial periodic structure. Such a construction justifies the interest in the study of higher dimensional periodic structures but creates at the same time a lot of questions about the relevance of the choice of the dimension and of the orientation of the subspace to project the structure. We do not enter in this very popular domain which has a lot of applications not only in the analysis of quasicrystals (fully recognized as an important class of physical systems by awarding the Nobel prize for their discovery in 2011) but in various different branches of physics and mathematics, including chaotic dynamical systems, singularity theory, etc. For an introduction to quasicrystals and related mathematical domains see [17, 24].

9.4 Lattice defects

The classification of the mesomorphic states of matter uses an idealization that the ordered phase of matter is extended indefinitely in space in order to be globally invariant under an allowed subgroup H of $E(3)$. This idealization is not bad if the real sample under study is large enough (as compared to the size of the unit cell) so that its symmetry can be recognized. But, in nature, samples are not only limited in size, but they also can be non-perfect, i.e. they can have defects.

Application to physically real objects of lattice theory is related to the description and classification of typical defects and boundaries. The first step in defect description should include the description of so called topologically stable defects, which persist in the medium even under small (local) deformation.

A very intuitive and visual description of possible defects in regular (periodic) physical materials (crystals, liquid crystals) is based on the "cut and glue" construction of defects for regular lattices.

We give below several examples of such defect constructions. The simplest defect is a vacancy which corresponds to removing one vertex of the lattice without qualitatively disturbing the surrounding (see Figure 9.5, left). This means that testing the lattice locally in any region outside of a small neighborhood of a vacancy we cannot notice the presence of the defect.

A more complicated defect, linear dislocation, is shown in Figure 9.5, center and right. To construct such a defect we remove (we can also insert) one ray of points (eventually several parallel rays) and glue the two boundaries of

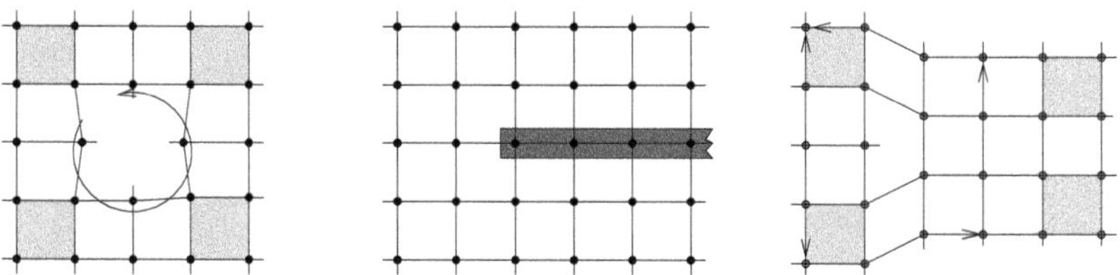

FIG. 9.5 – Lattice with a vacation (left). Construction of a linear dislocation (center). Lattice with a linear dislocation (right).

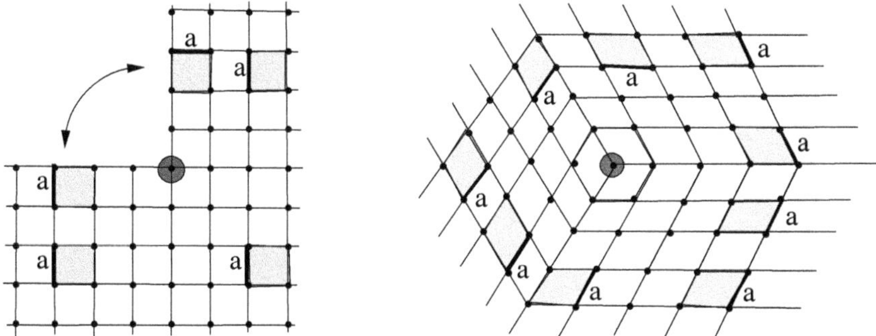

FIG. 9.6 – Construction of the rotational disclination by removing the solid angle $\pi/2$ shown on the left picture.

the cut by a parallel translation along the transversal direction. For the two-dimensional lattice the linear dislocation results in a point (codimention-2) defect. But the evolution of the elementary cell along a closed path surrounding the defect does not modify the elementary cell. In order to characterize the linear dislocation we can introduce the Burgers vector which characterizes what happens with the closed contour chosen on the initially perfect lattice after constructing the dislocation.

The next important defect of the regular lattice is the rotational disclination. We can get it by removing (or adding) an angular wedge from the regular lattice and then joining the two boundaries by rotating them. Examples of such a construction of $\pi/2$ and π rotational disclinations are shown in Figures 9.6 and 9.7. The effect of the evolution of the elementary cell along a closed path surrounding the rotational disclination consists in rotation of the elementary cell by an angle associated with rotational disclination.

We need to distinguish rotational disclination from the angular dislocation shown in Figure 9.8. Angular dislocation is less typical as a defect of real crystals but it turns out to be of primary importance in integrable dynamical

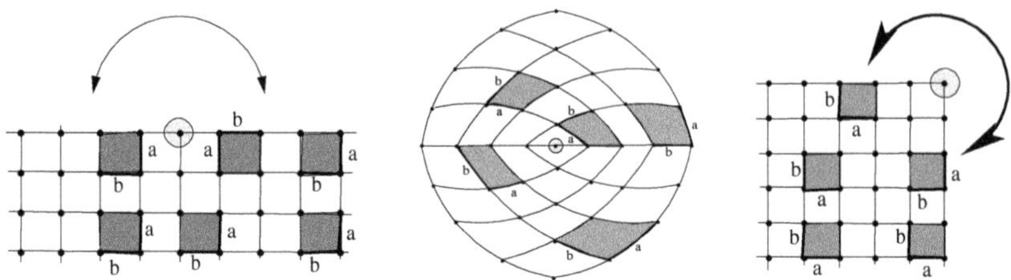

FIG. 9.7 – Construction of the rotational disclination by removing the solid angle π (Left) and $3\pi/2$ (Right). The reconstructed lattice after removing the π solid angle (center).

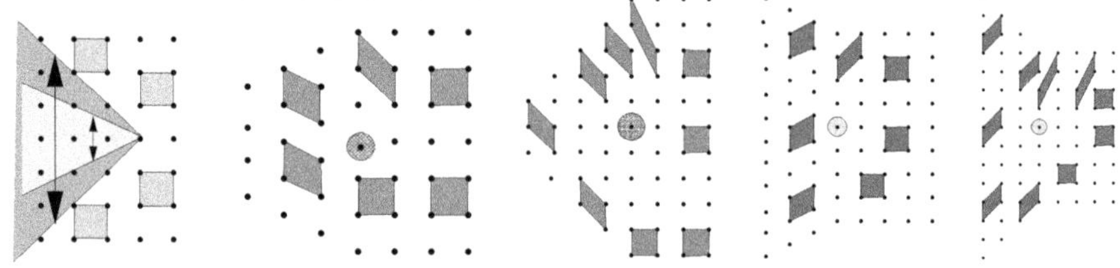

FIG. 9.8 – Construction of the angular dislocation by removing or adding one of the solid angles shown on the left picture. Reconstructed lattices after removing or adding small or large sectors are shown together with transport of the elementary cell along a closed path around the defect on the reconstructed lattice. The identification of boundaries after removing or adding a solid angle is done by the parallel shift of lattice points in the vertical direction.

systems as a defect of regular lattices associated with focus-focus singularities (see next subsection).

9.5 Lattices in phase space. Dynamical models. Defects.

Lattices appear naturally not only in the configuration space, as localized positions of atoms or more complicated particles. We turn now to dynamical systems, in particular to Hamiltonian systems. The basic object of our study is the phase space formed by conjugated position and momentum variables. The notion of integrable classical system leads to the appearance of toric fibrations.

The most evident appearance of lattices is associated with quantization of classical Hamiltonian integrable systems. Integer values of actions correspond to quantum states forming local lattices of quantum states.

Integrable problem in classical mechanics and corresponding quantum problems are very special to be associated directly with concrete physical systems. Certain qualitative features of integrable classical problems inherited also by quantum systems remain valid after small deformation because of their topological origin. This justifies the study of integrable systems from the point of view of further analysis of generic (non-integrable) systems.

To see the relation of regular lattices and their defects to dynamical systems we can start with one-degree of freedom problem. The Hamiltonian system describing the motion of a particle in a one dimensional potential can be imagined for simplicity as harmonic or slightly anharmonic oscillator.

Near the minimum of the potential the classical phase portrait shown in Figure 9.9, e can be topologically described as a system of circles (Figure 9.9, a), fibered over an interval and a singular fiber, a point, associated with the boundary point, the minimum. The corresponding system of quantum levels is a sequence of points which can be deformed to a regular one-dimensional lattice (with a boundary) associated to the harmonic oscillator (9.9, b). Small deformations of the one-dimensional problem cannot change qualitatively neither classical fibration, nor the lattice of quantum states. Qualitative modification of classical fibration is related with bifurcation of the phase portrait associated with the appearance of new stationary points on the energy surface. Subfigure 9.9, f shows qualitatively new phase portrait after the bifurcation associated with the formation of two new stationary points and the separatrix. Classical fibration 9.9, c now has a singular fiber (associated with a separatrix) and three regular regions, associated with locally defined lattices.

Completely integrable classical Hamiltonian for a two-degree of freedom system can be represented by its two-dimensional energy momentum (EM) map each regular point of which is associated with a regular T^2 fiber. The corresponding quantum system is characterized by the joint spectrum of two mutually commuting integrals of motion. In the case of the two-dimensional isotropic harmonic oscillator (see Figure 9.10) two integrals of motion can be chosen as the energy E, which is the eigenvalue of the Hamiltonian H and the projection of angular momentum m which is the eigenvalue of L_z.

$$H = \frac{1}{2}(p_1^2 + q_1^2) + \frac{1}{2}(p_2^2 + q_2^2), \tag{9.1}$$

$$L_z = p_1 q_2 - p_2 q_1. \tag{9.2}$$

Their joint eigenvalues form a regular two-dimensional lattice bounded by two rays.

Along with special fibers associated to boundary lines of the energy-momentum map, it is possible that integrable fibrations have also singular fibers inside the energy momentum map. Typical images of energy momentum

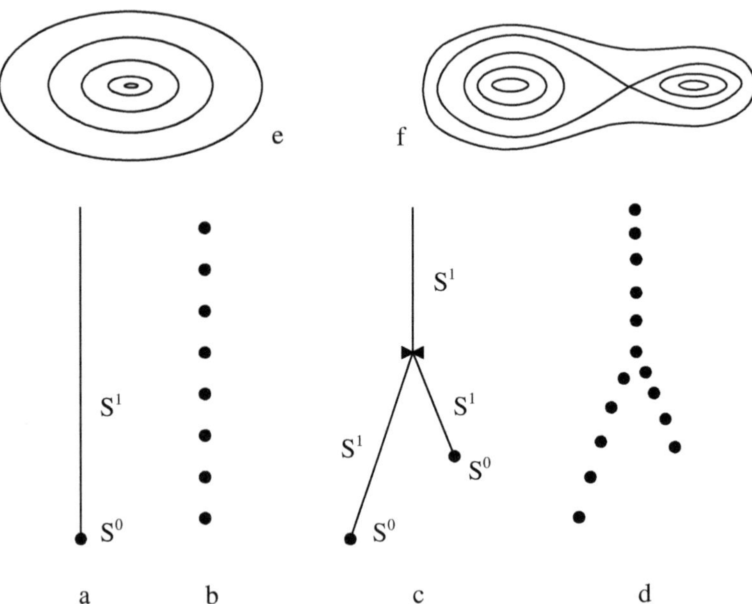

FIG. 9.9 – Classical and quantum bifurcations for the one degree of freedom system. Situations before (a,b,e) and after (c,d,f) the bifurcation are shown. (a) Energy map for a harmonic oscillator type system. Inverse images of each point are indicated. (b) Quantum state lattice for a harmonic oscillator type system. (c) Energy map after the bifurcation. Inverse images of each point are indicated. (d) Quantum state lattice after bifurcation represented as composed of three regular parts glued together. (e) Phase portrait for a harmonic oscillator type system. Inverse images are S^1 (generic inverse image) and S^0 - inverse image for a minimal energy value. (f) Phase portrait after bifurcation.

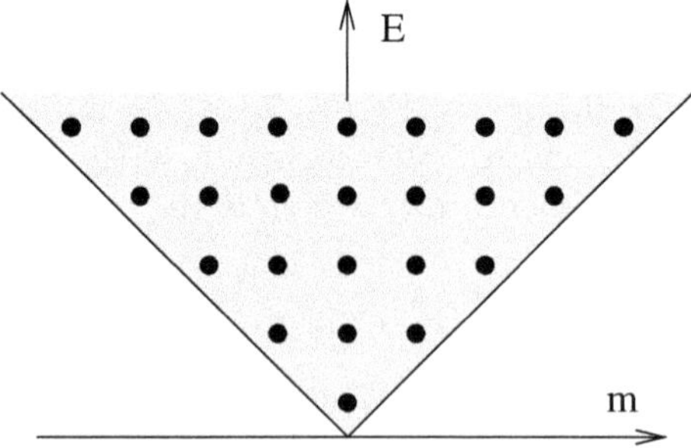

FIG. 9.10 – Joint spectrum of two commuting operators (9.1, 9.2) together with the image of a classical EM map for a two-dimensional isotropic harmonic oscillator.

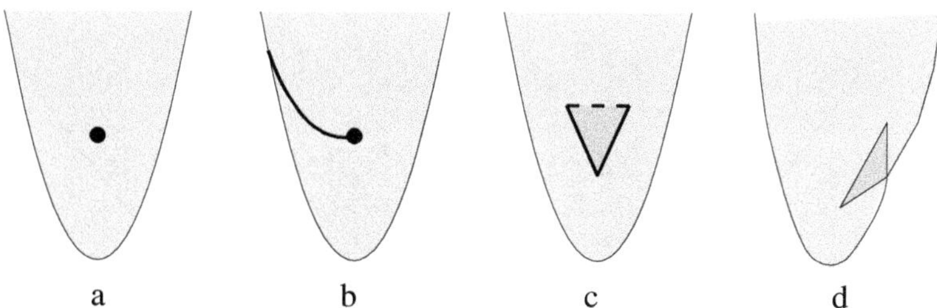

FIG. 9.11 – Typical images of the energy momentum map for completely integrable Hamiltonian systems with two degrees of freedom in the case of: (a) integer monodromy, (b) fractional monodromy, (c) nonlocal monodromy, and (d) bidromy. Values in the lightly shaded area lift to single 2-tori; values in darkly shaded area lift to two 2-tori.

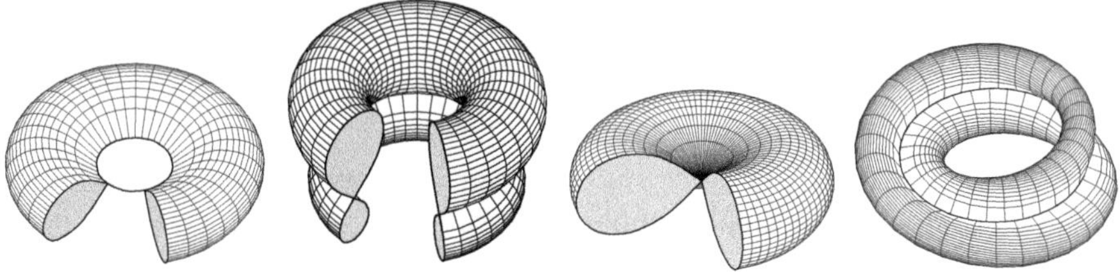

FIG. 9.12 – Two dimensional singular fibers in the case of integrable Hamiltonian systems with two degrees of freedom (left to right): singular torus, bitorus, pinched and curled tori. Singular torus corresponds to critical values in Figure 9.11 (c,d), (ends of the bitorus line). Bitorus corresponds to critical values in 9.11 (c,d), which belong to the singular line (fusion of two components). Pinched torus correspond to the isolated focus focus singularity in Figure 9.11 (a). Curled torus is associated with the critical values at the singular line in Figure 9.11 (b), (fractional monodromy).

maps possessing singular fibers for the two-degree-of-freedom Hamiltonian systems are shown in Figure 9.11. Visualization of singular fibers is given in Figure 9.12. The presence of singular fibers can be considered for classical fibration as a singularity which naturally influences the regular character of the fibration. For corresponding quantum systems regular regions of classical fibration correspond to locally regular lattices of common eigenvalues of mutually commuting operators. Singular fibers result in formation of defects of lattices of common eigenvalues.

For integrable systems with two-degrees of freedom the simplest codimension two singularity of the energy momentum map is the so called focus-focus point associated with a pinched torus. Its manifestation on the joint spectrum lattice for the corresponding quantum problem is shown in Figure 9.13 on the

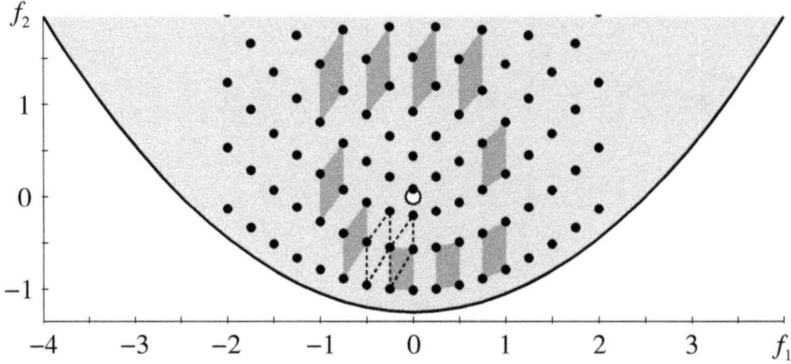

FIG. 9.13 – Joint spectrum of two commuting operators together with the image of the classical EM map for the resonant $1 : (-1)$ oscillator [80]. Quantum monodromy is seen as a result of transportation of the elementary cell of the quantum lattice along a closed path through a non simply connected region of the regular part of the image of the EM map.

example of the resonant $1 : (-1)$ oscillator. The two mutually commuting integrals of motion for this example are given by

$$f_1 = \frac{1}{2}(p_1^2 + q_1^2) - \frac{1}{2}(p_2^2 + q_2^2), \tag{9.3}$$

$$f_2 = p_1 q_2 + p_2 q_1 + \frac{1}{4}(p_1^2 + q_1^2 + p_2^2 + q_2^2)^2. \tag{9.4}$$

It is clear that outside a small neighborhood of a codimension-2 defect the lattice of common eigenvalues remains regular, i.e. it can be transformed (within a local simply connected region of the image of the energy momentum map) to a simple square lattice by an appropriate choice of variables (local actions). At the same time the existence of a singularity imposes that along a closed path surrounding the singularity the unique choice of action variables does not exist. The evolution of the elementary cell of the local lattice along a path surrounding the singularity leads to a new choice of local action variables. Transformation between initial and final choices of local action variables is named a quantum monodromy. The type of quantum monodromy depends on the type of singularity of integrable classical fibration. The simplest singularity of classical integrable fibration, i.e. singly pinched torus, corresponds to transformation of the basis of the elementary cell of the quantum lattice by the matrix M

$$M = \begin{pmatrix} 1 & 0 \\ 1 & 1 \end{pmatrix}, \tag{9.5}$$

which is defined up to the $SL(2, Z)$ transformation.

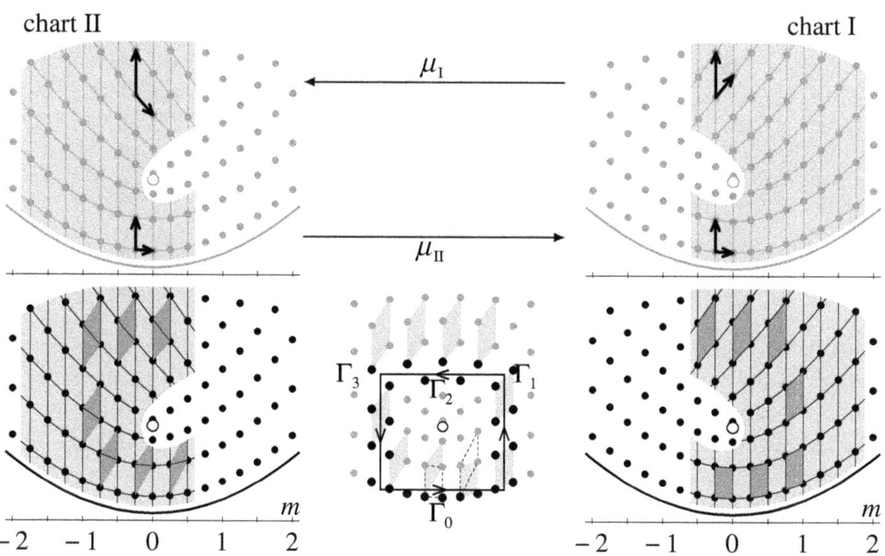

FIG. 9.14 – Two chart atlas which cover the quantum lattice of the 1 : (−1) resonant oscillator system represented in Figure 9.13. Top plots show the choice of basis cells and the gluing map between the charts. Bottom plots show the transport of the elementary cell (dark grey quadrangles) in each chart. Central bottom panel shows the closed path Γ and its quantum realization (black dots) leading to nontrivial monodromy (compare with Figure 9.13).

A possible choice of two overlapping simply connected charts with associated evolution of elementary cells for each chart is used in Figure 9.14 to explain the appearance of quantum monodromy for a lattice with a defect. Among the different possible visualizations of such simple-monodromy defect the most natural is that represented in Figure 9.15. Its construction is similar to that used for the "angular dislocation defect" shown in Figure 9.8. The idea of the construction of the defect is as follows. We cut from the regular lattice a wedge shown in Figure 9.14, left, and identify points on the two boundary rays of the cut. The wedge is chosen in such a way that the number of removed points from the lattice is a linear function of the integral of motion. After identification of the boundaries of the cut the reconstructed lattice remains regular except in the neighborhood of a singular point and is characterized by a quantum monodromy matrix (9.5).

Along with codimension-2 singularities classical fibrations for integrable dynamical systems have codimension-1 singularity lines. Such singularity is associated, for example, with a curled torus (see Figure 9.12) and can be studied on a concrete example of the two-dimensional nonlinear 1 : (−2) resonant oscillator.

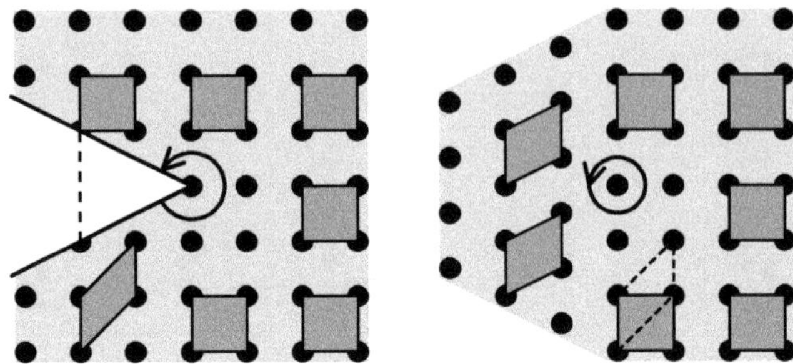

FIG. 9.15 – Construction of the 1:(−1) lattice defect starting from the regular Z^2 lattice. The solid angle is removed from the regular Z^2 lattice and points on the so obtained boundary are identified by vertical shifting. Dark grey quadrangles show the evolution of an elementary lattice cell along a closed path around the defect point.

Two integrals of motion for this problem are given by

$$f_1 = \frac{\omega}{2}(p_1^2 + q_1^2) - \frac{2\omega}{2}(p_2^2 + q_2^2) + R_1(q,p), \tag{9.6}$$

$$f_2 = Im[(q_1 + ip_1)^2(q_2 + ip_2)] + R_2(q,p). \tag{9.7}$$

Here R_i are higher order terms which ensure the compactness of the subspaces with fixed energy. The corresponding image of the energy momentum map together with the lattice of the joint quantum spectrum are shown in Figure 9.16.

The new qualitative feature which appears with this example is the possibility to define a generalization of quantum monodromy in case when the closed path on the image of the energy-momentum map crosses the line of singularities. The construction of the defect by the cutting and gluing procedure of a regular lattice is shown in Figure 9.17. The key point now is the possibility to go to sublattice (of index two in this concrete case) and to show that it is possible to define what happens with the elementary cell when crossing the line of singularities. At the same time being in regular region, it is possible to return to the original elementary cell. The monodromy matrix written in this case for an elementary initial cell includes fractional entries. That is why the corresponding qualitative feature was named fractional monodromy.

9.6 Modular group

In order to see the relation between lattices and functions of complex variables let us remember that an elliptic function is a function f meromorphic

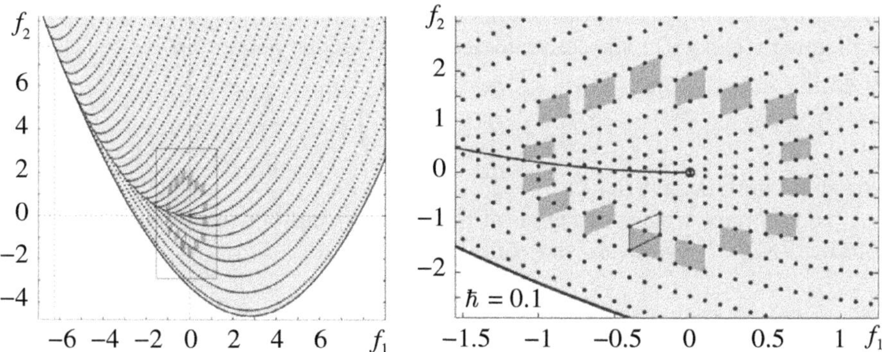

FIG. 9.16 – Joint quantum spectrum for two-dimensional nonlinear 1 : (−2) resonant oscillator [80]. The singular line is formed by critical values whose inverse images are curled tori shown in Figure 9.12. In order to get the unambiguous result of the propagation of the cell of the quantum lattice along a closed path crossing the singular line, the elementary cell is doubled.

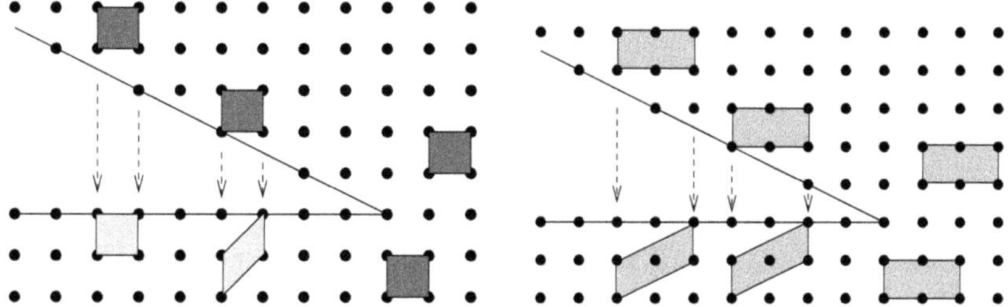

FIG. 9.17 – Representation of a lattice with a 1 : 2 rational defect by cutting and gluing. Left: The elementary cell goes through cut in an ambiguous way. The result depends on the place where the cell crosses the cut. Right: Double cell crosses the cut in an unambiguous way.

on $\mathbb{C}$ for which there exist two non-zero complex numbers ω_1 and ω_2 with $\omega_1/\omega_2 \notin \mathbb{R}$, such that $f(z) = f(z+\omega_1)$ and $f(z) = f(z+\omega_2)$ for all z. Denoting the "lattice of periods" by $\Lambda = \{m\omega_1 + n\omega_2 \mid m, n \in \mathbb{Z}\}$, it follows that $f(z) = f(z+\omega)$ for all $\omega \in \Lambda$. The complex numbers (ω_1, ω_2) generating the period lattice are defined up to $SL(2, Z)$ transformation, like quadratic forms or bases of two-dimensional lattices. Note, that for two-dimensional real lattices the group describing the transformation of bases is often extended by including reflections. In such case the group is $GL_2(\mathbb{Z})$, which includes integer 2 by 2 matrices with determinant ± 1. In complex analysis the holomorphic transformations includes only those with positive determinant, whereas transformations with negative determinant are anti-holomorphic. This means that under holomorphic transformations a pair of complex numbers (vectors) α_1, α_2 will generate exactly the same lattice as the lattice generated by ω_1, ω_2 if and only if

$$\begin{pmatrix} \alpha_1 \\ \alpha_2 \end{pmatrix} = \begin{pmatrix} a & b \\ c & d \end{pmatrix} \begin{pmatrix} \omega_1 \\ \omega_2 \end{pmatrix} \tag{9.8}$$

for some matrix in $SL_2(\mathbb{Z})$.

To compare with more general $GL(2,R)$ group, let us consider the action of the $GL(2,R)$ group on the complex plane $z \in \mathbb{C}$

$$z \in \mathbb{C}, \quad g = \begin{pmatrix} a & b \\ c & d \end{pmatrix} \in GL(2,R); \quad g \cdot z = \frac{az+b}{cz+d} \in \mathbb{C}. \tag{9.9}$$

We verify easily that

$$-I_2 \cdot z = z, \quad \Im(g \cdot z) = \frac{\det g}{|cz+d|^2} \Im(z). \tag{9.10}$$

These transformations show that the upper half part, $\mathcal{H}$, of the complex plane is invariant under transformations by $SL_2(\mathbb{R})$ matrices with a positive determinant. If we apply transformation with a negative determinant, the imaginary part of the complex number changes the sign.

In order to study the rational transformations of the upper half complex plane $\mathcal{H}$, which leave the period lattice invariant we need to be restricted to the $SL_2(\mathbb{Z})$ group rather than for a larger $GL_2(\mathbb{Z})$ one. Moreover, the element $-I_2 = \begin{pmatrix} -1 & 0 \\ 0 & -1 \end{pmatrix}$ from $SL_2(\mathbb{Z})$ acts trivially on $\mathcal{H}$. Thus, we can conclude that in fact it is the group $PSL_2(\mathbb{Z}) = SL_2(\mathbb{Z})/(\{\pm 1\}$ that acts. The subgroup $\{\pm 1\}$ is the center of the image of $SL_2(\mathbb{Z})$ in $PSL_2(\mathbb{Z})$.

The name modular group is reserved for the group

$$G = SL_2(\mathbb{Z})/\{\pm 1\},$$

which is the image of the group $SL_2(\mathbb{Z})$ in $PSL_2(\mathbb{R})$. But sometimes the discrete subgroup $SL_2(\mathbb{Z})$ of the group $SL_2(\mathbb{R})$ is also named a modular group.

The interest in the study of the lattices and modular group action on the upper half of the complex plane is related to the use of it as a model of hyperbolic space.

It is quite instructive to describe the fundamental domain of the modular group action on the upper half part of the complex plane and to compare the action of the modular group on the complex plane with the action of the $SL_2(\mathbb{Z})$ group on the cone of quadratic forms studied in chapter 6 in relation to the two-dimensional lattice classifications.

The choice of the fundamental domain of the modular group action is shown in Figure 9.18, where several images of the chosen fundamental domain under the modular group action are also shown. Special care should be taken for the indicated boundary of the fundamental domain in order to ensure that

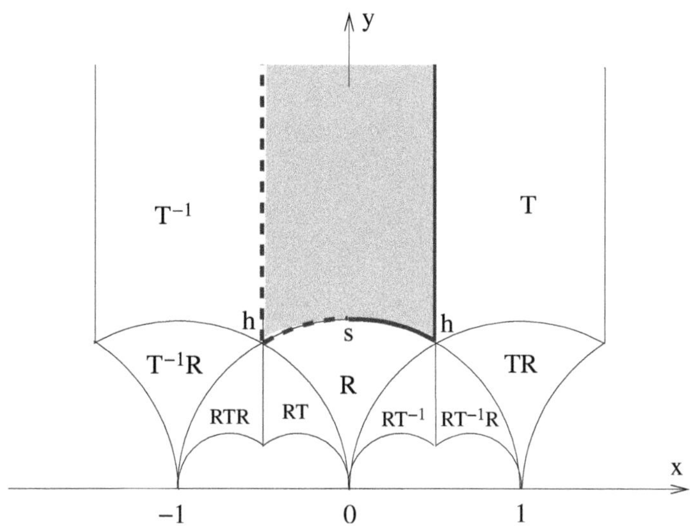

FIG. 9.18 – The fundamental domain of the modular group action on the upper half complex plane.

only one point from each orbit of the modular group action on the upper half of the complex plane is included in the fundamental region.

To describe the fundamental domain F let us represent it as a union of two subdomains $F = F^{(1)} \cup F^{(2)}$, where

$$F^{(1)} := \{z \in \overline{\mathbb{C}} : 0 \leq \Re z \leq \frac{1}{2}, \quad |z| \geq 1\}, \tag{9.11}$$

$$F^{(2)} := \{z \in \mathbb{C} : -\frac{1}{2} < \Re z < 0, \quad |z| > 1\}. \tag{9.12}$$

Here $\overline{\mathbb{C}}$ is the extended complex plane, note that ∞ is included in $F^{(1)}$ but not in $F^{(2)}$. The fundamental domain shown on Figure 9.18 by shading has boundaries marked by solid lines and boundaries marked by dashed lines. Only solid lines are included in the definition of the fundamental domain. To see the topology of the fundamental domain we need to identify two vertical boundaries and two halves of the circular boundary. The result is the topological sphere.

To see in more details the action of the modular group on the upper half complex plane let us introduce two generator of the $SL_2(\mathbb{Z})$ group.

Let

$$U = \begin{pmatrix} 1 & 1 \\ 0 & 1 \end{pmatrix}; \quad V = \begin{pmatrix} 0 & -1 \\ 1 & 0 \end{pmatrix}; \tag{9.13}$$

The corresponding mapping associated with the introduced action of $SL_2(\mathbb{Z})$ on the complex plane are given by

$$Uz = z + 1, \quad Vz = -1/z. \tag{9.14}$$

Let us note further that

$$U^k = \begin{pmatrix} 1 & k \\ 0 & 1 \end{pmatrix}; \quad V^2 = -I; \quad (VU)^3 = -I. \tag{9.15}$$

This means that from the point of view of the $SL_2(\mathbb{Z})$ group, V is a generator of a subgroup of order four, VU is a generator of a subgroup of order six.

But returning to the group of mappings on the upper half complex plane, i.e. to the $PSL_2(\mathbb{Z})$ group and taking into account the mentioned earlier fact that $-I$ acts trivially on the upper half complex plane (i.e. belongs to the center), we can say that the mapping V has order two and the mapping U has order three.

To characterize the fundamental domain we need to describe the stabilizers of different points belonging to the fundamental domain, i.e. find different strata of the group action. It can be checked [85] that all points have the trivial stabilizer except for point i denoted s on Figure 9.18, points $z = e^{\pi i/3}$, and $z = s^{2\pi i/3}$ denoted respectively as h and h' on Figure 9.18 and the ∞ point of the extended complex plane. Point i has a stabilizer generated by the element V, i.e. the stabilizer of point i is a group of order two. Two points $z = e^{\pi i/3}$, and $z = s^{2\pi i/3}$ belong to the same orbit. Their stabilizers are conjugate and generated by VU or by UV. The order of stabilizer is three. The ∞ point is invariant under the so called parabolic subgroup generated by element U. The corresponding discrete subgroup has infinite order. So finally we can say that the space of the orbits of the modular group action on the extended upper half complex plane $\mathcal{H}^*$ is a topological sphere with one point belonging to the stratum with the stabilizer being the group of order two and one point belonging to the stabilizer of order three and one point with stabilizer of infinite order.

9.7 Lattices and Morse theory

Many important physical characteristics of periodic crystals depend on the number and positions of stationary points of continuous functions defined on the Brillouin zone (see section 8.4). For three-dimensional crystals the Brillouin zone is a three-dimensional torus stratified by the action of the point symmetry group of the crystal. Morse theory is an appropriate mathematical tool which allows us to relate the number of stationary points of a smooth function with the topology of the space on which this function is defined. In the presence of symmetry additional restrictions on the number and position of stationary points follow from group action, in particular, from the existence of zero-dimensional strata formed by critical orbits which are stationary points for any invariant smooth function. In this section we illustrate the application of Morse theory to the description of the minimal possible system of stationary points for functions invariant with respect to point groups of 14 three-dimensional Bravais classes.

9.7.1 Morse theory

We start with short reminder of Morse theory. Let us consider a smooth real valued function f on a real compact manifold M of dimension d. If in a local coordinate system $\{x_i\}$, $1 \leq i \leq d = \dim M$, defined in a neighborhood of a point $m \in M$ the function f satisfies equations

$$\frac{\partial f}{\partial x_i} = 0; \quad \det \frac{\partial^2 f}{\partial x_i \partial x_j} \neq 0, \tag{9.16}$$

of vanishing gradient and non-vanishing determinant of the Hessian, we say that f has a non degenerate extremum at m. By a change of coordinates $\{x_i\} \mapsto \{y_i\}$ in a neighborhood of m the function can be transformed into $f = \sum_i \varepsilon_i y_i^2$ with $\varepsilon_i = \pm 1$. The number of "minus" signs is independent of the coordinate transformation. It is called the Morse index μ of this non degenerate extremum: for instance $\mu = 0$ for a minimum, $\mu = d$ for a maximum, and the intermediate values correspond to the different types of saddle points. By a small generic deformation all stationary points can be made non degenerate. A function on M with all its extrema non degenerate is called a Morse function. The essence of Morse theory is the relations between the numbers c_k of extrema of Morse index k and the topological invariants of the manifold M, its Betti numbers. The Betti number b_k is defined as the rank of the k-th homology group of M. Intuitively b_k is the maximal number of k-dimensional submanifolds of M which cannot be transformed into one another or into a submanifold of smaller dimension. For instance for the sphere S_d of dimension d, $b_0 = b_d = 1$ and all the other b_k vanish. More generally one has the Poincaré duality: $b_k = b_{d-k}$. The information about Betti numbers can be written in a form of a Poincaré polynomial $P_M(t)$ of a manifold M

$$P_M(t) = \sum_{i=0}^{d} b_k t^k; \quad d = \dim(M); \quad \text{e.g.} \quad P_{S_d}(t) = 1 + t^d. \tag{9.17}$$

The Poincaré polynomial of a topological product of manifolds is the product of the Poincaré polynomials of the factors. For instance, a d-dimensional torus is the topological product of d circles. This gives the Betti numbers for the d-dimensional torus T_d:

$$T_d = S_1^d \Rightarrow P_{T_d}(t) = (1+t)^d \Rightarrow b_k(T_d) = \binom{d}{k}. \tag{9.18}$$

For a compact manifold M the system of Morse relations consists of one equality

$$\sum_{k=0}^{d}(-1)^{d-k}(c_k-b_k)=0 \quad \Leftrightarrow \quad \sum_{k=0}^{d}(-1)^{d-k}c_k=\sum_{k=0}^{d}(-1)^{d-k}b_k \stackrel{\text{def}}{=} \chi(M), \tag{9.19}$$

where $\chi(M)$ is the Euler Poincaré characteristic, and the system of inequalities

$$\sum_{k=0}^{\ell}(-1)^{\ell-k}(c_k-b_k)\geq 0, \quad 0\leq \ell < d, \tag{9.20}$$

which can be simplified to a more crude form $c_k \geq b_k$. These simplified inequalities are not equivalent to Morse inequalities (9.20) but give lower bounds to the number of extrema of a Morse function.

For functions defined on the Brillouin zone (i.e. for a torus) for $d=2$ and $d=3$ the relations (9.19), (9.20) become

$$\begin{aligned} d=2, \quad & c_0-c_1+c_2=0, \quad c_0\geq 1\leq c_2, \quad c_0+1\leq c_1\geq c_2+1, \\ d=3, \quad & c_0-c_1+c_2-c_3=0, \quad c_0\geq 1\leq c_3, \quad c_1\geq c_0+2, \quad c_2\geq c_3+2. \end{aligned} \tag{9.21}$$

Thus for the two-dimensional torus the minimal number of stationary points for a Morse function cannot be smaller than four, whereas for the three-dimensional torus the minimal number is eight.

9.7.2 Symmetry restrictions on the number of extrema

In the presence of symmetry acting on a manifold all stationary points belonging to the same orbit of the group action naturally have the same Morse index. Moreover for invariant functions all orbits isolated in their strata should be formed by stationary points. Such orbits are named *critical orbits.* These stationary points are fixed (their position does not vary under small deformation of the Morse function). Thus in the presence of symmetry it is quite useful to find first all critical orbits and then verify if some other stationary points should exist in order to satisfy Morse inequalities.

Before passing to the application of the Morse analysis for functions defined on the Brillouin zone for different point symmetry groups we consider two simpler examples for a function defined on the two-dimensional sphere in the presence of symmetry. In the case of the O_h group action on the sphere (see section 4.5.1, Figure 4.20) there are three critical orbits: one consists of 6 points (stabilizer C_{4v}), another of 8 points (stabilizer C_{3v}), and the third one is formed by 12 points (stabilizer C_{2v}). We have 26 fixed stationary points among which points with C_{4v} and C_{3v} stabilizers should be stable, i.e. to be maxima or minima and cannot be saddles. One can easily verify that six maxima/minima, eight minima/maxima and 12 saddles satisfy

Morse inequalities and consequently the minimal number of stationary points for a Morse function on the sphere in the presence of O_h symmetry is 26.

As another example let us study the Morse function of the sphere in case of the C_{2h} point group action. There is only one critical orbit of the C_{2h} group action consisting of two points (see section 4.5.1, Figure 4.17). These two points should have the same Morse index. In order to construct a Morse function with the minimal number of stationary points it is necessary to add two orbits of two points located on C_h stratum. Positions of stationary points on one-dimensional stratum are not fixed and the distribution of stationary points among these three orbits is arbitrary. The only condition imposed by the Morse relation for the function invariant under C_{2h} action and possessing the minimal possible number of stationary points is the existence of two equivalent minima, two equivalent maxima and two equivalent saddles.

As a crystallographic application we give here the list of critical orbits and the minimal number of stationary points for functions defined on the Brillouin zone (three-dimensional torus) in the presence of point symmetry group action for 14 Bravais classes. The results of the analysis are represented in the form of Table 9.2 taken from [72]. For each of the 14 Bravais classes given in the first column we list in columns 2-6 all critical orbits classified by their k-values. Eight points corresponding to $k = 0$ (one point) and to $2k = 0$ (seven points) are critical for all Bravais classes. Under the presence of symmetry seven points associated with the $2k = 0$ form orbits consisting of one or several equivalent points. The numbers of critical points in each individual orbit of the symmetry group action are shown in column 3. For points with higher local symmetry (i.e. for $nk = 0$ with $n = 3, 4, 6$) columns labeled by $nk = 0$ indicate the number of critical points within the corresponding orbit of the group action. The points between [] have to be maxima or minima. The column labeled "nb" gives the minimal possible number of stationary points for each Bravais class. This number for two Bravais classes, namely for $Fmmm$ and $Im\overline{3}m$, is larger than the number of stationary points associated with critical orbits. These additional stationary points which are obliged to exist for the Morse functions are indicated explicitly as $n_{\text{crit}} + n_{\text{non-crit}}$.

Finally, the last four columns give the possible distribution of stationary points into subsets of stationary points with a given Morse index. Several lines give alternative distributions for the simplest Morse type functions, i.e. for Morse type functions with the minimal number of extrema.

TAB. 9.2 – List of the critical orbits on the Brillouin zone for the action of point symmetry group G of the 14 Bravais classes and the numbers and Morse indices of extrema of G-invariant functions with the minimum number of stationary points. Columns "nk" give the number of critical points satisfying the $nk = 0$ condition. See text for further details.

Bravais class	0	$2k$	$4k$	$3k$	$6k$	nb	0, 3	1, 2	2, 1	3,0
$P\bar{1}$	1	1,1,1,1,1,1,1				8	1	1+1+1	1+1+1	1
$P2/m$	1	1,1,1,1,1,1,1				8	1	1+1+1	1+1+1	1
$C2/m$	1	1,1,1,2,2				8	1	1+2	1+2	1
$Pmmm$	1	1,1,1,1,1,1,1				8	1	1+1+1	1+1+1	1
$Cmmm$	1	1,1,1,2,2				8	1	1+2	1+2	1
$Fmmm$	1	1,1,1,4				8+2	1	4	1+1+2	1
							1+1	4	1+2	1
$Immm$	1	1,2,2,2	2			10	1	2+2	2+2	1
							2	2+2	1+2	1
$P4/mmm$	1	1,1,1,2,2				8	1	1+2	1+2	1
$I4/mmm$	1	1,2,4	2			10	1	4	2+2	1
							2	4	1+2	1
$R\bar{3}m$	1	1,3,3				8	1	3	3	1
$P6/mmm$	1	1,3,3		2	2	12	1	2+3	2+3	1
							2	2+3	1+3	1
							2	1+3	1+3	2
							3	2+3	1+2	1
$Pm\bar{3}m$	[1]	1,3,3				8	1	3	3	1
$Fm\bar{3}m$	[1]	3,4	6			14	1	3	6	4
							1	4	6	3
$Im\bar{3}m$	[1]	1,6	[2]			10+6	1	6	1+6	2
							2	6	6	1+1

Appendix A

Basic notions of group theory with illustrative examples

We give in this appendix basic group-theoretical definitions used in the main body of the book.

Group

A group G is a set with a composition law: $\circ \in M(G \times G, G)$, which is associative:

$$\forall g, h, k \in G, \quad (g \circ h) \circ k = g \circ (h \circ k),$$

which has a neutral element e:

$$\forall g \in G, \quad e \circ g = g = g \circ e$$

and every element has an inverse one:

$$\forall g \in G, \quad \exists g^{-1}, \quad g^{-1} \circ g = e = g \circ g^{-1}.$$

There are two usual notations for the group law and the neutral element.

For the group operation the sign $+$ is used and for the neutral element 0 is used. Examples: the additive group of integers $\mathbb{Z}$, of real or complex numbers, $\mathbb{R}$ or $\mathbb{C}$, the additive group M_{mn} of $m \times n$ matrices with real (respectively, complex) elements. This notation is generally restricted to *Abelian groups*, i.e. the groups with a commutative law: $a + b = b + a$. The inverse element of a is denoted in this convention by $-a$ and is called the opposite.

For the group operation the sign multiplication, $\times$, is used (often this sign is simply omitted). The neutral element is noted $\mathbf{1}$ or $\mathbf{I}$. Examples: the multiplicative groups $\mathbb{R}^{\times}$, $\mathbb{C}^{\times}$; the n-dimensional linear groups $GL_n(\mathbb{R})$, $GL_n(\mathbb{C})$, i.e. the multiplicative groups of the $n \times n$ matrices on $\mathbb{R}$ or $\mathbb{C}$ with non-vanishing determinant. The inverse element of g is denoted by g^{-1}.

In all examples we have just given, the groups have an infinite or continuous number of elements. An example of a *finite group* is $\mathcal{S}_n$ the group of permutation of n objects, formed of $n!$ elements. The number of elements of a finite group G is named the order of the group and is denoted by $|G|$.

Subgroup

When a subset $H \subset G$ of elements of G forms a group (with the *composition law of G restricted to H*), we say that H is a subgroup of G and we shall denote it by $H \leq G$ (this is not a general convention) or by $H < G$ when we want to emphasize that H is a strict G-subgroup, i.e. H is a subgroup of G and $H \neq G$. Note that from $A \leq B$ and $B \leq G$ it follows that $A \leq G$.

Examples[1]: The subset $U(n)$ of matrices of $GL_n(\mathbb{C})$ which satisfy $m^* = m^{-1}$ is a subgroup of $GL_n(\mathbb{C})$; it is called the n-dimensional unitary group.

In particular $U(1) < \mathbb{C}^\times$. Note also that $GL_n(\mathbb{Z}) < GL_n(\mathbb{R}) < GL_n(\mathbb{C})$.

Since the determinant of the product of two matrices is the product of their determinants, in a group of matrices the matrices of determinant one form the subgroup which is often referenced as "special": for $GL_n(\mathbb{Z})$, $GL_n(\mathbb{R})$, $GL_n(\mathbb{C})$, $U(n)$ we denote them respectively by $SL_n(\mathbb{Z})$, $SL_n(\mathbb{R})$, $SL_n(\mathbb{C})$, $SU(n)$.

Another general example of a subgroup is the one generated by one element. Let $g \in G$ and consider its successive powers: $g, g^2, g^3, \ldots$ The order of g is the smallest integer such that $g^n = I$. If no such n exists, we say that g is of infinite order. When g is of finite order n the subgroup generated by g is formed of distinct powers of g; it is called a cyclic group of order n and it is usually denoted by Z_n or C_n. For example $e^{2\pi ik/n} \in \mathbb{C}^\times$ generates the cyclic group

$$Z_n \equiv C_n = \left\{ e^{2\pi ik/n}, 0 \leq k \leq n-1 \right\} < U(1) < \mathbb{C}^\times .$$

Note that the intersection of subgroups of G is a G-subgroup. Generally, the union of subgroups is not a subgroup.

The important example of the orthogonal group $O(n)$ can be introduced as

$$O(n) = U(n) \cap GL_n(\mathbb{R}) < GL_n(\mathbb{C}), \quad SO(n) = SU(n) \cap SL_n(\mathbb{R}) < SL_n(\mathbb{C}).$$

Note that the matrix elements of $O(n)$ are real and those of $U(n)$ are complex.

It is useful to have a complete list of the finite subgroups of $O(2)$. The matrices of $O(2)$ of determinant ± 1 are rotations $r(\theta)$ and reflections $s(\phi)$ through the axis of azimuth ϕ:

$$r(\theta) = \begin{pmatrix} \cos\theta & -\sin\theta \\ \sin\theta & \cos\theta \end{pmatrix}, \quad \theta (\mathrm{mod}\ 2\pi); \tag{A.1}$$

$$s(\phi) = \begin{pmatrix} \cos(2\phi) & \sin(2\phi) \\ \sin(2\phi) & -\cos(2\phi) \end{pmatrix}, \quad \phi (\mathrm{mod}\ \pi). \tag{A.2}$$

They satisfy the following relations:

$$r(\theta)r(\theta') = r(\theta+\theta'); \quad s(\phi)s(\phi') = r(2(\phi-\phi')); \tag{A.3}$$

$$r(\theta)s(\phi) = s(\theta/2+\phi) = s(\phi)r(-\theta). \tag{A.4}$$

[1] We denote by $m^\top$ the transpose of the matrix m, i.e. $(m^\top)_{ij} = m_{ji}$ and by m^* the Hermitian conjugate of m, i.e. $(m^*)_{ij} = \overline{m}_{ji}$, the complex conjugate of m_{ji}.

TAB. A.1 – Multiplication table of the group $\mathcal{S}_3$. The elements are given as permutation matrices and also by their cycle decomposition. One sees from this table that the alternate group $\mathcal{A}_3 = \{I, (123), (132)\}$ formed by odd permutations is a subgroup.

$\mathcal{S}_3$	$\begin{pmatrix} 1 & 0 & 0 \\ 0 & 1 & 0 \\ 0 & 0 & 1 \end{pmatrix}$ I	$\begin{pmatrix} 0 & 0 & 1 \\ 1 & 0 & 0 \\ 0 & 1 & 0 \end{pmatrix}$ (123)	$\begin{pmatrix} 0 & 1 & 0 \\ 0 & 0 & 1 \\ 1 & 0 & 0 \end{pmatrix}$ (132)	$\begin{pmatrix} 0 & 1 & 0 \\ 1 & 0 & 0 \\ 0 & 0 & 1 \end{pmatrix}$ (12)	$\begin{pmatrix} 0 & 0 & 1 \\ 0 & 1 & 0 \\ 1 & 0 & 0 \end{pmatrix}$ (13)	$\begin{pmatrix} 1 & 0 & 0 \\ 0 & 0 & 1 \\ 0 & 1 & 0 \end{pmatrix}$ (23)
I	I	(123)	(132)	(12)	(13)	(23)
(123)	(123)	(132)	I	(13)	(23)	(12)
(132)	(132)	I	(123)	(23)	(12)	(13)
(12)	(12)	(23)	(13)	I	(132)	(123)
(13)	(13)	(12)	(23)	(123)	I	(132)
(23)	(23)	(13)	(12)	(132)	(123)	I

In particular:

$$r(\theta)r(-\theta) = I, \quad (s(\phi))^2 = I; \tag{A.5}$$

$$s(\phi)r(\theta)s(\phi) = r(-\theta); \quad r(\theta)s(\phi)r(\theta)^{-1} = s(\phi + \theta). \tag{A.6}$$

We denote by C_n the n-element group formed by rotations $r(2\pi k/n)$, $0 \leq k \leq n-1$. When $n > 2$ the n reflections $s(\phi + \pi k/n)$, $0 \leq k \leq n-1$ form with C_n a non commutative group of $2n$ elements that we denote by $C_{nv}(\phi)$[2], $\phi(\text{mod } 2\pi/n)$.

$$C_n = \{r(2\pi k/n), \quad 0 \leq k \leq n-1\} \tag{A.7}$$

$$C_{nv}(\phi) = C_n \cup \{s(\phi + \pi k/n), \quad 0 \leq k \leq n-1\}. \tag{A.8}$$

The C_n, $n \in \mathbb{N}_+$ form the complete (countable) list of finite subgroups of $SO(2)$. The $C_{nv}(\phi)$ are a continuous infinity of finite subgroups of $O(2)$. They are the symmetry groups of the regular n-vertex polygons.

Every finite group can be considered as a subgroup of permutation group $\mathcal{S}_n$ when n is large enough. The permutation group itself $\mathcal{S}_n$ is a subgroup of $O(n) < U(n)$ when its permutation $1, 2, 3, \ldots, n \mapsto i_1, i_2, \ldots, i_n$ is represented by the matrix p_{ij} with elements $p_{1i_1} = p_{2i_2} = \ldots = p_{ni_n} = 1$ and all the other elements are zero.

An example of $\mathcal{S}_3$ group is detailed in Table A.1 where the multiplication table of six elements of $\mathcal{S}_3$ is given using the representation of elements by permutation matrices and by their cycle decomposition.

The group $C_{3v}(\phi)$ has the same multiplication table as $\mathcal{S}_3$ when the following correspondence (bijection) between the elements of the two groups is made:

$$I \leftrightarrow I, \quad r(2\pi/3) \leftrightarrow (123), \quad r(4\pi/3) \leftrightarrow (132),$$

$$s(\phi) \leftrightarrow (12), \quad s(\phi + 2\pi/3) \leftrightarrow (23), \quad s(\phi + 4\pi/3) \leftrightarrow (31).$$

The subgroup C_{3v} of the $O(2)$ is the symmetry group of an equilateral triangle. The representation of the group $\mathcal{S}_3$ by permutation matrices corresponds to

[2] Alternative widely used notation is D_n.

orthogonal transformation of the three dimensional space leaving invariant the line carrying the vector of coordinates $(1, 1, 1)$, so it leaves invariant the plane orthogonal to it. In that plane, $\mathcal{S}_3$ permutes the vertices $(1, -1, 0)$, $(0, 1, -1)$, $(-1, 0, 1)$ of an equilateral triangle. This construction can be extended to the symmetry group of the $n - 1$ dimensional simplex (regular tetrahedron in 3-dimensions). This symmetry group of orthogonal transformations is also the permutation group of its n vertices.

$GL_n(\mathbb{Z})$ is another important example of groups. For the multiplication in Z an integer $n \neq 1$ has no inverse; an $n \times n$ matrix with integer elements has an inverse with integer elements if and only if its determinant is ± 1. Since the product of two integer matrices (i.e. matrices with integer elements) is an integer matrix, the integer matrices with determinant ± 1 form a group $GL_n(\mathbb{Z})$; $(GL_1(\mathbb{Z}) = Z_2 \equiv \{\pm 1\})$.

Lattice - notion from the theory of partially ordered sets.

The set of subgroups of a group is an example of a *lattice.* A lattice is a partially ordered set such that for any two given elements, x, y there exists a unique minimal element among all elements z such that $z > x$ and $z > y$, and similarly for any two given elements x, y, there exists a unique maximal element among all elements z' such that $z' < x$ and $z' < y$. For any two subgroups x, y belonging to the lattice of subgroups the unique minimal subgroup z among all z such that $z > x$ and $z > y$ is the subgroup generated by the union of x and y. The intersection of two given subgroups x and y is the maximal subgroup among all subgroups z' such that $z' < x$ and $z' < y$. In particular, a lattice of subgroups of a group G has unique minimal and maximal elements. The group G is the maximal element and the trivial subgroup $\{I\}$ is the minimal one.

Cosets

Let $H < G$. The relation among the elements of $G : x \in yH$ is an equivalence relation; it is reflexive: $x \in xH$; symmetric: $x \in yH \;\Leftrightarrow\; y \in xH$; transitive: $x \in yH,\ y \in zH \;\Rightarrow x \in zH$. The equivalence classes are called *cosets.* We denote by $G : H$ the quotient set, i.e. the set of cosets.

Note that each coset has the same order (number of elements) as H. For gH, the left multiplication by g of the elements of H defines one-to-one (bijective) correspondence between the two cosets H and gH. This gives the relation

$$|G| = |G : H||H|. \tag{A.9}$$

The order of the quotient $|G : H|$ is also called the *index of the subgroup H* in G. This proves the Lagrange theorem (the oldest theorem in group theory proven even before Galois had introduced the notion "group"):

Theorem. For the finite group G, the order of a subgroup divides the order of group, $|G|$.

Invariant subgroup

When defining cosets, we could have pointed out that we were using left cosets; similarly we can introduce right cosets Hx. In general $xH \neq Hx$. When left and right cosets are identical, H is named an *invariant subgroup*[3]. We will also write this property $H \lhd G$.

$$H \lhd G \stackrel{\text{def}}{=} H \leq G. \quad \forall g \in G, \quad gH = Hg. \tag{A.10}$$

Evidently, *every subgroup of an Abelian group is invariant.*

Every non-trivial group has two invariant subgroups: $\{1\}$ and G itself. If there are no other invariant subgroups, the group G is named *simple*. The set of invariant subgroups forms a lattice (sublattice of the subgroup lattice). Beware that $K \lhd H$, $H \lhd G$ does not imply $K \lhd G$.

Quotient group

When $K \lhd G$ is an invariant subgroup there is a natural group structure on $G : K$; indeed $gK \circ hK = (gK)(hK) = gKhK = (gh)K$, i.e. the multiplication of cosets is well defined. We call this group the *quotient group*[4] of G by K and we denote it by G/K. Since the determinant of a matrix is invariant by conjugacy by an invertible matrix, the "special" subgroups are invariant. The corresponding quotient groups are

$$\begin{aligned} &GL_n(\mathbb{C})/SL_n(\mathbb{C}) = \mathbb{C}^\times; \quad GL_n(\mathbb{R})/SL_n(\mathbb{R}) = \mathbb{R}^\times; \\ &U(n)/SU(n) = U(1); \quad O(n)/SO(n) = Z_2. \end{aligned}$$

Note that an index 2 subgroup is always an invariant subgroup because left and right cosets coincide automatically.

Double cosets

A generalization of the cosets is the notion of the *double cosets* which defines the following equivalence relation between elements of G.

When $H < G, K < G$, $x \in HyK$ is an equivalence relation between $x, y \in G$. Indeed let $x = h(x)yk(x)$ with $h(x) \in H$, $k(x) \in K$, then $y = h(x)^{-1}xk(x)^{-1}$; if moreover $y = h(y)zk(y)$, then $x = h(x)h(y)zk(y)k(x) \in HzK$. We will denote by $H : G : K$ the set of H-K-double cosets of G. Note that $H : G : K \neq K : G : H$.

When either H or K is an invariant subgroup of G, then the HK is a subgroup of G and the double cosets are either left or right coset of HK. Indeed assume $H \lhd G$, then $HaK = aHK$.

Conjugacy classes

Two elements $x, y \in G$ are conjugate if there exists a $g \in G$ such that $y = gxg^{-1}$. Conjugacy is an equivalence relation among the elements of a

[3] An often used synonym is *normal subgroup*.

[4] Sometimes this is called the "factor" group.

group. A group is therefore a disjoint union of its conjugacy classes. Note that gh is conjugate to hg. The elements of a conjugacy class have the same order. In physical language the conjugate elements correspond to symmetry operations equivalent with respect to the symmetry group.

Examples.

1) $U(n)$ or $SU(n)$. Any unitary matrix can be diagonalized by conjugacy with unitary matrices which can be chosen of determinant 1. So in $U(n)$, the unitary matrices with the same spectrum, i.e. with the same set of eigenvalues, or same characteristic polynomial, form a conjugacy class of $U(n)$; indeed, by conjugacy in $U(n)$ a unitary matrix can be brought to diagonal form and by conjugacy with a permutation matrix (which are also unitary) the eigenvalues can be put in a chosen order, e.g. in increasing values.

2) $GL_n(\mathbb{C})$ or $SL_n(\mathbb{C})$. The situation is different: by conjugacy in $SL_n(\mathbb{C})$ one can put the eigenvalues of a matrix of the group in a given order along the diagonal; however if there are degenerate eigenvalues, the matrix might not be diagonalizable, but one can put it in Jordan form (some 1's on the first diagonal above the main diagonal). For example, for $SL_2(\mathbb{C})$, the conjugacy classes can be labeled by the trace t of the matrix $t = z + z^{-1}$ (where z and z^{-1} are the eigenvalues) when $t \neq \pm 2$ because for $t = \pm 2$ the eigenvalues are degenerate. Among the matrices with trace $t = \pm 2$, the matrices $\pm I$ form each a conjugacy class; the other matrices of trace 2 form one conjugacy class since they are equivalent to $\begin{pmatrix} 1 & 1 \\ 0 & 1 \end{pmatrix}$. The other matrices of trace -2 form another conjugacy class which contains $\begin{pmatrix} -1 & 1 \\ 0 & -1 \end{pmatrix}$. This example also shows that the characteristic polynomials of the matrices of $GL_n(\mathbb{C})$ or $SL_n(\mathbb{C})$ are not sufficient to label the conjugacy classes.

3) $O(2)$. The equations (A.6) shows that all reflections $s(\phi)$ form a unique conjugacy class of $O(2)$ while each pair of rotations and its inverse, $r(\theta)$ and $r(-\theta)$, form one conjugacy class.

4) $SO(n)$. The matrices of $SO(n)$ have in general non real eigenvalues (which form pairs of complex conjugate phases); so they cannot be diagonalized by conjugacy with real matrices. However they can be put in the form of diagonal 2×2 blocks when n is even, and each block is a matrix $r(\theta)$ defined in equation (A.1); when n is odd there is also a single 1 (which can be placed at the end of the diagonal). So conjugate matrices of $SO(n)$ have same set of the $\pm\theta$'s (the rotation angles). For instance, conjugate classes of the 3-dimensional rotation group $SO(3)$ contain all rotations with the same single rotation angle (in absolute value modulo π).

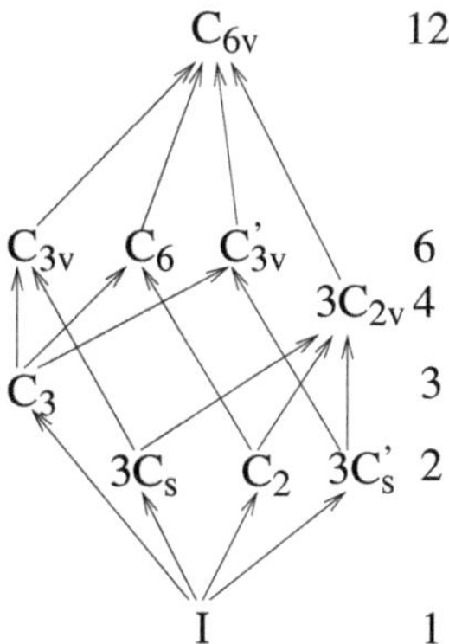

FIG. A.1 – Partially ordered set of conjugacy classes of subgroups of the C_{6v} group.

5) $\mathcal{S}_n$. To describe the conjugacy classes of $\mathcal{S}_n$ one needs the notion of partition. A set of integers whose sum is n form a partition of n; the partition is defined by these integers independently of their order. The conjugacy classes of $\mathcal{S}_n$ are labeled by the partitions of n, corresponding to the decomposition of a permutation into k cycles $\{c_i, 1 \leq i \leq k\}$ of length $l(c_i)$ with $\sum_i l(c_i) = n$.

Partially ordered set of conjugacy classes of subgroups

Two G-subgroups H, H' are said to be conjugate if there exists $g \in G$ such that $H' = gHg^{-1}$. This is an equivalence relation among subgroups of a group. We denote by $[H]_G$ the conjugacy class of H. When a subgroup is alone in its conjugacy class, this is an invariant subgroup.

We have seen that the subgroups of a group form a lattice. Beware that this is not generally true for the set of subgroup conjugacy classes. We can say that partial order $[H]_G \prec [K]_G$ between conjugacy classes of subgroups exists when $\exists x \in G$, $xHx^{-1} < K$ but there is no element of G which conjugates K into a strict subgroup of H. Equivalently, if a G-subgroup H cannot be conjugate to one of it strict subgroup, then the conjugacy relation among G-subgroups is compatible with the partial order of the subgroup lattice. It is clear that there exists a natural partial order on the set of conjugacy classes of finite (respectively, finite index) subgroups of a group G. Figure A.1 gives an example.

Center

An element which commutes with every element of a group G forms a conjugacy class by itself (this is always the case of the identity). These elements form a subgroup called the *center* of G and often denoted by $C(G)$.

The center is an invariant subgroup of G. If the group A is Abelian, $C(A) = A$. Examples:

$$C\left(GL_n(\mathbb{C})\right) = \mathbb{C}^\times I, \quad C\left(GL_n(\mathbb{R})\right) = \mathbb{R}^\times I,$$
$$C\left(U(n)\right) = U(1)I, \quad C\left(O(n)\right) = Z_2 I.$$

Similarly :

$$C\left(SL_n(\mathbb{R})\right) = \begin{cases} Z_2 I & \text{for } n \text{ even} \\ \{I\} & \text{for } n \text{ odd} \end{cases} \quad C\left(SO(n)\right) = \begin{cases} C(SL_n(\mathbb{R})) & \text{for } n > 2 \\ SO(2) & \text{for } n = 2 \end{cases}$$

Centralizers, Normalizers.

The *centralizer* of $X \subset G$ is the set of elements of G which commute with **every** element of X; this set is a G-subgroup. We denote it by $C_G(X)$. If X coincides with G, the centralizer becomes the center of G: $C(G) = C_G(G)$.

The normalizer of $X \subset G$ is a G-subgroup

$$N_G(X) = \{g \in G,\ gXg^{-1} = X\}. \tag{A.11}$$

Note that $C_G(X) \lhd N_G(X)$, i.e. the centralizer of X is an invariant subgroup of the normalizer of X.

From the definition of the normalizer, when $H \leq G$, the normalizer $N_G(H)$ is the largest G-subgroup such that $H \lhd N_G(H)$. For instance if $N_G(H) = G$, then $H \lhd G$ is the invariant subgroup of G.

Homomorphism

A group *homomorphism* or, shorter, a *group morphism* between the groups G, H is a map $G \xrightarrow{\rho} H$ compatible with both groups laws

$$G \xrightarrow{\rho} H, \quad \rho(xy) = \rho(x)\rho(y), \quad \rho(1) = 1 \in H. \tag{A.12}$$

This implies

$$\rho(x^{-1}) = \rho(x)^{-1}. \tag{A.13}$$

A morphism of a group G into the groups $GL_n(\mathbb{C})$, $GL_n(\mathbb{R})$, $U(n)$, $O(n)$ respectively is called a n-dimensional (complex, real) linear, unitary, orthogonal representation of G,

The *image* of the morphism $G \xrightarrow{\rho} H$ is denoted by Im ρ. It is a subgroup of H, Im $\rho \leq H$, which includes images of all elements of G.

The *kernel* Ker ρ of the morphism $G \xrightarrow{\rho} H$ is the set $K \in G$ which is mapped on $I \in H$. Ker ρ is an invariant subgroup of G. There is an important relation between image and kernel: Im $\rho = G/$Ker ρ.

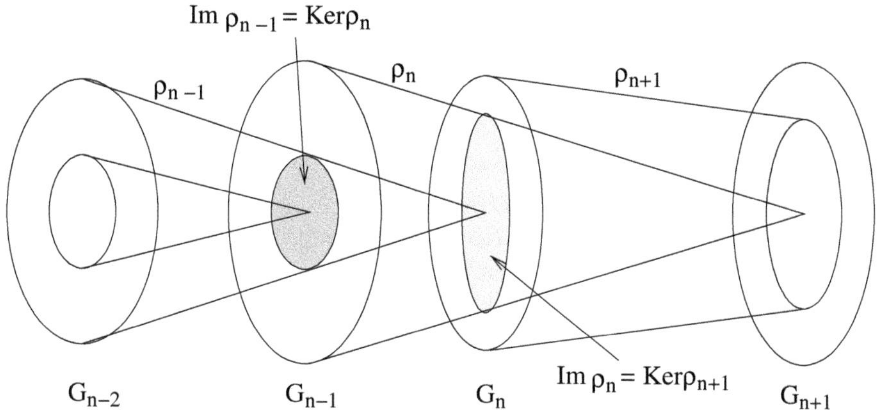

FIG. A.2 – Schematic representation of an exact sequence of homomorphisms of groups.

Sequence of homomorphisms

Let us consider the following sequence of homomorphisms of Abelian groups:

$$G_1 \xrightarrow{\rho_1} G_2 \xrightarrow{\rho_2} \cdots G_{n-1} \xrightarrow{\rho_{n-1}} G_n \xrightarrow{\rho_n} G_{n+1} \cdots \tag{A.14}$$

Such a construction, named complex, is quite useful to relate topological and group-theoretical properties. If for all n we have Im ρ_{n-1} = Ker ρ_n, the sequence in (A.14) is an *exact sequence of homomorphisms.*

Examples: If $H \lhd G$ and $G \xrightarrow{\rho} G/H$ we can write

$$1 \rightarrow H \xrightarrow{i} G \xrightarrow{\rho} G/H \rightarrow 1, \tag{A.15}$$

where $H \xrightarrow{i} G$ is the injection map, i.e. $\forall x \in H$, $i(x) = x \in G$. An exact sequence of this type is named *short exact*. The part of the diagram $1 \rightarrow H \xrightarrow{i} G$ means that $1 \rightarrow H$ is the injection of the unit into H and Ker $i = 1$, i.e. i is injective. The fact that ρ is surjective is expressed by $G \xrightarrow{\rho} G/H \rightarrow 1$, For any homomorphism ρ there is always a short exact sequence

$$1 \rightarrow \text{Ker } \rho \xrightarrow{i} G \xrightarrow{\rho} \text{Im } \rho \rightarrow 1, \tag{A.16}$$

Isomorphism

A bijective morphism ρ is called an *isomorphism.* In other words if Ker $\rho = I_G$, and Im $\rho = H$ than ρ is an isomorphism and $G \sim H$. When we want to classify groups, this will be done up to isomorphism, except if we precise explicitly a more refined classification. Often when we write about "abstract groups" we mean an isomorphism class of groups.

Examples:

For every prime number p there is (up to isomorphism) only one group of order p: this is $\mathbb{Z}_p$.

There are exactly two non-isomorphic groups of order 4.

Automorphism

An isomorphism from G to G is called an *automorphism* of G. The composition of two automorphisms is an automorphism. Moreover there exists the identity automorphism I such that $\forall g \in G$, $I(g) = g$, and every automorphism has an inverse. Thus the automorphisms of G form a group, Aut G. The conjugation by a fixed element $g \in G$ induces a G-automorphism: $(gxg^{-1})(gy^{-1}g^{-1}) = g(xy^{-1})g^{-1}$ which is an "inner" automorphism. The set of inner automorphisms forms a subgroup of Aut G that we denote by In Aut G. Note that the elements $c \in C(G)$ of the center of G induce the trivial automorphism I_G, so we have the exact sequence:

$$1 \to C(G) \to G \xrightarrow{\theta} \text{In Aut } G \to 1,$$

An automorphism which is not inner, is called outer automorphism. We note that In Aut G in an invariant subgroup of Aut G.

Making groups from groups

Given two groups G_1, G_2, one can form a new group, $G_1 \times G_2$, the *direct product* of G_1 and G_2: the set of elements of $G_1 \times G_2$ is the product of the set of elements of G_1 and of G_2, i.e. the set of ordered pairs: $\{(g_1, g_2), g_1 \in G_1, g_2 \in G_2\}$, the group law is

$$(g_1, g_2)(h_1, h_2) = (g_1 h_1, g_2 h_2).$$

When $G_1 \neq G_2$, $G_1 \times G_2 \neq G_2 \times G_1$, but they are isomorphic, i.e. $G_1 \times G_2 \sim G_2 \times G_1$.

Given a morphism $Q \xrightarrow{\theta} K$ one defines the *semi-direct product* as the group whose elements are the pairs $(k, q), k \in K, q \in Q$, and the group law is [using $q \cdot k$ as a short for $(\theta(q))(k)$]:

$$(k_1, q_1)(k_2, q_2) = (k_1 q_1 \cdot k_2, q_1 q_2). \tag{A.17}$$

Here we denote this semi-direct product by $K \rtimes Q$.

Examples:

The semi-direct product of $\mathbb{R}^n$ and $GL_n(\mathbb{R})$ is called the affine group

$$Aff_n(R) = \mathbb{R}^n \rtimes GL_n(\mathbb{R}). \tag{A.18}$$

Similarly one can define the complex affine group: $Aff_n(C) = \mathbb{C}^n \rtimes GL_n(\mathbb{C})$.

The Euclidean group E_n is the semi-direct product of $\mathbb{R}^n$ and orthogonal group

$$E_n = \mathbb{R}^n \rtimes O(n) \leq Aff_n(R).$$

Group extensions

Given two groups K, Q, a very natural problem is to find all groups E such that $K \triangleleft E$ and $Q = E/K$. E is called an extension of Q by K. The extension can be represented by a diagram

$$1 \to K \to E \to E/K \to 1,$$

which is not an exact sequence. The main problem is to classify different extensions up to equivalence.

The semi-direct product (and its particular case, the direct product) are particular examples of an extensions. But in the general case of an extension E of Q by K there is no subgroup of E isomorphic to the quotient Q.

Such an example is given by $SU(2)$ as an extension of $SO(3)$ by Z_2:

$$1 \to Z_2 \to SU(2) \to SO(3) \to 1.$$

$SU(2)$ is the group of two-by-two unitary matrices of determinant 1. Its center Z_2 has two elements, the matrices 1 and -1. These matrices are the only square roots of the unit. The three-dimensional rotation group $SO(3)$ is isomorphic to $SU(2)/Z_2$. This group has an infinity of square roots of 1: the rotations by π around the arbitrary axis. So $SO(3)$ is not a subgroup of $SU(2)$.

Appendix B

Graphs, posets, and topological invariants

The purpose of this appendix is to give a minimal required system of definitions and mathematical constructions needed to understand and to follow the discussion of the visualization of lattices by graphs and calculation of corresponding topological invariants introduced in Chapter 6, section 6.7.

We start by several intuitively evident but important definitions.

A *graph* $G = (V, E)$ consists of a finite set V of vertices (nodes) and a finite set E of edges. Every edge $e \in E$ consists of a pair of vertices, u and v, called its *endnodes*. We will denote the edge e by uv. Two vertices are said to be *adjacent* if they are joined by an edge. We will mainly consider here *simple* graphs, i.e. graphs in which every edge has distinct endnodes (no loops) and no two edges have the same two endnodes (no parallel or multiple edges). When every two nodes in G are adjacent, the graph G is said to be a *complete graph*. The complete graph on n nodes is usually denoted by K_n.

Let $G = (V, E)$ be a graph. A graph $H = (W, F)$ is said to be a *subgraph* of G if $W \subset V$ and $F \subset E$. Given an edge $e \in E$ in G, $G \backslash e := (V, E \backslash e)$ is called the graph obtained from G by deleting e.

Contracting an edge $e := uv$ in G means identifying the endpoints u and v of e and deleting the parallel edges that may be created while identifying u and v. G/e denotes the graph obtained from G by contracting the edge e. For an edge set $F \subseteq E$, G/F denotes the graph obtained from G by contracting all edges of F (in any order).

Two graphs $G = (V, E)$ and $G' = (V', E')$ are said to be *isomorphic*, $G \simeq G'$, if there exists a bijection $f : V \to V'$ such that $uv \in E \Leftrightarrow f(u)f(v) \in E'$.

A graph is said to be *connected* if, for every two nodes $u, v \in G$, there exists a path in G joining u and v. The *rank* of the graph is the number of nodes minus the number of connected components.

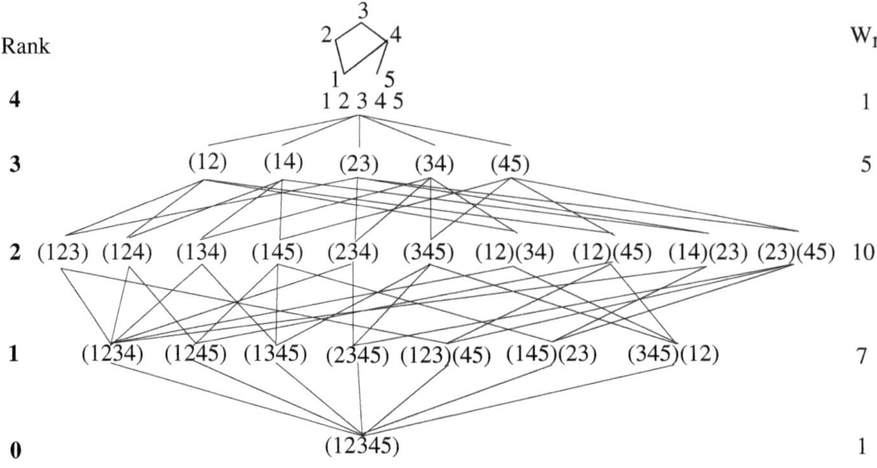

FIG. B.1 – Ranked partially ordered set of contractions for a graph representing the **14.28**-0 4-d lattice. Only fused vertices are shown for contracted subgraphs. Lines symbolize the partial order imposed by contractions.

For a given simple connected graph we can construct all graphs which can be obtained from the initial graph by one or several contractions of edges. The set of so obtained graphs form a partially ordered set (see definition of partially ordered set or *poset* in appendix A). An example of such a partially ordered set *(Poset)* is shown in Figure B.1. In Figure B.1 only the initial graph is shown. Graphs obtained by contractions of the initial graph are represented just by labels of nodes which were fused during the contraction. The rank is well defined for all contracted graphs. Contracting one edge decreases rank by one. Thus we obtain the so called ranked partially ordered set of contractions, P. There are some number of topological invariants which can be introduced for the ranked partially ordered set P.

The simplest invariant is the number of elements of rank k of the poset P. This invariant is named a simply indexed Whitney number of second kind, W_k. More complicated invariants are the doubly indexed Whitney numbers, W_{ij} of the second kind. In order to introduce them we need to study subsets of P with rank $r = i$ and with rank $r = j$. Whitney number W_{ij} of the second kind gives the number of pairs $\{(x^i, x^j) : x^i \leq x^j\}$ of elements of P which satisfy the order relation. If i and j are neighboring integers, the corresponding Whitney number is just equal to the number of contraction lines between neighboring rows of a poset. It is clear that simply indexed Whitney numbers are a special case of doubly indexed ones, namely $W_{0,j} = W_j$.

From Figure B.1 we immediately find the doubly indexed Whitney numbers which can be represented in the form of a triangular table

Graph	W_{00}	W_{i1}	W_{i2}	W_{i3}	W_{i4}
	1	7	10	5	1
		7	24	22	7
			10	20	10
14.28-0				5	5
					1

which shows some equivalence between W_{ij} values which remains valid for a wide class of graphs.

The formal definition of the doubly indexed Whitney numbers of the second kind can be written as follows

$$W_{ij}(P) = \left|\{(x^i, x^j) : x^i \leq x^j\}\right|, \tag{B.1}$$

where $|S|$ means the cardinality of the set S, i.e. the number of elements in the set.

The construction of the Whitney numbers of the first kind is based on the preliminary introduction of the Möbius function for a ranked partially ordered set. To be maximally concrete we restrict ourselves always to posets of contractions for a simple connected graph, which is one of the subgraphs of a complete graph K_n. We start by calculating values of the Möbius μ-function for all elements of the poset P. To find these values we use $\mu(g, g) = 1$ for the initial graph g. Next, for $b \neq g$ we calculate $\mu(b, 0)$ as a sum

$$\mu(b, 0) = -\sum_{g \geq c > b} \mu(c, g), \tag{B.2}$$

over all c which are partially ordered with respect to b and are strictly greater than b. The result of this calculation is illustrated in Figure B.2, upper right subfigure; it gives a system of simply indexed Whitney numbers of the first kind, $w_i = w_{0i}$. Generalization to doubly indexed Whitney numbers of the first kind is similar to what we have done for Whitney numbers of the second kind.

To calculate doubly indexed Whitney numbers of the first kind w_{ki} we need to analyze only the sub-poset of the initial poset taking into account the elements with the rank not exceeding $r - k$ where r is the rank of the initial graph, and calculate μ-values for this sub-poset. The lower left sub-figure of B.2 visualizes the "neglected" part of the initial poset by using dashed lines. For this sub-poset we calculate $\mu(b)$ values with respect to the rank 2 level. This explains why for all $b = (ij)$ elements now $\mu(b) = 1$ and $w_{11} = 4$. Going to the lower rank $r = 1$ we see that the value of $\mu(123)$, for example, should be calculated with respect to the level with the rank equal 2 and we have only three contributions from $(12), (13), (23)$ elements which all equal -1.

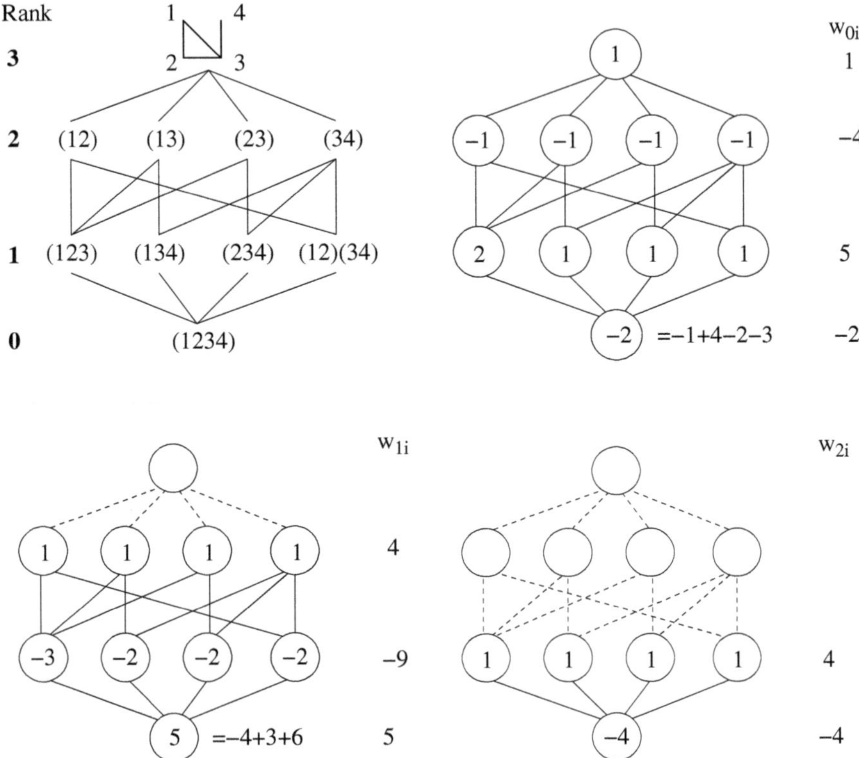

FIG. B.2 – A ranked partially ordered set of contractions for a graph representing the **8.12** 3-d lattice. The upper left subfigure shows initial graph and for each contraction step indicates fused vertices for contracted graphs. Lines symbolize the partial order imposed by contractions. The upper right subfigure reproduces the same poset and for each element a shows the value of the Möbius function $\mu(a, 0)$. On the right of the poset the values $w_{0i} = w_i$ of the Whitney numbers of the first kind are given which are the sum of μ-values for all elements of the same rank. In a similar way the lower left and right subfigures illustrate calculation of the Whitney numbers of the first kind for w_{1i} and for w_{2i}.

Completing calculations for all elements of rank 1 we get $w_{12} = -9$ and in a similar way we get $w_{13} = 5$.

Finally, to calculate Whitney number w_{2i} we need to study only the sub-poset of the initial poset taking into account the elements with the rank not exceeding 1 and calculate μ-values for this sub-poset. Lower right sub-figure of B.2 visualizes the "neglected" part and indicates corresponding $\mu(b)$ values and w_{2i}.

TAB. B.1 – Doubly indexed Whitney numbers of the first and second kind for several graphs corresponding to some of four-dimensional zonotopes.

Graph	w_{00}	w_{i1}	w_{i2}	w_{i3}	w_{i4}	$\sum_j w_{ij}^+$	W_{01}	W_{02}	W_{03}
	1	-4	6	-4	1	$16 = N_0$	4	6	4
		4	-12	12	-4	$32 = N_1$	4	12	12
			6	-12	6	$24 = N_2$		6	12
8.16-0				4	-4	$8 = N_3$			4
					1				
	1	-5	10	-10	4	$30 = N_0$	10	10	5
		5	-20	30	-15	$70 = N_1$	10	30	30
			10	-30	20	$60 = N_2$		10	20
20.30-0				10	-10	$20 = N_3$			5
					1				
	1	-6	13	-12	4	$36 = N_0$	6	11	6
		6	-24	30	-12	$72 = N_1$	6	24	24
			11	-24	13	$48 = N_2$		11	24
12.36-12				6	-6	$12 = N_3$			6
					1				
	1	-6	15	-17	7	$46 = N_0$	11	15	6
		6	-30	48	-24	$108 = N_1$	11	42	36
			15	-42	27	$84 = N_2$		15	30
22.46-0				11	-11	$22 = N_3$			6
					1				
	1	-7	19	-23	10	$60 = N_0$	12	17	7
		7	-36	60	-31	$134 = N_1$	12	49	44
			17	-49	32	$98 = N_2$		17	36
24.60-12				12	-12	$24 = N_3$			7
					1				
	1	-10	35	-50	24	$120 = N_0$	15	25	10
		10	-60	110	-60	$240 = N_1$	15	75	7
			25	-75	50	$150 = N_2$		25	60
30.120-60				15	-15	$30 = N_3$			10
					1				

Table B.1 gives Whitney numbers of the first and of the second kind for several four-dimensional zonotopes. Face numbers for corresponding zonotopes are given along with Whitney numbers of the first kind because of the simple relation between face numbers and Whitney numbers of the first kind. Namely, for subgraphs of K_{r+1} representing zonotopes we have [56]

$$\sum_{j=k}^{r} w_{kj}^{+} = N_k, \tag{B.3}$$

where $w_{kj}^{+} = |w_{kj}|$ and N_k are the number of k-faces of the zonotope associated with the graph.

Only $W_{i1}, W_{i2} W_{i3}$ are shown in table B.1 for the Whitney numbers of the second kind. The rest of the table can be easily reconstructed taking into account the symmetry of the table, namely $W_{ii} = W_{i4}$ and $W_{44} = W_{00} = 1$.

We note also that the singly indexed Whitney numbers of the first kind are the coefficients of the chromatic polynomial. The chromatic polynomial

$$P_G(t) = \sum_{k=0}^{r} w_{0k} t^{r+1-k} \tag{B.4}$$

shows how many different coloring of graph nodes are possible with t colors with the restriction on adjacent nodes to be of different color.

For more details on relevant material see [1, 20, 9].

Appendix C

Notations for point and crystallographic groups

The notation used for symmetry groups varies depending on the science domain and on the class of groups used in applications.

Point symmetry groups of three-dimensional space are the most widely used in different concrete applications. In fact, they are not the abstract groups but their representations in three-dimensional Euclidean space.

There are seven infinite families of groups and seven exceptional groups. The different notations for these groups are given in Table C.1.

We characterize shortly these groups here using Schoenflies notation.

The seven infinite series of point groups are:

C_n - group of order n generated by rotation over $2\pi/n$ around a given axis; $n = 1, 2, 3, \dots$ C_1 is a trivial, "no symmetry" group. In the limit $n \to \infty$ we get the $C_\infty = SO(2)$ group.

S_{2n} - group of order $2n$ generated by rotation-reflection over π/n around a given axis; $n = 1, 2, \dots$ For n-odd, the group S_{4k+2} is often noted as $C_{2k+1,i}$. i.e. as an extension of the C_{2k+1} group by inversion. In particular $S_2 = C_i$. In the limit $n \to \infty$ we get the $C_{\infty h}$ group.

C_{nh} - Group of order $2n$ obtained by extension of C_n by including reflection in plane orthogonal to the symmetry axis. $n = 1, 2, \dots$ For $n = 1$, the notation $C_s \equiv C_{1h}$ is used. In the limit $n \to \infty$ we get the $C_{\infty h}$ group.

C_{nv} - Group of order $2n$ obtained by extension of C_n by adding reflection in plane including the symmetry axis. $n = 2, 3, \dots$ In the limit $n \to \infty$ we get the $C_{\infty v} = O(2)$ group.

D_n - Group of order $2n$ obtained by extension of C_n by including symmetry axes of order two orthogonal to the C_n axis. $n = 2, 3, \dots$ In the limit $n \to \infty$ we get the D_∞ group.

D_{nd} - Group of order $4n$ obtained by extension of D_n by including reflection in the symmetry plane containing the C_n axis, but not containing orthogonal C_2 axes. $n = 2, 3 \dots$ In the limit $n \to \infty$ we get the $D_{\infty h}$ group.

TAB. C.1 – Different notations for point groups.

Schoenflies	C_n	S_{2n}	C_{nh}	C_{nv}	D_n	D_{nd}	D_{nh}
ITC (even n)	n	$\overline{(2n)}$	n/m	nmm	$n22$	$\overline{(2n)}2m$	n/mmm
ITC (odd n)	n	$\overline{n}$	$\overline{(2n)}$	nm	$n2$	$\overline{n}m$	$\overline{(2n)}2m$
Conway	nn	$n\times$	$n*$	$*nn$	$n22$	$2*n$	$*n22$
Schoenflies	T	T_d	T_h	O	O_h	I	I_h
ITC	23	$\overline{4}3m$	$m\overline{3}$	432	$m\overline{3}m$	235	$m\overline{3}\overline{5}$
Conway	332	$*332$	$3*2$	432	$*432$	532	$*532$

D_{nh} - Group of order $4n$ obtained by extension of D_n by including reflection in the symmetry plane orthogonal to the C_n axis and containing all orthogonal C_2 axes. $n = 2, 3\ldots$ In the limit $n \to \infty$ we get the $D_{\infty h}$ group.

The seven exceptional groups:

T - A group of order 12 contains all rotational symmetry operations of a regular tetrahedron.

T_d - Group of order 24. The symmetry group of a regular tetrahedron.

T_h - Group of order 24 obtained by extension of group T by adding an inversion symmetry operation.

O - A group of order 24 contains all rotational symmetry operations of a regular octahedron (or cube).

O_h - Group of order 48. The symmetry group of a regular octahedron (or cube).

I - A group of order 60 contains all rotational symmetry operations of a regular icosahedron (or dodecahedron).

I_h - Group of order 120. The symmetry group of a regular icosahedron (or dodecahedron).

C.1 Two-dimensional point groups

There are two families of finite two-dimensional point groups.

A C_n, group of order n, is generated by rotation over $2\pi/n$. $n = 1, 2, 3, \ldots$

Another family of groups is the extension of C_n by reflection in line passing through the rotation axis. There is no universal notation for groups in this family. D_n or C_{nv} notation is used because of obvious correspondence with notation for three-dimensional point groups.

Two continuous two-dimensional point groups $SO(2)$ and $O(2)$ can be described equally as C_∞ and D_∞ ($C_{\infty v}$) groups respectively.

C.2 Crystallographic plane and space groups

For the notation for two- and three-dimensional crystallographic groups we simply refer to the International Tables of Crystallography [14] or to any basic book on crystallography.

C.3 Notation for four-dimensional parallelohedra

We give here correspondence between different notations used for four-dimensional lattices. Delone was the first to give in 1929 a list of 51 combinatorial types of four-dimensional lattices. In [41] he gave figures of three-dimensional projections for all 51 found types, numerated consecutively by numbers from 1 to 51. For each of these 51 types he gave also numbers of facets of each type and in cases when several polytopes have the same numbers, he added information which allows us to make a distinction between different polytopes. In 1973 Shtogrin [87] found one combinatorial type missed by Delone. We give in Tables C.2 and C.3 characterization of all 52 types. Column "Delone" gives numbering used by Delone in [41] together with his description of the set of facets in the form used by Delone, namely: $(n_1)_{k_1} + (n_2)_{k_2} + \dots$ where $(n_i)_{k_i}$ gives the number n_i of facets with k_i 2-faces. The combinatorial type discovered by Shtogrin is denoted as **St**.

In the tables we refer also to two types of notations used by Engel [11, 49, 53]. A short notation indicates the number of facets and uses consecutive numbers $1, 2, \dots$ to label polytopes within the subset of polytopes with the same number of facets. The more detailed notation uses symbol $\mathbf{N_f}.\mathbf{N_v}$-$n_6$ where $\mathbf{N_f}$ is the number of facets, $\mathbf{N_v}$ is the number of vertices, and n_6 is the number of hexagonal 2-faces. When such labeling is insufficient, a full description uses 2-subordinate and 3-subordinate symbols $K_{\alpha_{n_\alpha}} \dots$ giving numbers K_α of 2-faces with n_α edges in case of the 2-subordinate symbol and numbers K_α of 3-faces with n_α 2-faces in the case of 3-subordinate symbol.

For zonohedral polytopes we give also the notation used by Conway [32] and slightly different but essentially the same notation used by Deza and Grishukhin [44] (see column DG). For non-zonohedral polytopes we do not use Conway notation which is based on a rather different principle and give notation used by Deza and Grishukhin for a zonotope contribution $Z(U)$ which allows us to write a non-zonohedral polytope as a Minkowski sum of $P_{24} = \mathbf{24}.\mathbf{24}$-0 and a zonotope $Z(U)$.

TAB. C.2 – Combinatorial types of four-dimensional zonohedral lattices. Correspondence between notations.

m	Engel	Engel (full)	Delone	Conway	DG
10	30-2	30.120-60	1	K_5	K_5
		$4_{90}6_{60}$; $8_{20}14_{10}$	$10_{14}+20_8$		
9	30-1	30.102-36	19	$K_{3,3}$	$K_{3,3}^*$
		$4_{108}6_{36}$; $6_{12}12_{18}$	$18_{12}+12_6$		
	28-4	28.96-40	4	K_5-1	$K_5-\mathbf{1}$
		$4_{90}6_{40}$; $6_6 8_{12}12_6 14_4$	$4_{14}+6_{12}+12_8+6_6$		
8	24-16	24.72-26	6	K_5-2	$K_5-\mathbf{2}$
		$4_{76}6_{26}$; $6_8 8_{10}12_4 14_2$	$2_{14}+4_{12}+10_8+8_6$		
	26-8	26.78-24	5	K_5-1-1	$K_5-2\times\mathbf{1}$
		$4_{92}6_{24}$; $6_8 8_8 12_{10}$	$10_{12}+8_8+8_6$		
7	16-1	16.48-16	8	K_4+1	$K_4+\mathbf{1}$
		$4_{48}6_{16}$; $6_6 8_8 14_2$	$2_{14}+8_8+6_6$		
	20-3	20.54-16	10	K_5-3	$K_5-\mathbf{3}$
		$4_{64}6_{16}$; $6_8 8_8 12_4$	$4_{12}+8_8+8_6$		
	22-2	22.54-12	11	C_{2221}	C_{2221}
		$4_{72}6_{12}$; $6_{16}12_6$	$6_{12}+16_6$		
	24-12	24.60-12	7	K_5-2-1	$K_5-\mathbf{1}-\mathbf{2}$
		$4_{86}6_{12}$; $6_{10}8_8 12_6$	$6_{12}+8_8+10_6$		
6	12-1	12.36-12	16	C_3+C_3	C_3+C_3
		$4_{36}6_{12}$; 8_{12}	12_8		
	14-2	14.36-8	13	K_4	K_4
		$4_{44}6_8$; $6_8 8_4 12_2$	$2_{12}+4_8+8_6$		
	20-2	20.42-6	12	C_{321}	C_{321}
		$4_{66}6_6$; $6_{12}8_6 12_2$	$2_{12}+6_8+12_6$		
	22-1	22.46-0	9	C_{222}	C_{222}
		4_{84}; $6_{16}12_6$	$6_{12}+16_6$		
5	10-1	10.24-4	17	C_3+1+1	$C_3+2\times\mathbf{1}$
		$4_{30}6_4$; $6_6 8_4$	4_8+6_6		
	14-1	14.28-0	15	C_4+1	$C_4+\mathbf{1}$
		4_{48}; $6_{12}12_2$	$2_{12}+12_6$		
	20-1	20.30-0	14	C_5	C_5
		4_{60}; 6_{20}	20_6		
4	8-1	8.16.0	18	$1+1+1+1$	$4\times\mathbf{1}$
		4_{24}; 6_8	8_6		

TAB. C.3 – Combinatorial types of four-dimensional lattices obtained as a sum $P_{24}+Z(U)$ of 24-cell $P_{24}=$ **24.24**-0 and a zonotope $Z(U)$. Correspondence between notations for polytopes and for zonotope contribution to the sum.

m	Engel	Engel (full)	Delone	$Z(U)$, [DG]
10	30-3	30.120-42	2	$K_5-\mathbf{1}$
		$4_{72}5_{36}6_{42}$; $6_6 8_2 10_{12} 12_6 14_4$	$4_{14}+6_{12}+12_{10}+2_8+6_6$	
	30-4	30.120-36	3	$K^*_{3,3}$
		$3_6 4_{54} 5_{54} 6_{36}$; $6_6 8_6 12_{18}$	$18_{12}+6_8+8_6$	
9	28-6	28.104-24	21	$K_5-2\times\mathbf{1}$
		$3_6 4_{52} 5_{54} 6_{24}$; $6_4 8_6 19_8 12_{10}$	$10_{12}+8_{10}+6_8+4_6$	
	28-5	28.104-30	20	$K_5-\mathbf{2}$
		$4_{70}5_{36}6_{30}$; $6_4 8_4 10_{14} 12_4 14_2$	$2_{14}+4_{12}+14_{10}+4_8+4_5$	
8	26-9	26.88-12	24	$K_5-\mathbf{1}-\mathbf{2}$
		$3_{12}4_{38}5_{60}6_{12}$; $6_2 8_{10} 10_8 12_6$	$6_{12}+12_{10}+6_8+4_6$	
	26-10	26.88-18	22	$K_5-\mathbf{3}$
		$3_6 4_{56} 5_{42} 6_{18}$; $6_2 8_8 10_{12} 12_4$	$4_{12}+12_{10}+8_8+2_6$	
	26-11	26.88-24	25	$K_4+\mathbf{1}$
		$4_{74}5_{24}6_{24}$; $6_2 8_8 10_{14} 14_2$	$2_{14}+14_{10}+8_8+2_6$	
	28-3	28.94-12	24	$K_5-\mathbf{1}-\mathbf{2}$
		$3_6 4_{60} 5_{54} 6_{12}$; $6_4 8_6 10_{12} 12_6$	$6_{12}+12_{10}+6_8+4_6$	
	28-2	28.94-18	23	C_{2221}
		$4_{78}5_{36}6_{18}$; $6_4 8_6 10_{12} 12_6$	$6_{12}+12_{10}+6_8+4_6$	
7	24-17	24.72-0	31	C_{222}
		$3_{24}4_{12}5_{72}$; $8_{18}12_6$	$6_{12}+18_8$	
	24-18	24.72-12b	30	$C_{221}+\mathbf{1}$
		$3_{12}4_{48}5_{36}6_{12}$; $8_{14}10_8 12_2$	$2_{12}+8_{10}+14_8$	
	24-19	24.72-12a	32	C_3+C_3
		$3_{12}4_{48}5_{36}6_{12}$; $8_{12}10_{12}$	$12_{10}+12_8$	
	24-20	24.72-24	33	K_4
		$4_{84}6_{24}$; $8_{16}10_6 14_2$	$2_{14}+6_{10}+16_6$	
	26-6	26.78-6	28	$C_{221}+\mathbf{1}$
		$3_{12}4_{52}5_{48}6_6$; $6_2 8_{10} 10_{12} 12_2$	$2_{12}+12_{10}+10_8+2_6$	
	26-7	26.78-12	27	C_{321}
		$3_6 4_{70} 5_{30} 6_{12}$; $6_2 8_{10} 10_{12} 12_2$	$2_{12}+12_{10}+10_8+2_6$	
	28-1	28.88-0	29	C_{222}
		$3_6 4_{72} 5_{54}$; $6_4 8_6 10_{12} 12_6$	$6_{12}+12_{10}+6_8+4_6$	

m	Engel	Engel (full)	Delone	$Z(U)$, [DG]
6	24-15	24.62-0	37	$C_4 + \mathbf{1}$
		$3_{24}4_{32}5_{48}$; $8_{18}10_{4}12_{2}$	$2_{12} + 4_{10} + 18_{8}$	
	24-13	24.62-6	38	$C_3 + 2 \times \mathbf{1}$
		$3_{18}4_{50}5_{30}6_{6}$; $8_{16}10_{8}$	$8_{10} + 16_{8}$	
	24-14	24.62-12	39	C_{221}
		$3_{12}4_{68}5_{12}6_{12}$; $8_{18}10_{4}12_{2}$	$2_{12} + 4_{10} + 18_{8}$	
	26-3	26.68-0	35	C_5
		$3_{18}4_{54}5_{42}$; $6_{2}8_{12}10_{12}$	$12_{10} + 12_{8} + 2_{6}$	
	26-4	26.68-6	34	$C_3 + 2 \times \mathbf{1}$
		$3_{12}4_{72}5_{24}6_{6}$; $6_{2}8_{12}10_{12}$	$12_{10} + 12_{8} + 2_{6}$	
	26-5	26.72-0	36	$C_4 + \mathbf{1}$
		$3_{12}4_{70}5_{36}$; $6_{2}8_{10}10_{12}12_{2}$	$2_{12} + 12_{10} + 10_{8} + 2_{6}$	
5	24-8	24.52-0	41	$4 \times \mathbf{1}$
		$3_{30}4_{40}5_{30}$; $8_{20}10_{4}$	$4_{10} + 20_{8}$	
	24-9	24.52-6	42	$C_3 + \mathbf{1}$
		$3_{24}4_{58}5_{12}6_{6}$; $8_{20}10_{4}$	$4_{10} + 20_{8}$	
	24-10	24.56-0c	43	$4 \times \mathbf{1}$
		$3_{24}4_{56}5_{24}$; $8_{16}10_{8}$	$8_{10} + 16_{8}$	
	24-11	24.56-0d	44	C_4
		$3_{24}4_{56}5_{24}$; $8_{18}10_{4}12_{2}$	$2_{12} + 4_{10} + 18_{8}$	
	26-2	26.62-0	40	$4 \times \mathbf{1}$
		$3_{18}4_{78}5_{18}$; $6_{2}8_{12}10_{12}$	$12_{10} + 12_{8} + 2_{6}$	
4	24-6	24.42-0	47	$3 \times \mathbf{1}$
		$3_{42}4_{36}5_{18}$; 8_{24}	24_{8}	
	24-5	24.42-6	**St**	C_3
		$3_{36}4_{54}6_{6}$; 8_{24}		
	24-7	24.46-0	48	$3 \times \mathbf{1}$
		$3_{36}4_{52}5_{12}$; $8_{20}10_{4}$	$4_{10} + 20_{8}$	
	26-1	26.56-0	45	$3 \times \mathbf{1}$
		$3_{24}4_{90}$; $6_{2}8_{12}10_{12}$	$12_{10} + 12_{8} + 2_{6}$	
3	24-3	24.36-0	49	$2 \times \mathbf{1}$
		$3_{54}4_{36}5_{6}$; 8_{24}	24_{8}	
	24-4	24.40-0	48	$2 \times \mathbf{1}$
		$3_{48}4_{52}$; $8_{20}10_{4}$	$4_{10} + 20_{8}$	
2	24-2	24.30-0	50	$\mathbf{1}$
		$3_{72}4_{24}$; 8_{24}	24_{8}	
1	24-1	24.24-0	51	24-cell
		3_{96}; 8_{24}	24_{8}	itself

Appendix D

Orbit spaces for plane crystallographic groups

A standard ITC [14] representation of 2D-plane crystallographic groups is given below along with corresponding description of orbit spaces (orbifolds) and with notation suggested by Conway which describes the topology and the singularity structure of orbifolds.

Orbifolds for five Bravais symmetry groups for two-dimensional lattices are discussed in section 4.5. Here we complete the discussion of orbifolds by treating all the rest of the 2D-symmetry groups.

Comments to figures are given here in parallel with explication of Conway notation [31, 33, 34].

The simplest group $p1$ (see Figure D.1) contains only translations and possesses only one type of orbit. By taking the elementary cell of the lattice formed by two independent translations we get a representation of a space of orbits as a parallelogram with respective points on the boundary being identified. After such identification we get that topologically the space of orbits for the $p1$ group is a torus. The presence of two nontrivial closed paths on the space of orbits (two generating circles for a torus) is manifested in the Conway notation as ◯. An interpretation of this notation ◯ is related to the fact that the torus can be obtained from a sphere by joining one handle.

Group $p2$ is discussed in 4.5.

The next example is the pm group (see Figure D.2). Along with translations the group pm contains reflection axes. There are two different axes which are not related by translation symmetry operation. Due to the presence of reflection axes the space of orbits has a boundary. For the pm group there are two inequivalent (by translation) boundaries formed by points belonging to orbits with stabilizer m. Each generic (principal) orbit with stabilizer $1 = C_1$ has two points in the elementary cell. Restricting to one point for each generic 1-orbit leads to the shaded region with the left and right boundary being identified. This identification leads to a cylinder as an orbifold. Two circular boundaries of this cylinder are formed by orbits with stabilizer m.

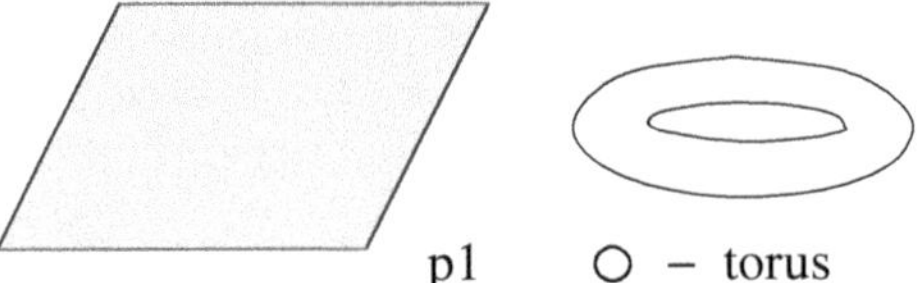

FIG. D.1 – Orbifold for the $p1$ crystallographic 2D-group.

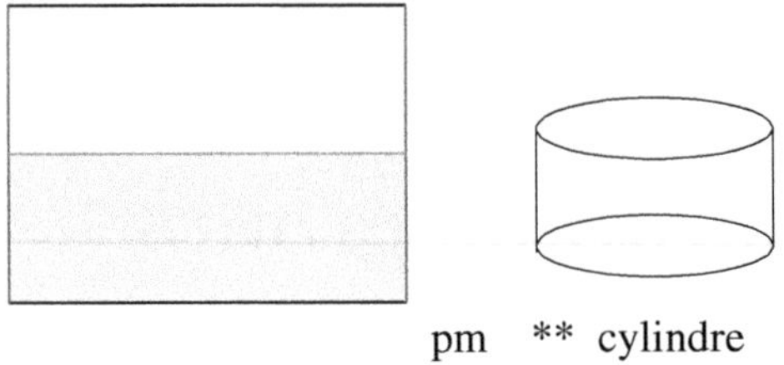

FIG. D.2 – Orbifold for the pm crystallographic 2D-group.

All other points represent generic orbits with stabilizer $1 = C_1$. The presence of a circular boundary is indicated in Conway notation by $*$. Consequently, the Conway notation for pm orbifold is $**$.

Group pg has only glide reflections (dash lines) in addition to two independent translations. All orbits are principal with the stabilizer 1. Due to glide reflection each elementary cell has two points from each orbit of the symmetry group action. The choice of one representative point from each orbit leads to the shaded region (see Figure D.3). Points on lower and upper boundaries should be identified because they are related by a translation. Points on left and right boundaries should be also identified but respecting the direction of the arrows. This identification is a result of transformation of the left boundary into the right boundary by the glide reflection (half-translation followed by a reflection). The resulting orbifold from the topological point of view is a Klein bottle. Conway notation for the Klein bottle is $\times\times$. Two signs $\times$ are used to indicate that the Klein bottle can be obtained from a sphere by adding two crosscaps.

The group cm contains reflection axes and glide reflection axes. Principal orbits (with stabilizer 1) have four points in the elementary cell. We can choose one point from each orbit by taking the shaded region of the elementary cell shown in Figure D.4. The boundary of this shaded region is formed by points belonging to orbits with stabilizer m (thick solid line). Two dash-dot lines forming the boundary of the triangle should be identified respecting the orientation of arrows because they are transformed one into another by a glide reflection. From the topological point of view the orbifold for the cm group is a Möbius band. The boundary of the band which is a topological circle is

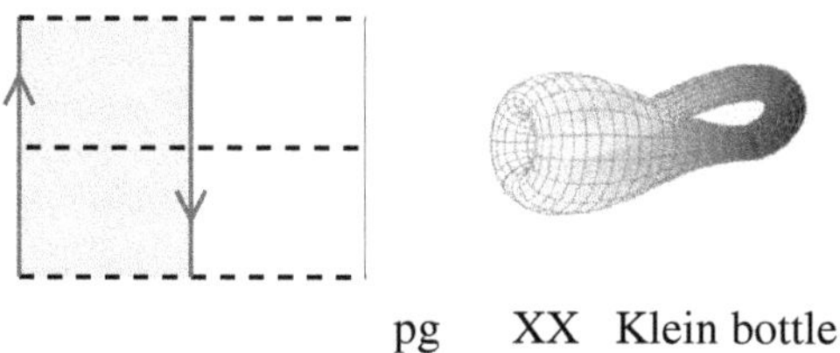

FIG. D.3 – Orbifold for the *pg* crystallographic 2D-group.

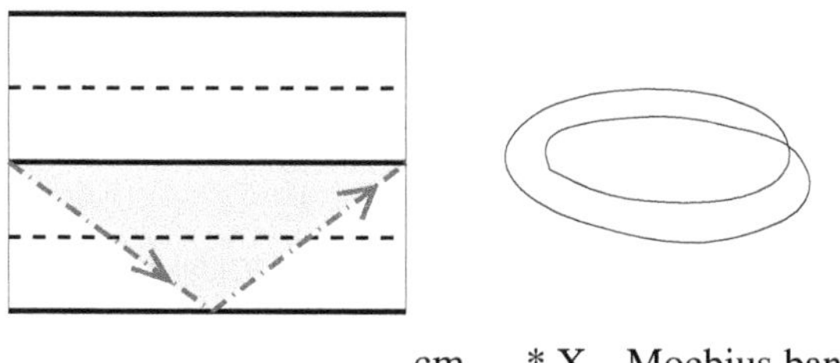

FIG. D.4 – Orbifold for the *cm* crystallographic 2D-group.

formed by orbits with stabilizer m. Conway notation for the orbifold is $*\times$. This notation indicates that there exists one boundary and one crosscap is glued to the demi-sphere to get the Möbius band.

The group *p2mm* is discussed in 4.5.

The group *p2mg* contains two independent rotation centers of order two and reflections in only one direction. It has also glide reflections (dash lines) whose axes are perpendicular to the reflection axes. The centers of rotation lie on glide reflection axes. There are two orbits with stabilizer $2 = C_2$ and each 2-orbit has two points in the elementary cell. Points with stabilizer m belong to reflection axes (solid lines). Each m-orbit has two points in one elementary cell. Each principal orbit has four points in the elementary cell. To form the space of orbits we choose the shaded rectangular region shown in figure D.5. Points situated on the glide reflection axes symmetrically with respect to the $2 = C_2$ axes should be identified. This gives the orbifold which from the topological point of view is a disk with a boundary formed by orbits with stabilizer m. Inside the disk there are two isolated orbits with stabilizer 2. Conway notation for an orbifold of *p2mg* group is $22*$. According to convention the orders of isolated rotation centers which do not belong to the boundary should be indicated before the boundary symbol, $*$.

The group *p2gg* contains two rotation centers of order two and glide reflections (dash lines) in two orthogonal directions. There are no reflections. The centers of rotation are not located on the glide reflection axes (see Figure D.6). The elementary cell contains two different orbits with stabilizer 2. Each orbit with stabilizer 2 has two representative points in the

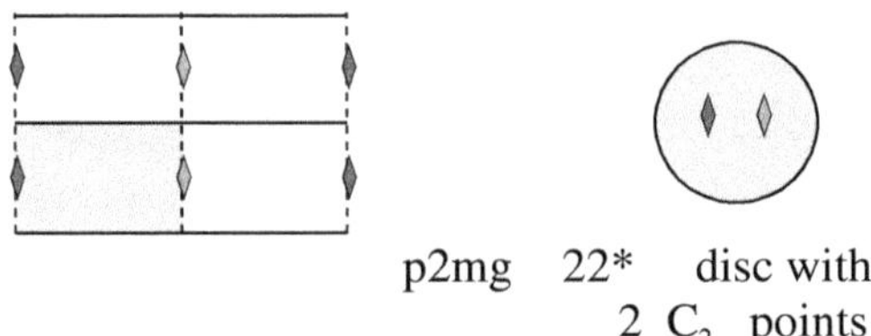

FIG. D.5 – Orbifold for the *p2mg* crystallographic 2D-group.

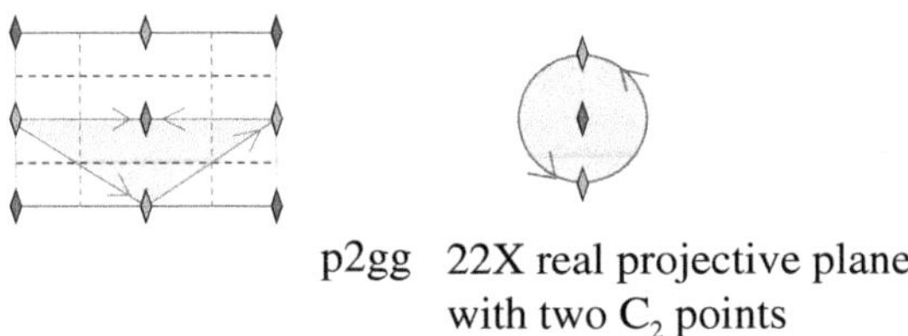

FIG. D.6 – Orbifold for the *p2gg* crystallographic 2D-group.

elementary cell. Each generic orbit with stabilizer 1 has four points in the elementary cell. The triangular shaded region shown in Figure D.6 contains one representative point from all orbits. Generic points on the boundary of a triangle should be identified under the action of 2-symmetry operations. Two half-sides of the base of the triangle can be first glued together. This leads to a subfugure D.6, right, which shows that the diametrically opposite points on the boundary of the disk should be identified. This is a standard construction of the real projective plane. Thus, the orbifold for the *p2gg* group is a real projective plane with two singular 2 points. The Conway notation for this orbifold is 22×. Initial 22 shows that there are two isolated 2 points. Symbol × indicates gluing of one crosscap. Remember that to get the real projective plane it is necessary to glue to a sphere one crosscup, whereas gluing two crosscaps leads to the Klein bottle.

The group *c2mm* is discussed in 4.5.

The group *p4* has two rotation centers of order four and one rotation center of order two. It has no reflection or glide reflections. Every principal orbit has four points in each elementary cell. The 2-orbit has two points in the elementary cell. Each of two C_4 orbits belonging to its own stratum (Wyckoff positions) has one point in the elementary cell. The shaded square shown in Figure D.7 includes one point from each orbit taking into account that points equivalent under C_4 rotation should be identified. The result of such identification is a sphere with three marked points, two C_4 (not conjugate) orbits and one C_2 orbit. The notation for the orbifold is 442 indicating the absence of the boundary and spherical topology.

The group *p4mm* is discussed in 4.5.

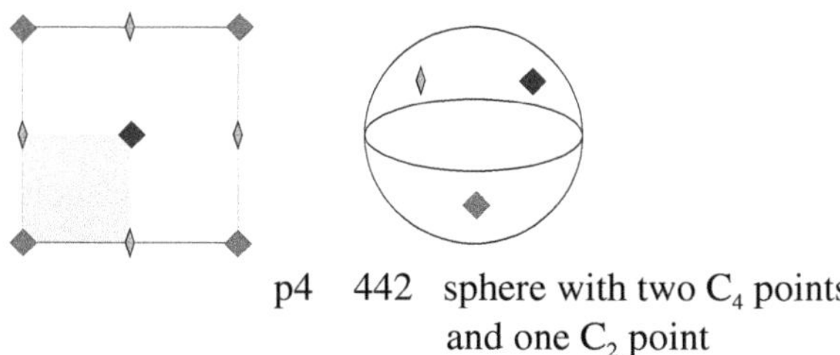

FIG. D.7 – Orbifold for the *p*4 crystallographic 2D-group.

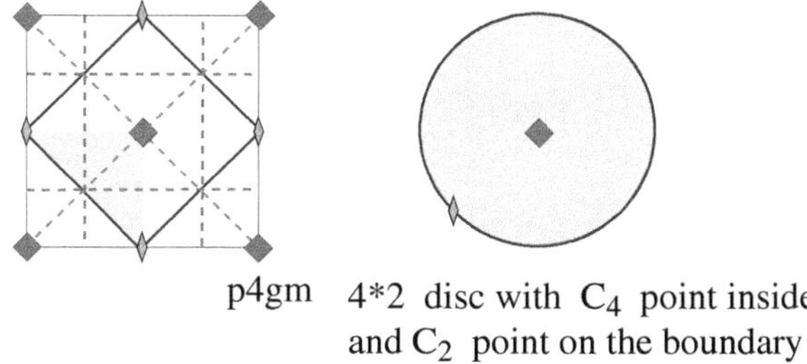

FIG. D.8 – Orbifold for the *p*4*gm* crystallographic 2D-group.

The group $p4gm$ has rotation centers of order four which form one orbit with stabilizer C_4. These rotation centers do not lie on reflection axes. The C_4 orbit has two representative points in each elementary cell. The group $p4gm$ has reflection axes in two orthogonal directions. There are rotation centers of order two which lie at the intersection of reflection axes. These order two rotation centers form one orbit with stabilizer $2m$. Each elementary cell contains two points belonging to the orbit with stabilizer $2m$. The group $p4gm$ has also two families of glide reflection axes - one in horizontal and vertical directions, the other at the angle of $\pi/4$ with these. Principal orbits (stabilizer C_1) has eight representative points in the elementary cell. Collecting one point from each orbit we get the space of orbits represented by a shaded triangle in figure D.8. Points on horizontal and vertical sides of this triangle should be identified due to action of C_4 rotation. Points belonging to the reflection axis (thick solid line in Figure D.8) form the boundary of the space of orbits together with one C_2-orbit on this boundary. Topologically the orbifold for the $p4gm$ group is a disc with one C_2 point on the boundary and one C_4 point inside. The Conway notation for this orbifold is $4{*}2$.

The group $p3$ has three different rotation centers of order three but no reflections or glide reflections. Each principal orbit with stabilizer $1 = C_1$ has three representative points in the elementary cell. Each of three orbits with

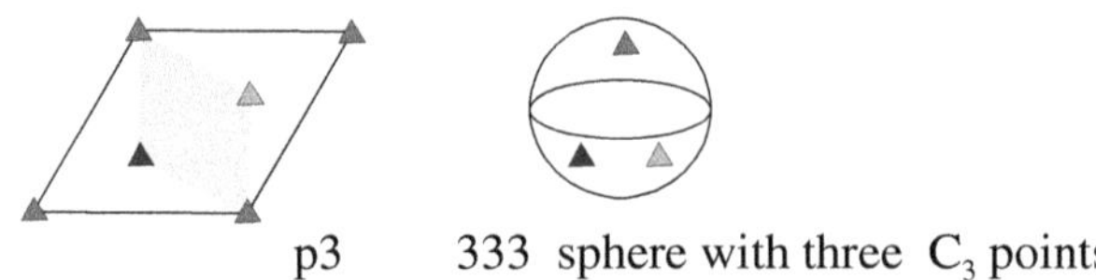

FIG. D.9 – Orbifold for the $p3$ crystallographic 2D-group.

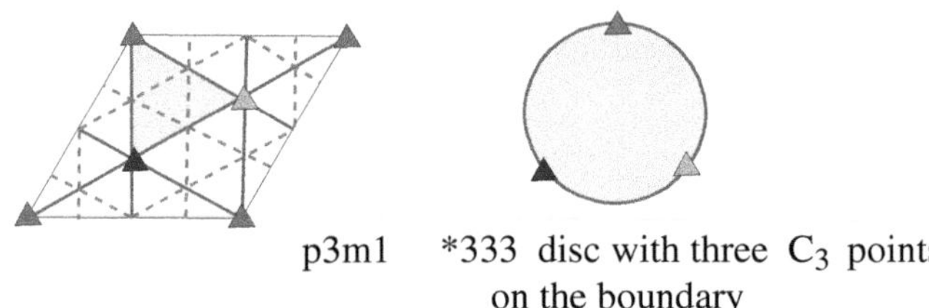

FIG. D.10 – Orbifold for the $p3m1$ crystallographic 2D-group.

stabilizer $3 = C_3$ forms its own stratum and has one representative point in the elementary cell. We can choose the domain of the elementary cell with one point from each orbit as shown in Figure D.9 by the shaded rhombus. The points on the sides of this rhombus equivalent with respect to order three rotation should be identified. This gives for the space of orbits from the topological point of view the sphere with three marked points being each representative of different strata with stabilizer $3 = C_3$. Conway notation for this orbifold is 333 indicating the absence of the boundary and existence of three singular points.

The group $p3m1$ has three different rotation centers of order three and reflection axes forming sides of an equilateral triangle. Each rotation center lies at the intersection of the reflection axes. There are additional glide reflections in three distinct directions whose axes are located halfway between adjacent parallel reflection axes. Each principal orbit has six representative points in the elementary cell. Orbits with stabilizer m (formed by points lying on reflection axes) have three representative points in the elementary cell. All these orbits belong to the same stratum. Finally, each of three orbits with stabilizer $3m$ has one representative point in the elementary cell and belongs to its own stratum. Collecting one point from each orbit we get the space of orbits represented in figure D.10 as a shaded triangle with its boundary. From the topological point of view the orbifold is a disc with three singular points at its boundary. The Conway notation for the orbifold of the group $p3m1$ is $*333$.

The group $p31m$ has two different types of rotation centers of order three. One rotation center of order three lies at the intersection of reflection axes forming an equilateral triangle. The stabilizer of the corresponding orbit is $3m$.

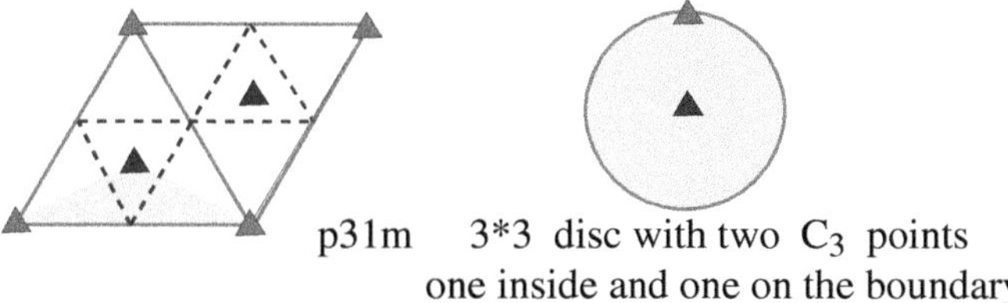

FIG. D.11 – Orbifold for $p31m$ crystallographic 2D-group.

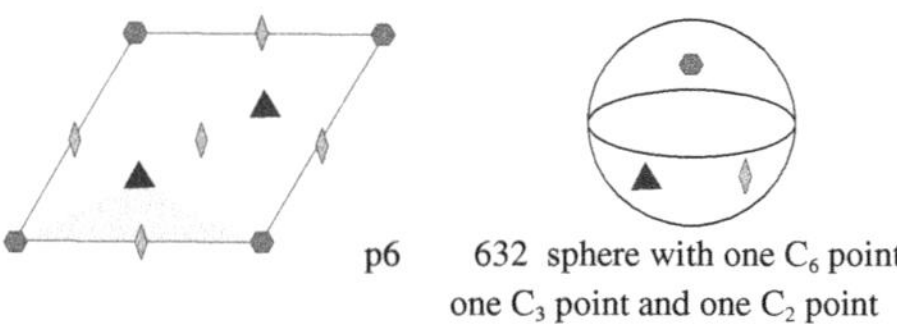

FIG. D.12 – Orbifold for the $p6$ crystallographic 2D-group.

This orbit has one representative point in the elementary cell. Another order three rotation center does not lie on the reflection axes and form an orbit with stabilizer C_3. The corresponding orbit has two representative points in the elementary cell. Points belonging to reflection axes form orbits with stabilizer m. Each such orbit has three representative points in the elementary cell. The group $p31m$ has also glide reflections in three distinct directions, whose axes are located halfway between adjacent parallel reflection axes. Principal orbits (with stabilizer C_1) have six points in the elementary cell. Taking one point from each orbit we form the shaded triangle (see Figure D.11) corresponding to the space of orbits after identification of points on the sides equivalent with respect to C_3 rotation. After such identification the orbifold becomes a topological disc with one singular point on the boundary and one C_3 point inside. The Conway notation of the orbifold is $3{*}3$.

The group $p6$ has one rotation center of order six, two rotation centers of order three and three rotation centers of order two. It has no reflection axes or glide reflection axes. Orbits with stabilizer $C_6 = 6$, $C_3 = 3$, $C_2 = 2$, and $C_1 = 1$ have respectively one, two, three, and six points in the elementary cell. The domain including one point from each orbit is shown by the shading in Figure D.12. To get the space of orbits we need to identify the points on the boundary of the shaded domain equivalent with respect to $C_2 = 2$ and $C_3 = 3$ rotations. After such identification the space of orbits becomes a sphere with three singular points corresponding to orbits with stabilizers $C_6 = 6$, $C_3 = 3$, and $C_2 = 2$ respectively. The notation for this orbifold is 632.

The group $p6mm$ is discussed in 4.5.

This completes the discussion of orbifolds for all plane crystallographic groups.

Appendix E

Orbit spaces for 3D-irreducible Bravais groups

There are three cubic Bravais groups $Pm\bar{3}m$, $Im\bar{3}m$, $Fm\bar{3}m$, corresponding to the same point symmetry group, O_h. We construct orbifolds for these three-dimensional irreducible Bravais groups to see the difference of group actions.

We start with the $Pm\bar{3}m$ group. Its elementary cell includes one point of the simple cubic lattice and is supposed to be of volume one. We use this primitive cell to represent different strata of the symmetry group action. Different strata (or systems of different Wyckoff positions according to ITC) are shown in Figure E.1.

There are two zero dimensional strata with stabilizer O_h, characterized as Wyckoff position a and b. The stabilizers of these two strata are not conjugate in the symmetry group of the lattice. Points a correspond to points forming the simple cubic lattice. Points b are situated in the center of the cubic cell formed by points a. Eight points of type a are shown in Figure E.1 but as soon as each point equally belongs to eight cells there is only one point a per cell.

There are also two zero-dimensional strata with stabilizer D_{4h} which are not conjugate in the symmetry group of the lattice. They are labeled as c and d (according to ITC). There are three positions of type c per cell and three positions of type d per cell. Each point of type c belongs to two cells whereas each point of type d belongs to 4 cells. That is why there are 6 points of type c and 12 points of type d drawn in Figure E.1.

There are six different one-dimensional strata of the $Pm\bar{3}m$ group action on the space. Two one-dimensional strata have as stabilizers two C_{4v} subgroups which are not conjugate in the lattice symmetry group. These two strata are shown on the same subfigure in Figure E.1. Each orbit of e (solid line) symmetry type has six points per primitive cell. Two points of each orbit are situated on each of three disconnected intervals shown by solid thick line. As soon as these solid lines are edges of the primitive cell and belong, in fact, to four cells, all other edges are equivalent and consequently belong to the

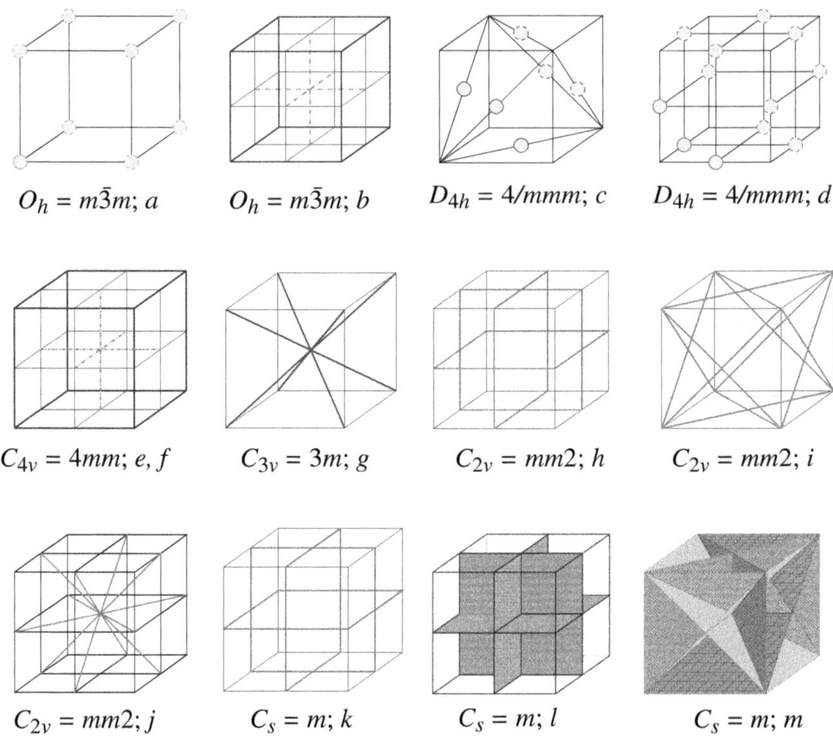

FIG. E.1 – Different strata for $Pm\bar{3}m$ Bravais group.

same stratum. Orbits of f type are situated inside the primitive cell on six intervals marked by the dash-dot line. One point of each orbit belongs to one of six equivalent intervals forming one stratum.

The C_{3v} stratum (type g of Wyckoff positions) consists of orbits having eight points per cell situated on the diagonals of primitive cell.

There are three non-conjugated C_{2v} strata (types h, i, j of Wyckoff positions). These strata are shown in three subfigures of Figure E.1. Each orbits has 12 points per cell for each of these three strata.

There are also three two-dimensional strata k, l, m. Each of them has $C_s \equiv m$ group as the stabilizer, but all of these three stabilizers are non-conjugate subgroups of the lattice symmetry group. The last three subfigures of E.1 show these strata. (Better visualization of m stratum can be done by using three rather than one subfigures. This is done for the $Fm\bar{3}m$ group in figure E.7, see three initial figures for stratum k.) Each orbit belonging to these strata has 24 points per cell.

At last, all points which do not belong to the mentioned above strata form generic stratum with trivial stabilizer $1 \equiv C_1$. It consists of orbits having 48 points per cell.

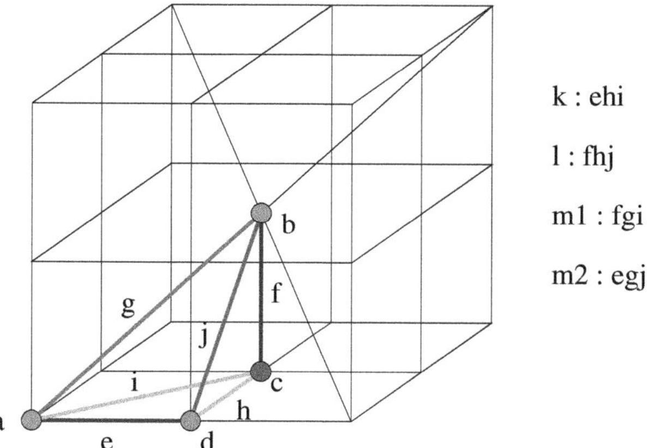

FIG. E.2 – Primitive cell and orbifold for the $Pm\bar{3}m$ three-dimensional Bravais group.

In order to construct the orbifold of the $Pm\bar{3}m$ group action we can choose a closed region (simplex) shown in Figure E.2. Its vertices are points from different zero-dimensional strata a, b, c, d. Six of its edges are also formed by points belonging to different one-dimensional strata: $e : a - d$; $f : b - c$; $g : a - b$; $h : c - d$; $i : a - c$; $j : b - d$. (Each edge is indicated by its two boundary vertices.) Among four faces, two belong to the same stratum of type m, namely, $m1 : fgi$ and $m2 : egj$ (the face is indicated by its three boundary edges). Each of two other faces belongs to its proper stratum: $k : ehi$, $l : fhj$. Internal points belong to generic stratum, n.

Topologically, the orbifold of the $Pm\bar{3}m$ group is a three-dimensional disk with all internal points belonging to the generic stratum and the boundary formed by 13 different strata.

Now we turn to the $Im\bar{3}m$ Bravais group. In order to have a cell whose symmetry coincides with the symmetry of the lattice, we are obliged to take a double cell which has volume 2 and includes two lattice points per cell. Different strata of the $Im\bar{3}m$ action are shown in Figure E.3. It is instructive to briefly compare the system of strata of $Im\bar{3}m$ with that of $Pm\bar{3}m$ by ignoring the difference in volumes of cells. The notation of strata by Latin letters follows again the notation of Wyckoff positions in ITC. Zero dimensional stratum a (stabilizer O_h) of $Im\bar{3}m$ includes points of strata a and b of $Pm\bar{3}m$. Zero dimensional stratum b of $Im\bar{3}m$ (stabilizer D_{4h}) includes points of strata c and d of $Pm\bar{3}m$. Zero dimensional strata c (stabilizer D_{3d}) and d (stabilizer D_{2d}) of $Im\bar{3}m$ are the new ones as compared to stratification imposed by $Pm\bar{3}m$.

One-dimensional stratum e of $Im\bar{3}m$ includes points belonging to e and f strata of $Pm\bar{3}m$. Stratum f (stabilizer C_{3v}) of $Im\bar{3}m$ action coincides with the stratum g of $Pm\bar{3}m$. The group $Im\bar{3}m$ has three one-dimensional strata

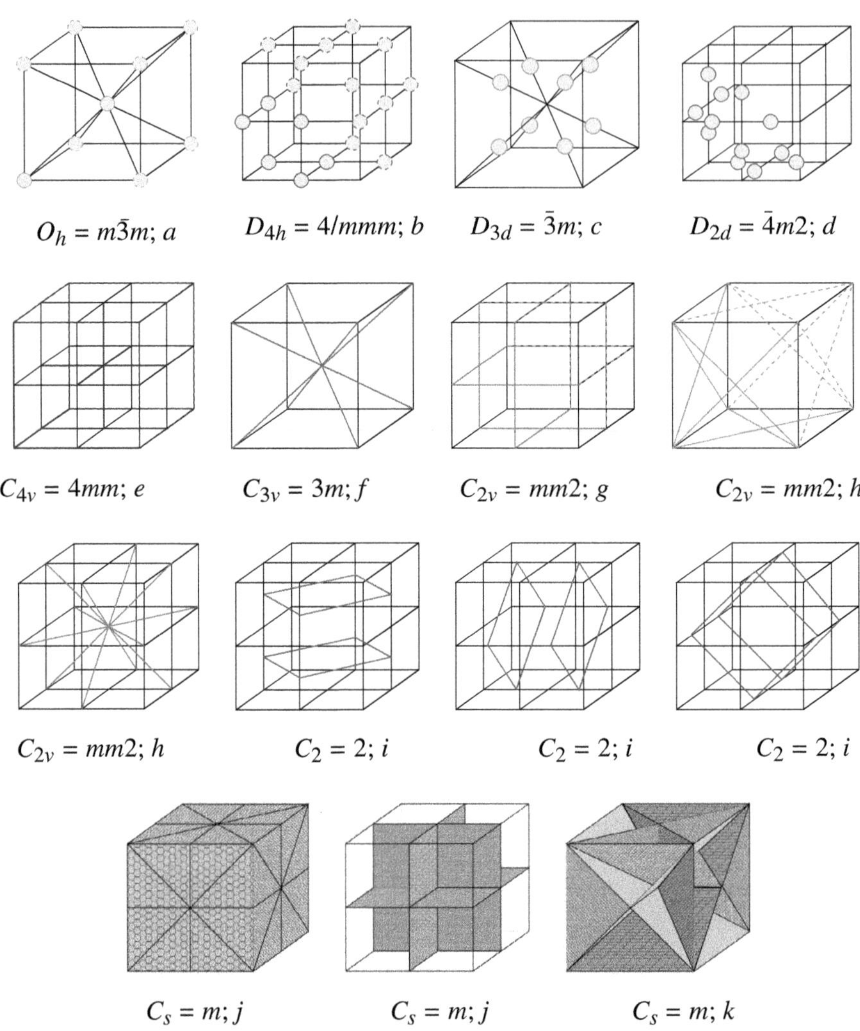

FIG. E.3 – Different strata for the $Im\bar{3}m$ Bravais group. In order to simplify visualization of the stratum d with stabilizer D_{2d} only translationally inequivalent points are represented.

g, h, i with stabilizer C_{2v}. The one-dimensional stratum g of $Im\bar{3}m$ action coincides with stratum h of $Pm\bar{3}m$. The one-dimensional stratum h of $Im\bar{3}m$ includes points of two strata i and j of $Pm\bar{3}m$. The one-dimensional stratum i of $Im\bar{3}m$ is a new one as compared to the stratification imposed by $Pm\bar{3}m$.

The two-dimensional stratum j (stabilizer C_s) of $Im\bar{3}m$ includes points belonging to two strata k and l of $Pm\bar{3}m$. The two-dimensional stratum k (stabilizer C_s) of $Im\bar{3}m$ reproduces stratum m of $Pm\bar{3}m$.

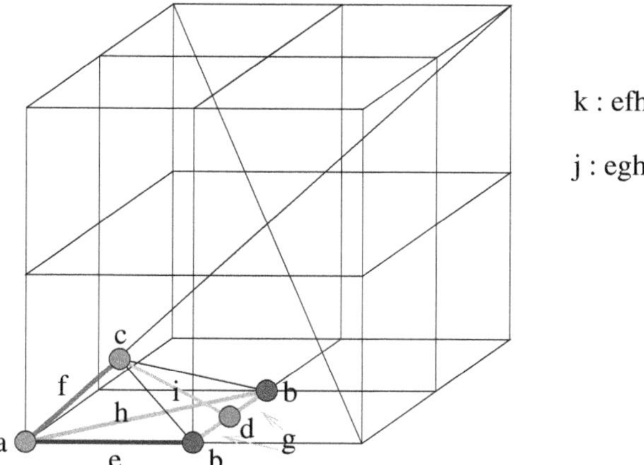

FIG. E.4 – Double cell and orbifold for the $Im\bar{3}m$ three-dimensional Bravais group.

In order to construct the orbifold and to take only one point from each orbit we can take the region shown in Figure E.4. The choice of this region coincides with the choice of the asymmetric unit suggested by ITC for the $Im\bar{3}m$ group. One should only additionally take into account the following important facts.

i) The two points marked by b in E.4 belong to the same stratum b and moreover to the same orbit with stabilizer D_{4h}. This can be easily seen because all points on the line *cd* have stabilizer C_2 and this C_2 rotation is obviously rotation around the *cd* line. This C_2 symmetry transformation unifies not only two points marked b into one orbit but also it acts on any point of the *bcb* triangular face of the chosen region. This indicates that pairs of respective points in two *cbd* triangles should be identified in order to construct the orbifold including only one point from each orbit. From the topological point of view the result of gluing two *cbd* triangles is the orbifold shown in Figure E.5. It can be represented as a three-dimensional body having the geometrical form of a double cone with two special points (c, d) at its apexes and two special points (a, b) on the equator. Moreover, all other points of the equator belong to two different (h and e) one-dimensional strata. Two other one-dimensional strata connect on the surface of double cone points a and c (stratum f) and points b and d (stratum g). Inside a double cone there is one more one-dimensional stratum i connecting points c and d. All other internal points belong to generic stratum l. Boundary points of the double cone which do not belong to the mentioned above zero-dimensional and one-dimensional strata form two two-dimensional strata. Stratum k consists of points of the upper part of the double cone boundary. This two-dimensional stratum has one-dimensional strata f, h, and e as its boundary. Stratum j consists of

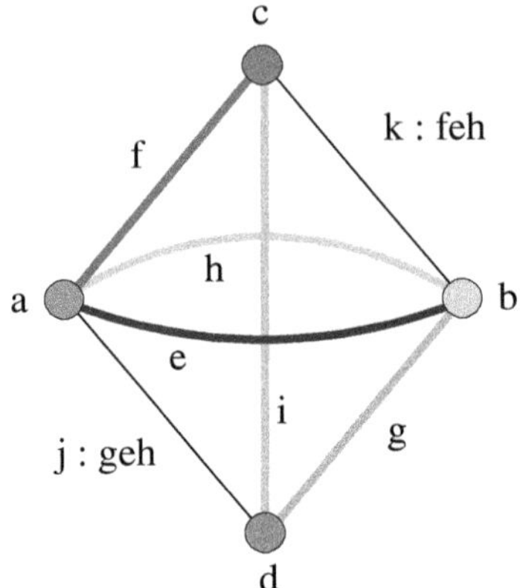

FIG. E.5 – Schematic representation of the orbifold for the $Im\bar{3}m$ three-dimensional Bravais group.

points of the lower part of the double cone boundary. This two-dimensional stratum has one-dimensional strata g, e, and h as its boundary.

In the case of the $Fm\bar{3}m$ Bravais group the choice of the cell respecting the O_h holohedry of the lattice leads to the quadruple cell of volume 4 as compared to the primitive cell. All zero-, one- and two-dimensional strata of the lattice symmetry group action on this quadruple cell are shown in Figures E.6, E.7.

The notation of strata by Latin letters follows again the notation of Wyckoff positions adapted in ITC. A zero dimensional stratum with stabilizer D_{2d} is shown in two subfigures in order to see better the location of all points. In this case one orbit includes 24 points per quadruple cell. In a similar way a one-dimensional stratum of type f (stabilizer C_{3v}) is represented in four sub-figures. Two of three non-conjugated in the lattice symmetry group strata with stabilizer C_{2v}, namely strata of type h and i, are also shown in two sub-figures. A two-dimensional stratum of type j (stabilizer C_s) is represented in two subfigures which coincide with figures of stratum j of Bravais group $Im\bar{3}m$. A two-dimensional stratum of type k (stabilizer C_s) is represented in six subfigures. Three of these subfigures reproduce figures of stratum k for the $Im\bar{3}m$ group or stratum m for the $Pm\bar{3}m$ group.

In order to construct the orbifold we need to take one representative point from each orbit. This can be done by restricting the quadruple cell to the region having tetrahedral geometry (see figure E.8) with coordinates of vertices

$a : \{0,0,0\};\ \ b : \{1/2,0,0\};\ \ c : \{1/4,1/4,1/4\};\ \ d : \{1/4,1/4,0\}.$

This choice coincides with the choice of the asymmetric unit for the $Fm\bar{3}m$ group made in ITC. The $Fm\bar{3}m$ orbifold is a topological three-dimensional disk. All its internal points belong to the generic C_1 stratum. The stratification

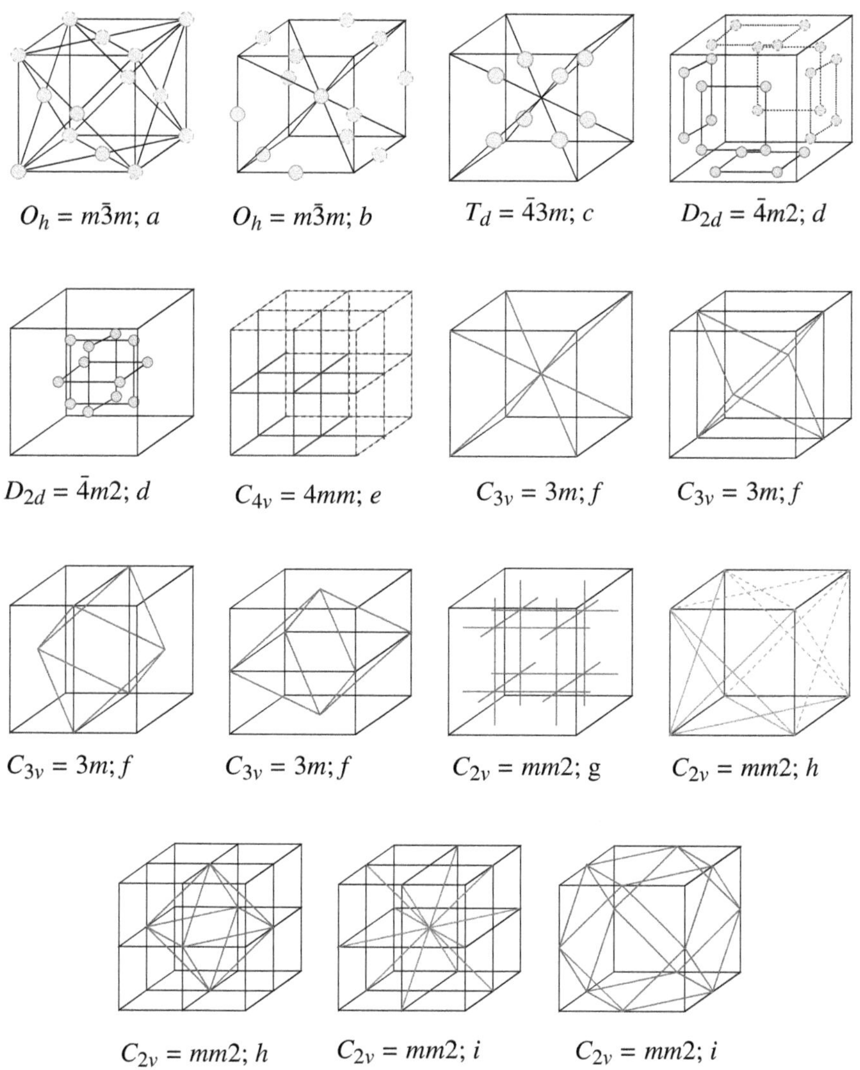

FIG. E.6 – Different zero- and one-dimensional strata for the $Fm\bar{3}m$ Bravais group.

of boundary is similar to the $Pm\bar{3}m$ orbifold. For $Fm\bar{3}m$ all four vertices belong to different zero-dimensional strata, but among six edges there are two belonging to the same stratum, and among four faces, three belong to the same stratum.

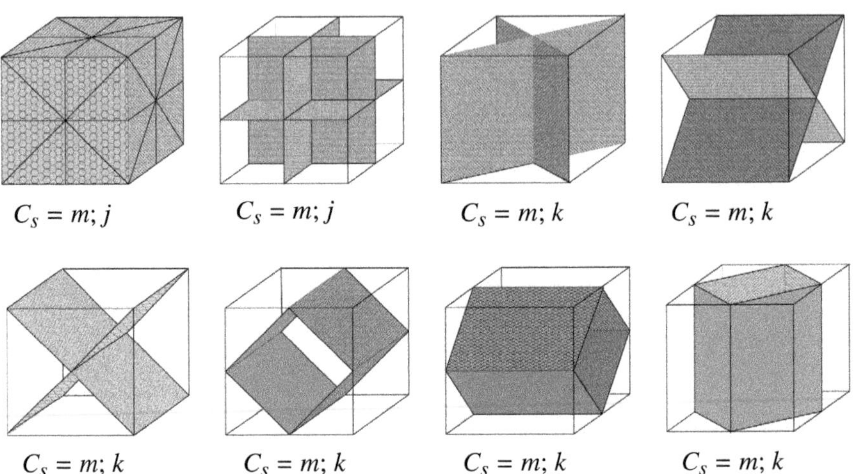

FIG. E.7 – Different two-dimensional strata for the $Fm\bar{3}m$ Bravais group.

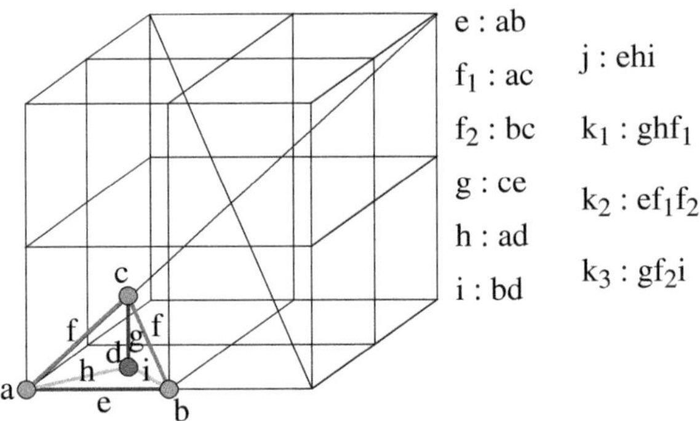

FIG. E.8 – Schematic representation of orbifold for the $Fm\bar{3}m$ three-dimensional Bravais group.

Bibliography

[1] M. Aigner, *Combinatorial theory*, Springer-Verlag. Berlin, 1979.

[2] A. Barvinok, *A Course in Convexity*, AMS, Providence, 2002.

[3] H. Brown, R. Bülow, J. Neubüser, H. Wondratschek, and H. Zassenhaus, *Crystallographic groups of four-dimensional space,* Wiley, New York, 1978.

[4] J.H. Conway, H. Burgiel, Ch. Goodman-Strauss, *The Symmetry of things*, A K Peters, Ltd. 2008.

[5] J.H. Conway and N.J.A. Sloane. *Sphere Packings, Lattices, and Groups,* Springer Verlag, New York, 1988.

[6] H.S.M. Coxeter. *Regular polytopes*, Dover, 3rd ed., 1973 (1st ed. Macmillan, 1947).

[7] H.S.M. Coxeter, *Introduction to Geometry*, Wiley and Sons, New York, 1969.

[8] B.N. Delone, *The St.Petersburg School of Number Theory.* AMS, 2005; translated from Russian edition, 1947.

[9] M. Deza, M. Laurent, *Geometry of cuts and metrics*, Springer Verlag, Berlin, 1997.

[10] P. Duval, *Homographies, Quaternions, and Rotations*, Clarendon Press, Oxford, 1964.

[11] P. Engel, *Geometric Crystallography*, Reidel, Dordrecht, 1986.

[12] E.S. Fedorov, *Nachala ucheniya o figurakh*, 1885. (In Russian.)

[13] B. Grünbaum and G.C. Shepard, *Tilings and Patterns,* W.H. Freeman, New York, 1987.

[14] *International Tables for Crystallography. Vol. A. Space Group Symmetry.* Hahn T. Ed. Kluwer, Dordrecht, 1996.

[15] J. Martinet, *Les réseaux parfaits des espaces Euclidiens*, Masson, Paris, 1996.

[16] M. Newman, *Integral Matrices,* Academic Press, New York, 1972.

[17] M. Senechal, *Quasicrystals and geometry.* Cambridge University Press, 1995.

[18] R.P. Stanley, *Combinatorics and Commutative Algebra,* Birkhäuser, Boston, 1996.

[19] *Symmetries in Nature. Scientific heritage of Louis Michel.* Eds. T. Damour, I. Todorov, B. Zhilinskii, World Scientific, 2014.

[20] K. Truemper, *Matroid decomposition.* Leibniz, Plano, Texas, 1998.

[21] W.P. Thurston, *Three-Dimensional Geometry and Topology,* Volume 1, (Princeton University Press, 1997).

[22] H. Weyl, *Symmetry.* Princeton Univ. Press, Princeton, 1952.

[23] G.M. Ziegler, *Lectures on polytopes.* Springer-Verlag, 1995.

Cited literature

[24] V.I. Arnol'd, Remarks on quasicrystallic symmetries. Physica D **33**, 21-25 (1988).

[25] I.A. Baburin, P. Engel, On the enumeration of the combinatorial types of primitive parallelohedra on E^d, $2 \leq d \leq 6$, Acta Cryst. A, **69**, 510-516 (2013).

[26] J. Bamberg, G. Cairns, D. Kilminster, The crystallographic restrictions, permutations, and Goldbach conjecture, Amer. Math. Month. **110**, 201-209 (2003).

[27] B.A. Bernevig, *Topological insulators and topological superconductors.* Princeton University Press, Princeton, 2013.

[28] L. Bieberbach, Ueber die Bewegungsgruppen der n-dimensionalen euklidischen Räume mit einem endlichen Fundamentalbereich, Göttinger Nachrichten (1910) 75-84; see also Ueber die Bewegungsgruppen der euklidischen Räume. Zweite Abhandlung. Die Gruppen mit endlichem Fundamentalbereich, Math. Ann. **72**, 400-412 (1912).

[29] A. Bravais, Mémoire sur les systèmes formés par des points distribués régulièrement sur un plan ou dans l'espace. Journal de l'Ecole Polytechnique, XIX, 1-128 (1850).

[30] E. Cartan, La géométrie des groupes simples. Annali Mat., **4**, 209-256 (1927).

[31] J.H. Conway, The orbifold notation for surface groups. In: Groups, Combinatorics and Geometry, London Mathematical Society Lecture Note Series **165**, Cambridge University Press, 1992, 438-447.

[32] J.H. Conway, The sensual (quadratic) forms. Math. Assoc. Amer., The Carus Math. Monographs, **26** (1997).

[33] J.H. Conway, O.D. Friedrichs, D.H. Huson, W.P. Thurston, On three-dimensional space groups. Contributions to Algebra and Geometry **42** (2001), 475-507.

[34] J.H. Conway, D.H. Huson, The orbifold notation for two-dimensional groups. Structural Chemistry, **13**, (2002) 247-257.

[35] J.H. Conway, N.J.A. Sloane. Low-dimensional lattices. VI. Voronoi reduction of three-dimensional lattices. Proc.R. Soc. Lond. A **436**, 55-68 (1992).

[36] J.H. Conway, N.J.A. Sloane. A lattice without a basis of minimal vectors. Mathematika, **42** (1995) 175-177.

[37] H.S.M. Coxeter. Discrete groups generated by reflections. Ann. Math., **85**, 588-621 (1934).

[38] V. Danilov, V. Grishukhin. Maximal Unimodular Systems of Vectors, Europ. J. Combinatorics, **20**, 507-526 (1999).

[39] B.L. Davies, R, Dirl, B. Goldsmith, Classification of subgroups of free Abelian groups, in "*Symmetry and Structural Properties of Condensed Matter*, T. Lulek, W. Florek, B. Lulek, eds., Proceedings of the 4th International SSPCM, Poznan, Poland, 1996; World Scientific, Singapore, 505-508 (1999).

[40] R. Decartes, *Traité sur la Lumière,* Paris, 1644.

[41] B.N. Delone (Delaunnay), Sur la partition régulière de l'espace à 4 dimensions. Bull. Acad. Sci. URSS, 79-110 (1929); ibid. 145-164.

[42] B.N. Delone, R.V. Galiulin, M.I. Shtogrin, "Bravais theory and its generalization to n-dimensional lattices". In Auguste Bravais: Collected Scientific Works, Nauka, Leningrad, 1974, Ch.1, sect. 5; English translation "On the Bravais types of lattices"... 1975, Plenum Publishing.

[43] B.N. Delone, N.P. Dolbilin, M.I. Shtogrin, R.V. Galilulin, A local criterion for regularity of a system of points. Reports of the Academy of Sciences of the USSR, **227** (in Russian); English translation: Soviet Math. Dokl. **17** No 2, 319-322 (1976).

[44] M. Deza, V.P. Grishukhin, More about the 52 four-dimensional parallelotopes. Taiwanese J. Math. **12**, 901-916 (2008).

[45] P.G.L. Dirichlet, Über die Reduction der positiven quadratischen Formen mit drei unbestimmten ganzen Zahlen, J. reine angew. Math., **40** 209-227 (1850); [Oeuvre Vl. II, p. 41-59].

[46] N.P. Dolbilin, J.C. Lagarias, M. Senechal, Multiregular Point Systems, Discrete and Computational Geometry, **20**, 477-498 (1998).

[47] P. Engel, Mathematical problems in modern crystallography, Comput. Math. Appl., **16**, 425-436 (1988).

[48] P. Engel, The enumeration of four-dimensional polytopes, Discrete Math. **91**, 9-31 (1991).

[49] P. Engel. On the symmetry classification of the four-dimensional parallelohedra. Zeitschrift für Kristallographie, **200**, 199-213 (1992).

[50] P. Engel. Investigations of parallelohedra in $\mathbb{R}^d$. *Voronoi's Impact on Modern Science,* **2**, eds. P. Engel, H. Syta, Institute of math., Kyiv, 1998, 22-60.

[51] P. Engel, The contraction types of parallelohedra in R^5, Acta Cryst. A **56**, 491-496 (2000).

[52] P. Engel, On the determination of a basis of facet vectors, Rendiconti del Circolo Matematico di Palermo, Serie II, Suppl. **70**, 269-278 (2002).

[53] P. Engel, L. Michel, M. Senechal. Lattice geometry, IHES/P/04/45, 280 p.

[54] R.M. Erdal. Zonotopes, dicings, and Voronoï's conjecture on parallelohedra. European Journal of Combinatorics **20**, 527-549 (1999).

[55] G. Friedel, Les états mésomorphes de la matière. Ann. Physique (Paris) **18**, 273-474 (1922).

[56] C. Greene, T. Zaslavsky, On the interpretation of Whitney numbers through arrangements of hyperplanes, zonotopes, non-Radon partitions, and orientations of graphs. Trans. Am. Math. Soc. **280**, 97-126 (1981).

[57] V.P. Grishukhin, Free and nonfree Voronoi polytopes. Math. Notes **80**. 355-365 (2006).

[58] V.P. Grishukhin, The Minkowski sum of a parallelotope and a segment, Sb.Math. **197**, 1417-1433 (2006).

[59] P.M. Gruber, S.S. Ryshkov, Facet-to-facet implies face-to-face, Europ. J. Comb. **10**, 83-84 (1989).

[60] C. Hermann, Kristallographie in Räumen beliebiger Diemensionzahl. I. Die Symmetrieoperationen, Acta Cryst., **2**, 139-145 (1949).

[61] H. Hiller, The crystallographic restrictions in higher dimensions, Acta Cryst., A **41**, 541-544 (1985).

[62] J. Kepler, Strena seu de nive sexangula, 1611.

[63] A. Korkin, G. Zolotareff, Sur les formes quadratiques. Math. Ann., **6**, 366-389 (1873).

[64] J. Kuzmanovich, A. Pavlichenkov, Finite groups of matrices whose entries are integers. Amer. Math. Monthly, **109**, 173-186 (2002).

[65] J.L. Lagrange, Recherches d'arithmétique. Oeuvre, Vol. III, Gauthier-Villars, Paris, 1773, 695-795.

[66] A.J. Mayer, Low dimensional lattices have a strict Voronoï basis. Mathematika, **42**, 229-238 (1995).

[67] P. McMullen, Convex bodies which tile the space by translation, Matematica **27**, 113-121 (1980).

[68] N.D. Mermin, The topological theory of defects in ordered media, Rev. Mod. Phys. **51**, 591-648 (1979).

[69] L. Michel, Invariance in quantum mechanics and group extensions. In *Group Theoretical Concepts and Methods in Elementary Particle Physics.* F. Gürsey edit., Gordon and Breach, New York (1964), 135-200. Reprinted in [19].

[70] L. Michel, The description of the symmetry of physical states and spontaneous symmetry breaking. In: *Symmetries and broken symmetries in condensed matter physics*, Proceedings of the Colloque Pierre Curie held at the Ecole Supérieure de Physique et de Chimie Industrielles de la Ville de Paris, Paris, September 1980, edit. N.Boccara, IDSET Paris (1981) 21-28; reprinted in [19], p.235.

[71] L. Michel, Symmetry defects and broken symmetry. Configurations. Hidden symmetry. Rev. Mod. Phys. **52**, 617-651 (1980).

[72] L. Michel, Extrema of P-invariant functions on the Brillouin zone. In: *Scientific Highlights in Memory of Léon Van Hove.* edit. F. Nicodemi, World Scientific, (1993) 81–108.

[73] L. Michel, Bravais classes, Voronoï cells, Delone symbols. In: *Symmetry and structural properties of condensed matter.* Proceedings of the third international school on theoretical physics, Zajączkowo, Poland, 1-7 September 1994, edit. T. Lulek, W. Florek, S. Wałcerz. World Scientific (1995) 279–316.

[74] L. Michel, Physical implications of crystal symmetry and time reversal, In: *Symmetry and structural properties of condensed matter.* (Proc. 4th Intern. School Theor. Phys. Zajaczkowo, Poland 29 August - 4 September 1996), edit. T. Lulek, W. Florek, B. Lulek, World Scientific (1997) 15-40.

[75] L. Michel, Editor, Symmetry, Invariants, Topology. Physics Reports, **341**, 1-395 (2001).

[76] L. Michel, J. Mozrzymas. Les concepts fundamentaux de la crystallographie. C. R. Acad. Sci. Paris, **308**, II 151-158 (1989).

[77] L. Michel, S.S. Ryshkov, M. Senechal. An extension of Voronoi's theorem on primitive parallelotopes. Europ. J. Combinatorics, **16**, 59-63 (1995).

[78] H. Minkowski, *Diophantische Approximationen*, Teubner, Leipzig, 1907; [Reprinted by Chelsea, New York, 1957].

[79] M. Morse, Relations between the critical points of a real function of n independent variables, Trans. Am. Math. Soc. **27**, 345-396 (1925).

[80] N.N. Nekhoroshev, D.A. Sadovskii, B. Zhilinskii, Fractional Hamiltonian monodromy. Ann. Henri Poincaré, **7**, 1099-1211 (2006).

[81] W. Plesken, T. Schulz, Counting crystallographic groups in low dimensions. Experimental mathematics, **9**, 407-411 (2000).

[82] G.-C. Rota, On the foundation of combinatorial theory, Z. Wahrsch. **2**, 340-368 (1964).

[83] S.S. Ryšhkov, E.P. Baranovskii, Repartitioning complexes in n-dimensional lattices (with full description for $n \leq 6$), in *Voronoï impact on modern Science*, P. Engel, H Syta, eds., Proc Inst. Math. Acad. Sci. Ikraine, vol. **21**, 115-124 (1998).

[84] L.A. Seeber, Untersushungen über die Eigenschaften der positiven ternären quadratischen Formen, Freiburg, 1831.

[85] J-P. Serre, *A course in arithmetic.* Springer, Berlin, (original French edition 1973).

[86] P.D. Seymour, Decomposition of regular matroids. J. Comb. Theory, Ser. B, **28**, 305-359 (1980).

[87] M.I. Shtogrin, Regular Dirichlet-Voronoï partitions for the second triclinic group. Proc. Steklov Inst. Math., **123** (1973).

[88] M.D. Sikirić, V. Grishukhin, A. Magazinov, On the sum of a parallelotope and a zonotope. European J. Combin. **42**, 49-73 (2014).

[89] B. Souvignier, Enantomorphism of crystallographic groups in higher dimensions with results in dimensions up to 6. Acta Cryst. A **59**, 210-220 (2003).

[90] R.P. Stanley, On the number of faces of centrally-symmetric simplicial polytopes. Graphs Combin., **3**, 55-66 (1987).

[91] F. Vallentin, Sphere coverings, lattices and tilings (in low dimensions). PhD, Technische Universität München, 2003.

[92] B.A. Venkov, On a class of Euclidean polytopes, (in Russian), Vestnik LGU, ser. Mat-fiz. khim., **9**, 11-31 (1954); reprinted in B.A. Venkov, Selected works, Nauka, Leningrad, 1981, pp.306-326.

[93] B.A. Venkov, On projection of parallelohedra, Math. Sbornik **49**, 207-224 (1959) (in Russian).

[94] G. Voronoï, Recherches sur les parallélloèdres primitifs. I. Propriétés générales des parallélloèdres. Crelle = J. Reine Angew. Math. **134**, 198-287 (1908).

[95] G. Voronoï, Recherches sur les paralléloèdres primitifs. II. Domaines de formes quadratiques correspondant aux differents types de paralléloèdres primitifs. Crelle = J. Reine Angew. Math. **136**, 67-181 (1909).

[96] H. Whitney, A set of Topological Invariants for Graphs. Am. J. Math. **55**, 231-235 (1933).

[97] H. Whitney, 2-Isomorphic Graphs. Am.J. Math. **55**, 245-254 (1933)

[98] E. Wigner, Über die Operation der Zeitumkehr inder Quantenmechanik. Nachr. Ges. Wiss. Göttingen Math.-Phys. K1. 546-559 (1932).

[99] O.K. Zhitomirskii. Verschärfung eines Satzes von Voronoi. Zhurnal Leningradskogo Math. Obshtestva **2**, 131-151 (1929).

Index

www.ingramcontent.com/pod-product-compliance
Ingram Content Group UK Ltd.
Pitfield, Milton Keynes, MK11 3LW, UK
UKHW021433280726
14060UKWH00001BA/54